The Plants of Bali Ngemba Forest Reserve, Cameroon

– A Conservation Checklist –

Yvette Harvey, Benedict John Pollard, Iain Darbyshire,
Jean-Michel Onana and Martin Cheek

Royal Botanic Gardens, Kew
National Herbarium of Cameroon

Published by the Royal Botanic Gardens, Kew

PLANTS PEOPLE
POSSIBILITIES

First published in 2004 by
Royal Botanic Gardens, Kew
Richmond, Surrey, TW9 3AB, UK
www.kew.org

ISBN 1-84246-075-7

Typeset by Yvette Harvey, Penelope Doyle, Iain Darbyshire and Benedict John Pollard (Introductory Chapters) and Yvette Harvey (Main Checklist).

Cover design by Jeff Eden, Media Resources,
Information Services Department,
Royal Botanic Gardens, Kew.

Printed in the UK by Cromwell Press

For information or to purchase all Kew titles please visit
www.kewbooks.com or email publishing@kew.org

Front cover: Bali Ngemba Forest Reserve from the boundary between grassland and forest on the eastern side of the site. Photo taken by Dave Roberts, April 2004.
Rear cover: Satellite image of Bali Ngemba Forest Reserve. Produced by Justin Moat.

CONTENTS

List of figures iv
Preface 1
Acknowledgements 3
New Name 7
Foreword 9
The Checklist Area 11
Vegetation 13
Geology, Geomorphology, Soils & Climate 27
Threats to Bali Ngemba Forest Reserve 28
Invasive, Alien & Weedy Plants 29
The Evolution of this Checklist 31
History of Botanical Exploration of Bali Ngemba Forest Reserve 32
Ethnobotany 43
The Birds of Bali Ngemba Forest Reserve 45
Endemic, Near-Endemic & New Taxa at Bali Ngemba 49
Validation of New Name 51
Figures 54
Red Data Plant Species of Bali Ngemba Forest Reserve 57
Bibliography 74
Read This First!: Explanatory Notes to the Checklist 76
Vascular Plant Checklist 81
Dicotyledonae 81
Monocotyledonae 133
Gymnospermae 151
Pteridophyta 151

LIST OF FIGURES

Fig. 1	The Bamenda Area	8
Fig. 2	The location of Bali Ngemba F.R.	10
Fig. 3	Vegetation map of Bali Ngemba F.R. showing the different Mapping Units.	12
Fig. 4	Vegetation map of Bali Ngemba F.R. and the surrounding area.	14
Fig. 5	*Magnistipula butayei* subsp. *balingembaensis* (Chrysobalanaceae).	50
Fig. 6	*Strombosia* sp. 1 (Olacaceae).	54
Fig. 7	*Vepris* sp. B (Rutaceae).	55
Fig. 8	*Scleria afroreflexa* (Cyperaceae).	56

PREFACE

We don't know for how long a forest has grown on the hills of the Bali Ngemba Forest Reserve, or how it has changed over the centuries. What we can easily see from our particular perspective in time is that in a few decades the forest has suffered serious degradation due to intense human pressure, including farming, man-made fires and introduction of exotic tree species, coupled with insufficient attention given to protection and management. Some people may have already written off this mere patch of 1100 hectares as having no particular value beyond the occasional harvesting of *Eucalyptus* logs for a few more years, until it is finally absorbed into the surrounding farmland. Who would have guessed that even the dedicated botanists who decided to give it a closer look would find no less than 39 plant species of special conservation interest, and at least 25 undescribed taxa? Being intermediate in most respects between the higher, cooler Kilum-Ijim forest to the north and the lower, moister Kupe-Bakossi forests to the south, who would have expected this degraded forest remnant to produce as many as 12 species not detected in the course of intensive searching by many of the same trained eyes in the latter, much larger sites?

One of the important lessons from this work, as well as the previous excellent work by the same researchers, is that if we think we know by now what lives in the forests of Cameroon, we should look again. In the case of this fascinating landscape called the Bamenda Highlands, we now know that each hill and valley, while being similar to the next one, may well contain something new and different.

While we continue to explore the biodiversity of these hills, it is already clear that virtually every remaining forest patch in the Bamenda Highlands, big or small, is probably important for the conservation of some of the extremely rare plants and animals that live scattered in those patches. Fortunately, in recent years there has been a growing interest among local communities, government agencies and international organisations in taking the necessary action to prevent destruction of these small but globally significant forests. This book should contribute to motivating and guiding all those concerned with their conservation.

Let us take the opportunity to give credit to those who had the foresight to create the Bali Ngemba Forest Reserve nearly a century ago, and let us not fail them by standing by while it is finally destroyed. Let us also honour the memory of those who for ages before hunted silently and gathered medicine in the shadows of the forest, never believing it could one day disappear.

John DeMarco
Project Leader, Gola Forest Programme
c/o Conservation Society of Sierra Leone,
PO Box 1292 Freetown,
Sierra Leone

Former co-manager, Bamenda Highlands Forest Project and Kilum Ijim Forest Project, Cameroon (1996-2003)

ACKNOWLEDGEMENTS

The following abbreviations are to be found in the text: ANCO (Apicultural and Nature Conservation Organisation (formerly North-West Beekeeping Association – NOWEBA)); BAT (British American Tobacco); BEPA (Boyo Environmental Protection Agency); BERUDEP (Belo Rural Development Programme); BHFP (Bamenda Highlands Forest Project); CRES (Conservation & Research for Endangered Species); ERUDEF (Environment and Rural Development Foundation); EW (Earthwatch); HNC (Herbier National Camerounais); HSBC (HSBC Bank Plc); KIFP (Kilum-Ijim Forest Project); LBG (Limbe Botanic Garden); MINEF (Ministry of the Environment and Forests); NEPA (Natural Environment Protection Agency); NOWEBA (North-West Beekeeping Association); ONADEF (National Office for the Development of Forestry); ORDEP (Organisation for Rural Development and Environmental Protection); RT (Rio Tinto Plc); SATEC (Sustainable Agriculture Technicians).

Firstly we thank those who facilitated our fieldwork at Bali Ngemba F.R.:

HRH the Fon of Bali, Dr Ganyonga III and his elders and chiefs are thanked for receiving delegations from our expeditions at his Palace, and for showing interest in the progress of our work. In particular, the observer from the Fon of Bali, Ba Doh Lawumma who joined our expedition in 2000.

The Bamenda Highlands Forest Project of BirdLife International (BHFP) at Bamenda, particularly co-manager John DeMarco, who supported the idea of doing a botanical survey of the Bali Ngemba Forest, who suggested using Mantum as a base and introduced us to its community, who arranged meetings with MINEF staff and their participation in the field, who provided BHFP and Kilum Ijim Forest Project (KIFP) staff to assist with the work and be botanically trained in the field, who co-ordinated the participation in our expeditions of representatives of many other NGOs. We also thank his co-manager Anne Gardner, their successor Michael Vabi, and deputy Grace Tima. We thank the following BHFP and KIFP staff members and volunteers, and the representatives of several Cameroonian forest NGOs who joined us in the forest:

2000: Ernest Keming (Ecomonitor, KIFP), Clement Toh (Ecomonitors, KIFP); Innocent Wultoff (BHFP); Grace Tima (Assistant Project Manager, BHFP); Clare Trinder (Environmental Education Officer, BHFP); Vincent Fomba (Staff forester BHFP/MINEF); Ndamsa, driver (BHFP); Rita Ngolan (BHFP); Delphine Wanduku (BHFP); Kenneth Tah (BHFP); Francis Njie (Bird Survey team, BirdLife International); Clare Wirmum (Mbiame Forest).

2001: Walters Cheso (Environmental Education Officer, BHFP); John DeMarco (co-Project Manager, BHFP); Clare Trinder (Environmental Education Officer, BHFP); Terence Suwinyi (BHFP); Kenneth Tah (BHFP); Innocent Wultoff (KIFP); Ernest Keming (KIFP); Rita Ngolan (KIFP); Clement Toh (Ecomonitors, KIFP).

2002: Cha Kenneth Yuh (senior bee-keeping trainer, BERUDEP); Florence Nkenya Azamah (senior forest technician, NEPA); Cletus Monju (field technician, SATEC); Marcellus Che (project execution manager, ORDEP).

2004: Terence Suwinyi (ANCO); Kenneth Tah (botanist, ANCO); Terence Atem (ERUDEF); Walters Cheso (formerly of BHFP).

We thank the staff of LBG who joined us on the following expeditions:

2001: Philip Nkeng (forester) and Raphael Kongor (Curator).
2004: Enow Kenneth.

The staff of MINEF at Bamenda and ONADEF for allowing us to conduct fieldwork inside the Forest reserve, and for assisting in the following expeditions:

2000: Kpoumie Chouaibou (Provincial Chief of Bureau, Protected areas, NW Province, ONADEF); Chief of Post Bali (MINEF); John Achu (Forest Guard, ONADEF); Vincent Fomba (Staff forester BHFP/MINEF);

2001: Patrick Fonji, Chief of Post Bali (MINEF); Waindah Nkemnya Mathias, observer for the Provincial Service of Forestry;

2002: Nangmo Yves Nestor (Delegate for Momo Division, MINEF); Kom Justin (Chief of Bureau, Bamenda Provincial Forestry Office, MINEF); Patrick Fonji (Chief of Post, Bali, MINEF).

Earthwatch funds allowed the employment of staff in Cameroon. We thank the following staff: Martin Etuge (Botanist: 2000; 2001; 2004); Freddy Epie (Assistant botanist: 2000); Emmanuel Ebong (Assistant botanist: 2000); Edmondo Njume (Assistant botanist: 2001; 2002; 2004); Martin Mbong (Administrator & Camp Manager: 2000; 2001; 2002); Ngolle Ngolle Hoffmann (Camp Manager: 2004); Robert Mesumbe (Apprentice Camp Manager:

2001); Henry Ngupi (Camp Assistant: 2002); Japhet Wain (Driver: 2000; 2001; 2002); Dansala Djire (Driver: 2001); Gilbert Tanyi (Driver: 2004).

We thank the people of Mantum and Bali who hosted us, particularly the families of Maurice Babila and Eric Nucam in whose compound we stayed on each of our visits. In addition, we are very grateful to the following for their assistance during our fieldwork:

Guides: Emmanuel Sama (2000); Maurice Babila (2000, 2002, 2004); Eric Nukam (2000, 2001, 2002, 2004); Peter Nkankano (2000, 2001, 2002, 2004); Henry Nsonfon (2001).
Cooks: Lilian Mankety Tanketi (2000, 2001, 2002, 2004); Patricia Lorga Tanketi (2000, 2002); Pamela Wolwonwo (2000); Patricia Lorga Tanketi (2001); Lovelyn Agem (2001); Sylvia Mbutigig Sakaah (2002, 2004).
Cleaner, clothes washing: Silvia Mbutigig Sakaah (2000, 2001); Lovelyn Agem (2002); Regina (assistant to Lovelyn Agem, 2002); Pamela Wolwonwo (2004).
Water Carrier: Yannick Njoya (2000); Vincent Tanketi (2001, 2002); Simon Tanketi (2002); Mdme Anna (2004).
Wood carrier: Laurence Tanketi (2000, 2001, 2002); Laurence Mbenbwo (2004).

At the Herbier National Camerounais, Yaoundé, we thank the leader, Gaston Achoundong, and the following, who have joined us in fieldwork at Bali Ngemba: Barthelemy Tchiengue (Researcher: 2004); Nicole Guedje (Researcher: 2004); Jean-Paul Ghogue (Technician, Leader HNC component: 2000); Fulbert Tadjouteu (Technician: 2000); Victor Nana (Technician: 2001; 2002); Felicité Nana (Secretary: 2000. Technician: 2002; 2004); Ela Ekeme Joseph (Accountant: 2002); Boniface Tadadjeu (Driver: 2000).

From the University of Yaoundé I, we thank the following for their assistance with fieldwork: Elvira Biye (Asst. Lecturer, Botany: 2000); Dr. Zapfack Louis (Senior Lecturer, University of Yaounde I and HNC Associate: 2002); Simo Placide (PhD. student under Dr L. Zapfack, joining the expedition to collect orchid taxa, particularly *Polystachya* and to collaborate with Dave Roberts: 2004).

Our fieldwork is not possible without the generous ongoing sponsorship and assistance from Earthwatch. While the Darwin Grant and Royal Botanic Gardens, Kew paid for National Herbarium expenses, 2000–2004, Earthwatch has provided all the other fieldwork funding in Bali Ngemba, apart from an RBG, Kew Overseas Fieldwork Committee grant in 2004. We are grateful to all the Earthwatch African Fellows and volunteers that gave their time to help us in the Forest Reserve (and their sponsor where appropriate):

2000: Kate Johnston (Australia); Julia Garcia (Spain); Joe Carson (USA); Hugh Guilbeau (USA); Pettula Nabukeera (Uganda: BAT); Raza Zulfiqar (Pakistan: BAT); Christo Van de Rheede (South Africa: BAT).
2001: Michael DeVille (USA); Agus Renggana (Indonesia: RT); Ian Rowe (Australia: RT); Mark Howson (UK: RT); Carola Girling (UK: EW Millennium Fellow); Elizabeth Merritt (UK: EW Millennium Fellow).
2002: (EW funded African Fellows): Alex Asase (Teaching assistant, Department of Botany, University of Ghana); Bakari Salim Mohamed (Assistant Regional Catchment Forest Manager, East Usambara Conservation Area Management Programme, Tanzania); Ignatius Malota (Senior Assistant Curator, National Herbarium & Botanic Gardens of Malawi, Malawi); Lazzarus Oketch Ojiek (Faculty of Forestry & Nature Conservation, Makerere University, P.O.Box 7052, Kampala, Uganda); Crentsil O'Rorke (Project Officer, Medicinal Plant Conservation Project, Aburi Botanic Gardens, Ghana).
2004: Adam Surgenor (UK); Zeynep Barlas (Turkey: HSBC); Armelle de Martrin (France: HSBC); Abdullah Shah (Pakistan: BAT).

A special word of thanks is given to Christo van de Rheede and Adam Surgenor who have graciously allowed us to use some of their plant portraits in this work. Thanks is also given to John Pool who gave access to his Spring 2004 images from Bali Ngemba and to Parmjit Bhandol for the images used in Plate 1(A–C).

Julian Laird, Gill Barker, Lucy Beresford-Stooke and Pamela Mackney of Earthwatch all helped greatly in recruiting various sorts of volunteers for us in Europe, as has Tania Taranovski in the U.S.A. We also thank Robert Llewellyn-Smith of the Earthwatch African Fellows Programme for providing us with botanists and other scientists from all over Africa to work with us.

In Douala we thank Constance and Valérie de Tailly of the British Consulate and the Brothers of La Procure Générale des missions Catholiques, for help and hospitality.

The Darwin Initiative grant to RBG, Kew has been the main single source of funds to our 'Conservation of the Plant Diversity of Western Cameroon' Project, from which this volume is one of the main outputs. We are extremely grateful for this assistance, without which the fieldwork, specimen identifications and species descriptions which form the basis of this book would not have been completed. A substantial part of the publication costs of this book were met by this grant. In particular we thank Valerie Richardson, Sarah Collins and Carrie Calhoun for their administration of this grant at DEFRA, as well as Des Bennett, financial accountant at RBG, Kew for assistance in managing the finances at Kew. The Edinburgh Centre for Tropical Forestry contracted by DEFRA to monitor the grant, are thanked for their constructive reviews of our project throughout its life.

At the Royal Botanic Gardens, Kew we thank those of our colleagues who joined in the fieldwork, George Gosline (2000), Laszlo Csiba (diploma student, 2001) and Parmjit Bhandol (Assistant Scientific Officer, 2001), Dave Roberts (orchid specialist, 2004) and Nina Rønsted (postgraduate researcher, Moraceae, 2004). The Publications Committee agreed to contribute towards the publication costs. John Harris, Ruth Linklater, Beth Lucas and Kate Hardwick are thanked for guidance on producing the camera-ready copy from which this book was produced, and for arranging scanning-in of images and design work. Chris Beard is thanked for her marvelous design of the plates.

At Kew we also thank Simon Owens, Keeper of the Herbarium, who has long supported our work in Cameroon, as has Daniela Zappi, Assistant Keeper for Regional teams, who has also championed our work. Eimear Nic Lughadha and Peter Crane have also been consistently supportive of our efforts in Cameroon.

Determinations of the specimens gathered in the course of our fieldwork were made by ourselves and by the following botanists, often world experts in their fields. We sincerely thank them all (in checklist order), and their institutions, for their work, often done without reservation, towards producing the names which are used in this book (acronyms follow Holmgren *et al.*).

We thank (in checklist order) F. Breteler (WAG) (Anacardiaceae), R. Becker (RNG) (Boraginaceae & Moraceae), M. Thulin (UPS) (Campanulaceae), C. Jongkind (WAG) (Combretaceae), M. Etuge (Dichapetalaceae & Costaceae), B. Sonké (University of Yaoundé I) (Rubiaceae), G. Achoundong (YA) (Violaceae), I. Nordal (O) (Anthericaceae), P. Boyce (Araceae), R. Faden (US) (Commelinaceae).

At Kew we thank (also in checklist order) K. Vollesen (Acanthaceae), C. Townsend (Amaranthaceae), G. Gosline (Annonaceae), H. Beentje (Apocynaceae & Compositae), D. Frodin (Araliaceae), D. Goyder (Asclepiadaceae) C. Sothers (Chrysobalanaceae), G. Prance (Chrysobalanaceae), N. Hind (Compositae), P. Wilkin (Convolvulaceae & Dioscoreaceae), Petra Hoffmann (Euphorbiaceae), G. Challen (Euphorbiaceae), P. Bhandol (Gesneriaceae), B. Mackinder (Leguminosae: Caesalpinioideae; Mimosoideae; Papilionoideae), R. Clark (Leguminosae: Papilionoideae), R. Polhill (Leguminosae: Papilionoideae, Loranthaceae & Viscaceae), B. Schrire (Leguminosae: Papilionoideae), B. Verdcourt (Leguminosae: Papilionoideae), E. Woodgyer (Melastomataceae), N. Rønsted (Moraceae), P. Green (Oleaceae), S. Dawson (Rubiaceae), T. Cope (Gramineae), D. Roberts (Orchidaceae), P. Cribb (Orchidaceae), M. Lock (Xyridaceae), P. Edwards (Lycopsida & Filicopsida).

A huge thanks goes to Stuart Cable of RBG., Kew. For the Mount Cameroon Project in 1997 and 1998, he originated the species database from which the checklist part of this book was produced. George Gosline subsequently developed the 'western Cameroon specimen' database, enabling us to print off data labels whilst still in the field, a real time-saver. He also developed the system by which data could be exported from the Access database using XML and XSLT into its Microsoft Word format. For meticulous data entry, Karen Sidwell, Suzanne White and Julian Stratton are also to be thanked. Without the database this volume would have taken considerably longer to produce.

Craig Hilton-Taylor, IUCN officer for Red Data, based at WCMC in Cambridge, is thanked for reviewing the Red Data assessments.

Finally, Penelope Doyle is to be thanked for typing some of Martin Cheek's Introductory Chapters (especially the long ones!).

NEW NAME

Benedict John Pollard
Herbarium, Royal Botanic Gardens, Kew, Surrey, TW9 3AE, U.K.

The following is published in this volume for the first time:

CHRYSOBALANACEAE

Magnistipula butayei De Wild. subsp. ***balingembaensis*** Sothers, Prance & B.J.Pollard p. 51

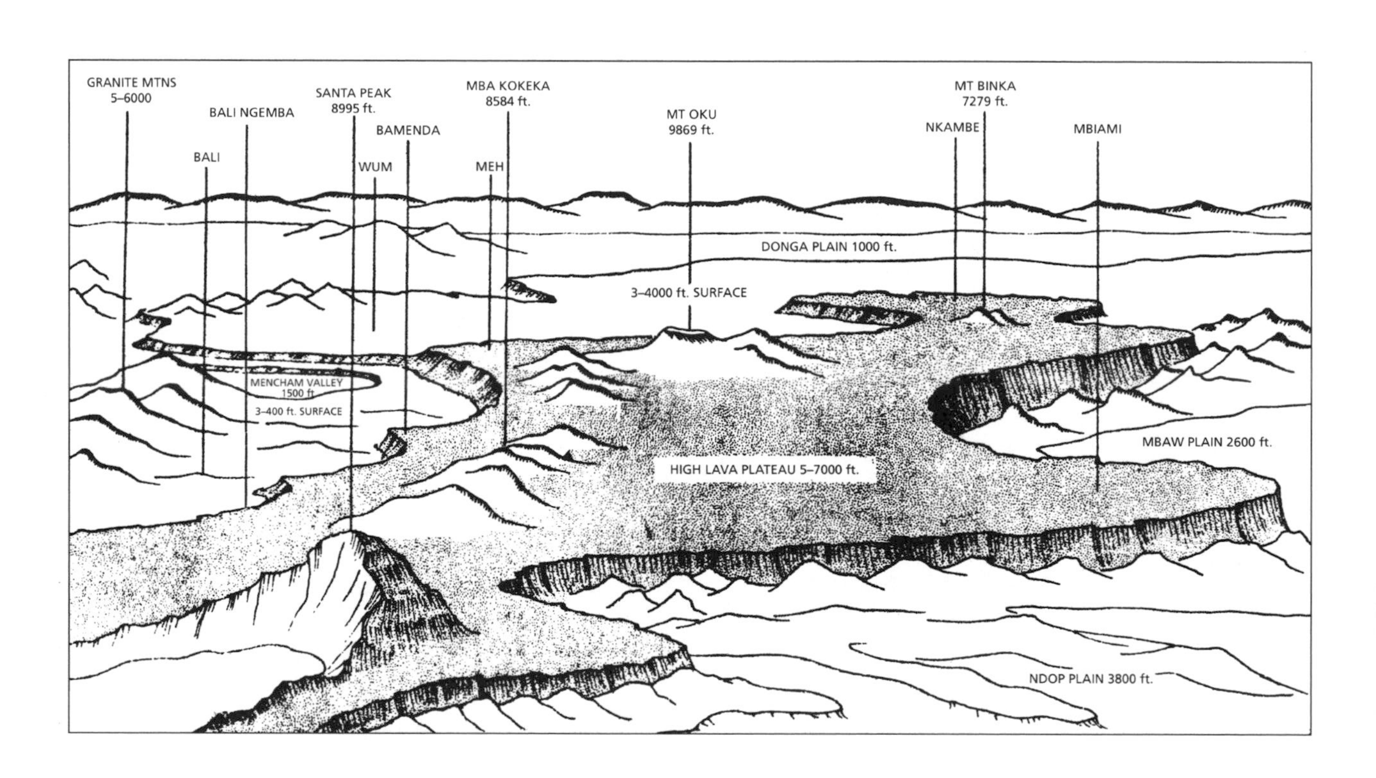

Fig. 1. The Bamenda Area. Reproduced with permission from Hawkins & Brunt (1965).

FOREWORD

Martin Cheek
Herbarium, Royal Botanic Gardens, Kew, Surrey, TW9 3AE, U.K.

Bali Ngemba Forest Reserve is located in the Bamenda Highlands of North West Province (Fig. 1), Cameroon, which themselves form part of the Cameroon Highlands. The Cameroon Highlands run from Bioko and Mt Cameroon in the south, northwards to Tchabal Mbabo and then continue eastwards as the Adamaoua area, following a geological fault line. It has been estimated that 96.5% of the original forest cover of the Bamenda Highlands above 1500 m altitude has been lost (Cheek *et al.* 2000).

It may seem curious that an obscure production Forest Reserve with an area of only 8 km^2, which is rarely mentioned in conservation circles and appears in no known directory of areas protected for conservation, should be the subject for a book such as this. The explanation is that Bali Ngemba is, in fact, of great biological and conservation importance. While about 70 km^2 of montane forest is preserved at Mt Oku and the Ijim Ridge (Kilum-Ijim) elsewhere in the Bamenda Highlands, that forest is almost entirely above 2000 m altitude. Bali Ngemba represents both the largest and the only officially protected block of forest below 2000 m alt. in the Bamenda Highlands. Nowhere else in the Bamenda Highlands displays such a continuous sequence of primary natural vegetation over such a large altitudinal range (1300–2200 m alt.), mainly in the submontane zone.

Moreover, these 8 km^2 harbour 38 Red Data (threatened with extinction) taxa, the densest concentration known in North West Province. Although Mt Oku and the Ijim Ridge are home to more Red Data species (56), these are dispersed over a much larger area. Moreover, Bali Ngemba's 38 Red Data species are set to increase as we analyse the latest collections. The species of this forest are so poorly worked out that 24 are new and unknown to science, all of which appear threatened, so that, when published, they will probably also be ascribed to the Red Data list.

So unique is the Bali Ngemba Forest Reserve that 12 species occur here and nowhere else in the world. A further 11 species are known from Bali Ngemba and only one other locality. These, and the other species of Bali Ngemba probably once occurred around the Bamenda Highlands and adjoining parts of the Cameroon Highlands, but have now been extirpated by mankind. If Bali Ngemba forest continues to be destroyed by wood-cutters and farmers as it is today, then, within a few years it is almost certain that these 12 species will be lost to the world, and many others will be significantly more threatened than before due to the loss of an important site. It is hoped that this book will bring to the attention of decision makers and the public as a whole, the importance of preserving intact the Bali Ngemba forest, for the Cameroon nation, and for the world.

This book is the fourth in a series of 'plant conservation checklists'. Its predecessors dealt with the plants of Mt Cameroon (Cable & Cheek 1998), of Mt Oku and the Ijim Ridge (Cheek *et al.* 2000) and of Kupe, Mwanenguba and the Bakossi Mts (Cheek *et al.* in press).

A major purpose of this book is to enable identification of the plant species within the checklist area, in particular those threatened with extinction – the highest priorities for conservation. To this end, details to aid the monitoring and management of each of the threatened taxa are given in a separate Red Data chapter, which includes illustrations for as many of the species involved as could be obtained. It is hoped that this information will aid the long-term survival of these threatened taxa.

Although the work presented here on the plants of Bali Ngemba is the most detailed and complete account available, it is not exhaustive. It is likely that yet more species await discovery in corners of the forest that we have not yet reached.

REFERENCES

Cable, S. & Cheek, M (1998). The Plants of Mount Cameroon, A Conservation Checklist. RBG, Kew, UK. lxxix + 198 pp.

Cheek, M., Onana, J.-M. & Pollard, B.J. (2000). The Plants of Mount Oku and the Ijim Ridge, Cameroon, A Conservation Checklist. RBG, Kew, UK. iv + 211 pp.

Cheek, M., Pollard, B.J., Darbyshire, I., Onana, J.-M. & Wild, C. (eds) (in press). The Plants of Kupe, Mwanenguba & the Bakossi Mts, Cameroon, A Conservation Checklist. RBG, Kew, UK.

Hawkins, P. & Brunt, M. (1965). The Soils and Ecology of West Cameroon. 2 Vols. FAO, Rome. 516 pp, numerous plates and maps.

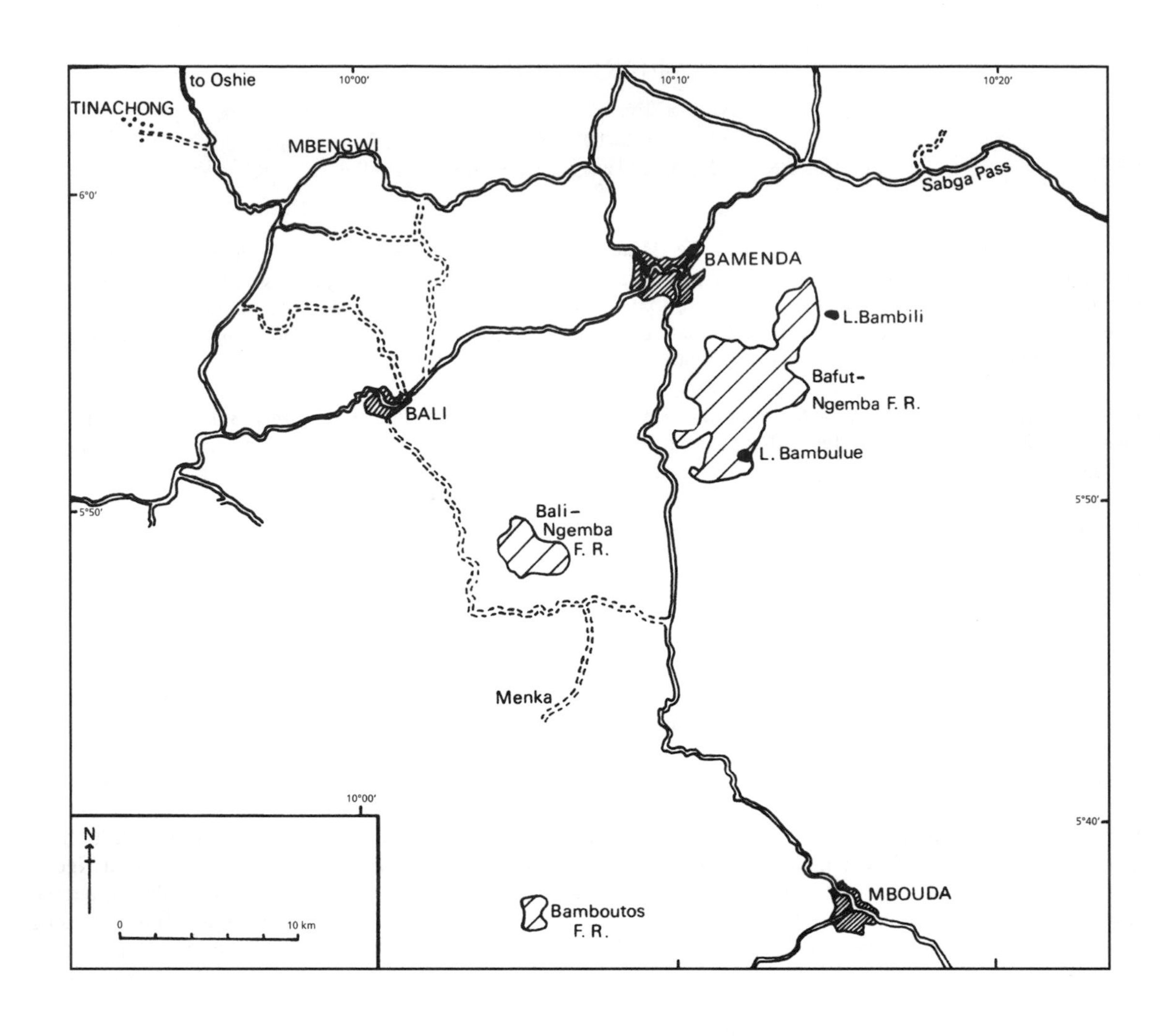

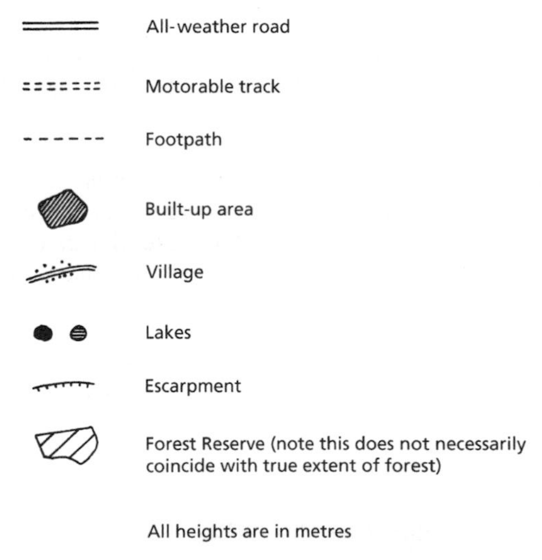

Fig. 2. The location of Bali Ngemba F.R. Reproduced with the permission of Birdlife International from Stuart (1986).

THE CHECKLIST AREA

Martin Cheek
Herbarium, Royal Botanic Gardens, Kew, Surrey, TW9 3AE. U.K.

The checklist area consists of the Bali Ngemba Forest Reserve in NW Province, Cameroon. It is the largest area of natural submontane forest known to survive in the Province and in the Bamenda Highlands as a whole. The forest is situated 5–10 km SE of Bali, a principal town in the province. The protected area has an altitudinal range of 1800–2200 m and lies at 5°49'N, 10°5'E (Fig. 2). The area of the forest reserve is usually given as 8 or 10 km^2.

No official government maps or documents concerning the reserve have yet been located by us, although these must surely exist. For this reason, information on the checklist area is derived from other sources. In practice, on the ground, there is little difficulty tracing the boundary of parts I am familiar with, e.g. the main point of access at Mantum. A cut-line is maintained, with farming or burnt grassland on one side and on the other the forest in the reserve; in addition, a line of *Cupressus lusitanica* forms part of the boundary.

The Forest Reserve is believed to be gazetted as a production reserve, i.e. for the production of timber. It was probably demarcated during the period of the British mandate between 1920–1960, since the earliest references to the reserve by name are the herbarium specimens of Ujor in May and June 1951. During the period of the British mandate, the current SW and NW Provinces of Cameroon were administered as an extension of Nigeria. Therefore the documents concerning the creation of the Forest Reserve may still be filed at Ibadan, the administrative centre of Nigerian forestry operations.

Bali Ngemba F.R. is mentioned in the survey of Cameroon Montane Forests by the precursor of BirdLife International (Stuart 1986) and also on their website (www.birdlife.org, BirdLife IBA factsheet CM016 'Bali Ngemba Forest Reserve', accessed Sept. 2004). But this is unusual. Despite its great biological significance, as revealed in this book, Bali Ngemba has been passed over in several national or provincial surveys of protected areas in Cameroon. For example Courade's (1974) map which shows other Forest Reserves in the Bamenda Highlands does not feature Bali Ngemba. Nor do works which treat protected areas or areas deserving of protection in Cameroon or West-Central Africa. Examples are: World Bank (1993), Collar & Stuart (1988); Gartlan (1989), MacKinnon & MacKinnon (1986).

Even in the BirdLife Survey of Stuart (1986), no botanical report for Bali Ngemba was given, whereas plant reports at some level were given for many other sites within the Cameroon Highlands.

Although the Bali Ngemba F.R. is featured on maps such as the 1:200,000 topographical map (Bafoussam sheet) and that accompanying the BirdLife survey (Stuart 1986, here reproduced as Fig. 2), the most detailed published map of the reserve available to us is that of Adolfe Tetsekoua reproduced in Chiron & Guiard (2001). This is believed to have been made as part of a University degree project and is reported (Babila pers. comm. 2000) to have the virtue of being made by surveying the forest reserve on the ground. This map, apart from being at a good scale, has the advantage of showing the courses of the several streams within the boundary of the reserve, and also of showing the area of the *Eucalyptus* plantation, and of one or two paths. However, several further important paths exist, but are not shown, nor are several major physical features helpful for navigation, such as 'Jerusalem Rock', nor are areas of habitation in the vicinity, such as Mantum or Pinyin. Most of the forest vegetation is shown as being degraded to some extent, whereas, in practice, extensive areas of reasonably intact tracts remain. The canopy and understorey, less so the herbaceous layer, are still intact over the greater part of the reserve.

REFERENCES

Chiron, G.R. & Guiard, J. (2001). Etude et conservation des orchidées de Bali Ngemba (Cameroun). Richardiana 1(4): 153–186.
Collar, N.J. & Stuart, S.N. (1988). Key Forests for Threatened Birds in Africa. International Council for Bird Preservation, Monograph No. 3. ICBP, Cambridge, UK. 102 pp.
Courade, G. (1974). Commentaire des cartes. Atlas Régional. Ouest I. ORSTOM, Yaoundé.
Gartlan, S. (1989). La Conservation des Écostystèmes forestièrs du Cameroun. IUCN Switzerland & UK. 186 pp.
MacKinnon, J. & MacKinnon, K. (1986). Review of the Protected Areas System in the Afrotropical Realm. IUCN Switzerland & UK. xviii + 259 pp.
Stuart, S.N. (ed.) (1986). Conservation of Cameroon Montane Forests. International Council for Bird Preservation, Cambridge, UK.
World Bank (1993). Ecologically Sensitive Sites in Africa. Vol. 1: Occidental and Central Africa. The World Bank, Washington, USA. 128 pp.

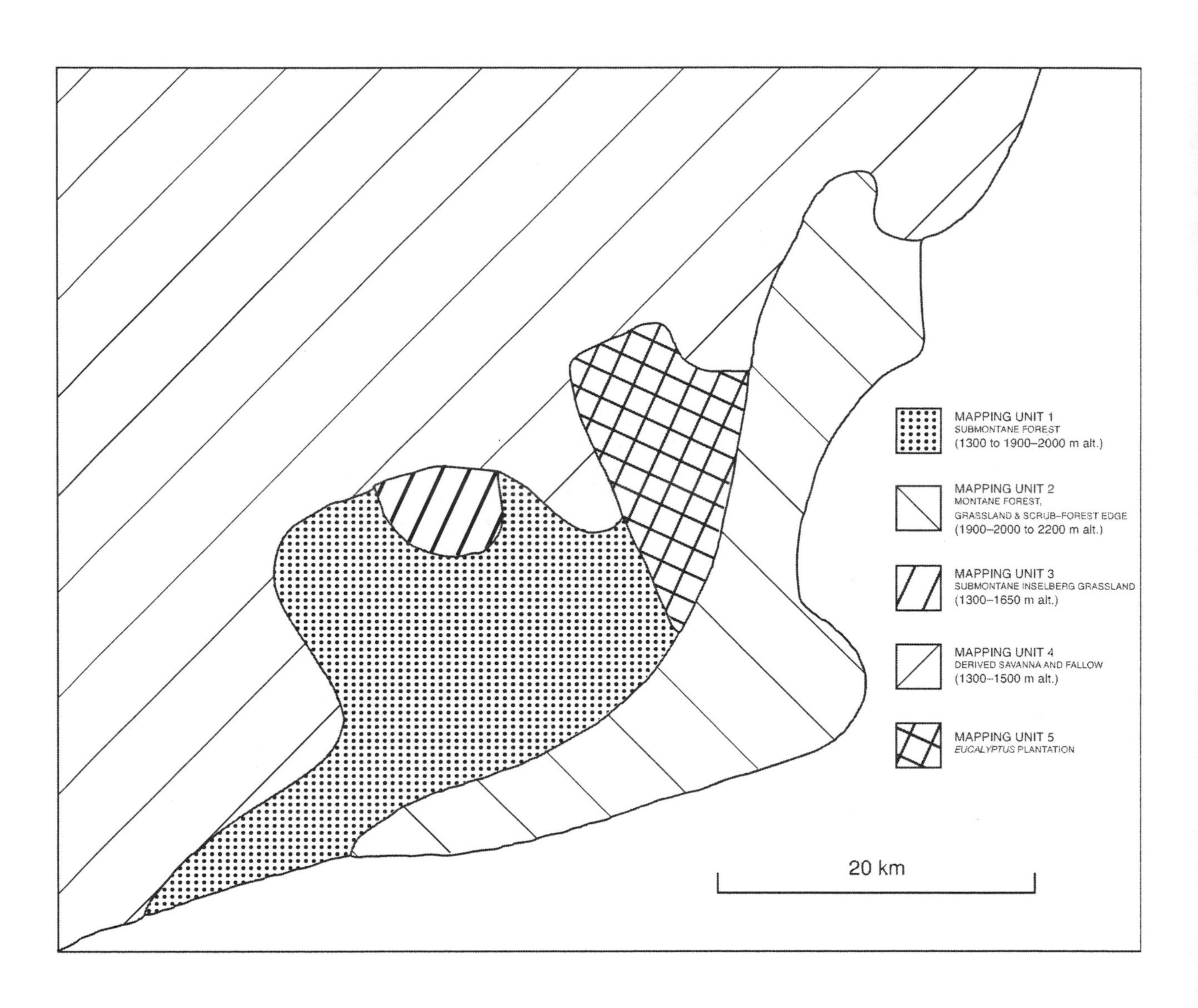

Fig. 3. Vegetation map of Bali Ngemba F.R. showing the different Mapping Units. Drawn by Martin Cheek.

VEGETATION

Martin Cheek
Herbarium, Royal Botanic Gardens, Kew, Surrey, TW9 3AE, U.K.

In this chapter the five vegetation types recognised in Fig. 3 (mapping units 1–5) are characterised; sampling by way of specimen collecting levels is gauged, the physiognomy and species composition of the vegetation is examined, endemic and threatened taxa and threats to the vegetation types are listed, and phytogeographical links are assessed.

PREVIOUS STUDIES

No on-the-ground studies of vegetational composition at Bali Ngemba appear to have been conducted prior to our own.

One previous provincial (Hawkins & Brunt 1965) and one national (Letouzey 1985) vegetation map of Cameroon show the forest patch corresponding to Bali Ngemba. The relevant parts of both these maps are reproduced in this book as Fig. 4 and Plate 2(C) respectively. Although the compilers of both maps did extensive botanical ground-truthing in the progress of their work, neither actually visited Bali Ngemba, but characterised the area of the forest from aerial photography and its vegetation type by extrapolation from observations made elsewhere in the Bamenda Highlands. This is regrettable because Bali Ngemba, despite some degradation, is still the best preserved, largest and most representative area of submontane forest known to survive in the Bamenda Highlands, so to interpret properly the much smaller fragments that occur elsewhere, an understanding of Bali Ngemba is essential.

The part of Hawkins & Brunt's (1965) map 10 (Fig. 4) that corresponds with Bali Ngemba shows a crescent of forest facing westwards. It is not subdivided further into vegetation types; in fact, throughout this map, forest is only shown as a single mapping unit. The accompanying text, however (Hawkins & Brunt 1965, 1: 208–214) shows that the authors recognised two forest types, based on altitude and divided by the 5,000 ft contour (=1500 m). Above this contour they recognise 'moist montane forest' and below it 'moist evergreen forest' (lowland evergreen forest). However, they deliberate at length (p. 213) about the possible existence of a 'transition forest' (between these two types). This is prompted by their reading of Lebrun's work in 1935, on altitudinal classification of the forests of Kivu and Ruwenzori, in which a distinct forest type was recognised as occurring between 3,300 ft and 5,250 ft, that is c.1000–1600 m.

It is clear that their 'transition forest' equates to what is referred to below, here and by other more recent workers in this part of Africa, as submontane forest. Hawkins and Brunt conclude with a statement that 'The few remnants of the original forest cover at lower altitudes in the Bamenda area, have not been sufficiently carefully studied to permit the suggestion that such a Transition Forest also occurred there.' Had they the opportunity to visit and study the forest at Bali Ngemba, they would have reached a very different conclusion. Instead, within this altitudinal range, they were only able to locate (p. 214) 'small stands of tall forest, protected for religious reasons, also remain ... and near Bambui village ... *Pterygota* ... 150–200 ft high.' The significance of the *Pterygota* is discussed below

Since Hawkins & Brunt's (1965) main geographical focus was what is today's NW Province of Cameroon, the larger part of their treatment of vegetation was devoted to grassland and savanna, since this dominates the province. They conclude that grassland and savanna in the Bamenda Highlands is derived, developing in areas which have been cleared of forest by man, the savanna species having migrated from lower altitudes. From their analysis it is clear that the forest of the type seen today at Bali Ngemba was formerly widespread and common in the Bamenda Highlands, occupying the 'low lava plateau' on which such large towns as Bali, Bamenda and Kumbo sit, and which today are among the most densely populated parts of Cameroon. It was probably this range of forest, also, that once covered the adjoining highland areas, in West Province, of the Bamilike people, as densely settled as NW Province and even more denuded of the original forest.

Before publishing his celebrated vegetation map of Cameroon in 1985, Letouzey released a series of other works on the subject through which it can be seen how his ideas developed. Amongst these was his monumental Étude Phytogéographique du Cameroun in which he first mentions Bali Ngemba (Letouzey 1968: 333). However, this reference is solely to the occurrence of *Podocarpus milanjianus* – a phytogeographically interesting gymnosperm – at Bali Ngemba. This record is based on the specimen collected by Ujor in 1951, which Letouzey probably came upon in FWTA 1 (Keay 1958: 32).

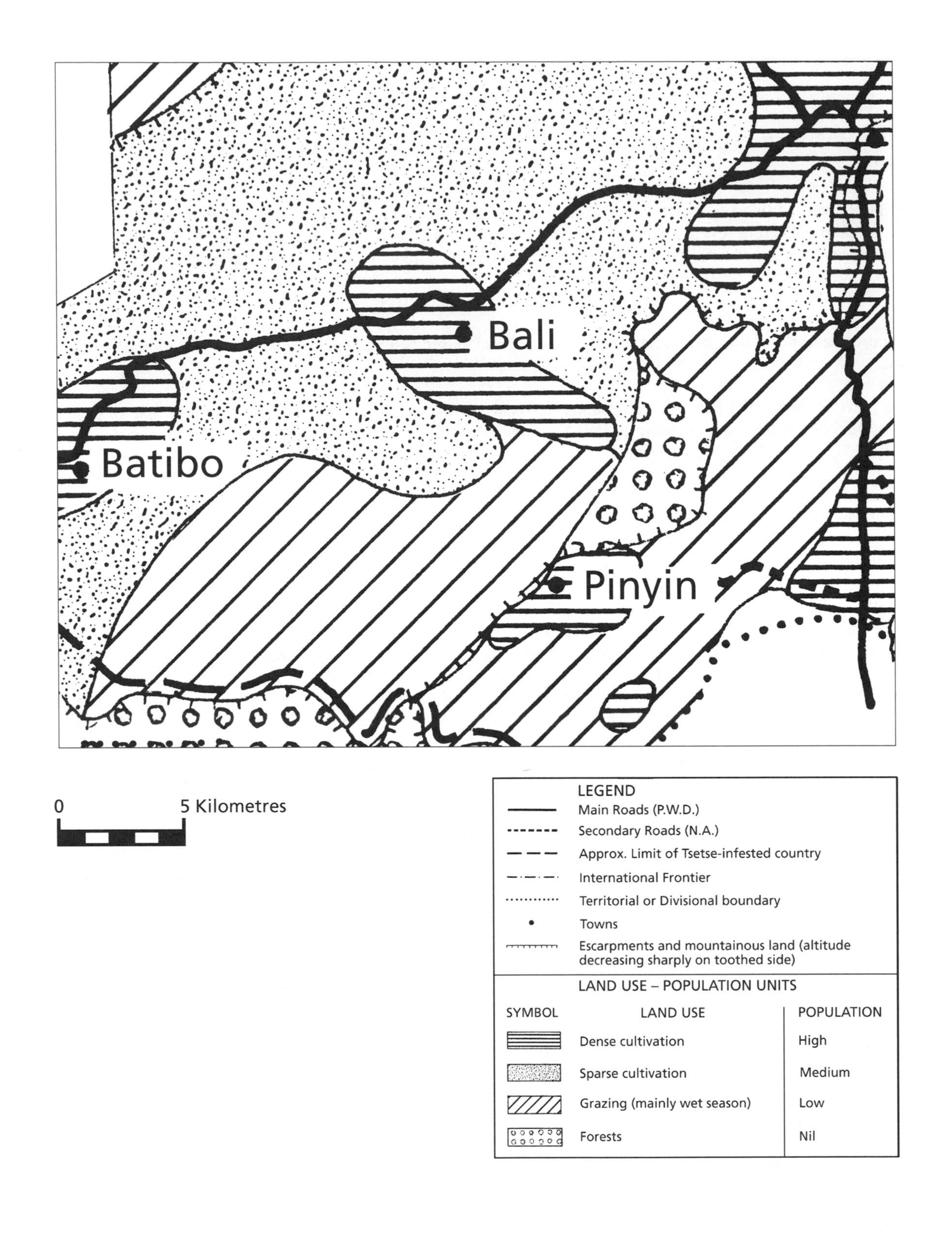

Fig. 4. Vegetation map of Bali Ngemba F.R. and the surrounding area. Reproduced with permission from Hawkins & Brunt (1965).

Letouzey's 1985 vegetation map of Cameroon classifies the forest corresponding with Bali Ngemba into two types:

- Submontane forest 800 to 1900–2000 m. alt. (Letouzey mapping unit 117)
- Montane forest 1900–2000 m. alt. and above (Letouzey mapping unit 108).

A third natural, but non-forest, vegetation type shown by Letouzey is his 'groupement saxicole divers' corresponding to the two grassy inselbergs inside the reserve boundary (his mapping unit 129; see also front cover). Surrounding the forest-inselberg block that corresponds with Bali Ngemba, Letouzey shows four anthropic types of vegetation:

- Mapping unit 113 extends from the E side of the forest northwestwards to Bamenda. This is characterised as *Sporobolus africanus* montane pasture (1600–2800 m) with gallery forest and has been characterised elsewhere in the Bamenda Highlands by Cheek *et al.* (2000). This grassland type is believed to result from repeated firing and cattle-tramping after forest clearance, and is generally species-poor and of low conservation value. Our survey did not include this vegetation type.
- Mapping unit 128 envelops most of the rest of the forest blocks. Letouzey characterises it as 'submontane grassland and farmbush, 1200–1800(–2000) m, more or less grazed and inhabited'. This is also believed to be derived from forest, but occurs at lower altitudes than *Sporobolus africanus* grassland. Part of the area mapped as this vegetation type by Letouzey was surveyed by us and is characterised below as 'submontane grassland and savanna'. It equates with the 'derived savanna' discussed by Hawkins & Brunt (1965: see above).
- Mapping unit 120 reaches the SW tip of the forest blocks. Letouzey characterises (translating informally) as domestic countryside with hedges 1000–2000 m. Our survey did not include this vegetation type.
- Mapping unit 121 abuts part of the NE forest boundary. Letouzey characterises it as domestic countryside without hedges or with dispersed enclosures, 1000–2000 m. Our survey did include some of this vegetation type, which is treated as mapping unit 4.

CLASSIFICATION OF VEGETATION TYPES: SUMMARY

The five vegetation types recognised in this treatment, together with their subdivisions, are as follows:

Mapping Unit 1: Submontane forest (1300 to 1900–2000 m alt.)

a) Lower submontane forest with *Pterygota* (1300–1500 m alt.)
b) Upper submontane forest with *Pterygota* (1500–1700 m alt.)
c) Extreme upper submontane forest (1700 to 1900–2000 m alt.)

Mapping Unit 2: Montane forest, grassland and scrub–forest edge (1900–2000 to 2200 m alt.)

a) Montane forest.
b) Montane grassland and scrub-forest edge.

Mapping Unit 3: Submontane inselberg grassland (1300–1650 m alt.)

Mapping Unit 4: Derived savanna and farm fallow (1300–1500 m alt.)

Mapping Unit 5: *Eucalyptus* plantation.

No lakes, which would be expected to have a separate vegetation type, are known from the checklist area. Several perennial streams run through the forest, exiting near Mantum, but no obligate rheophytic species are known, nor has any distinct aquatic or stream-side community been identified. Rockfaces do occur in the vicinity, and are likely to harbour a distinct vegetation type, but this has not been investigated.

LIMITATIONS OF CHARACTERISATION

A limitation of the treatments of the forest types that follow lies in their characterisation. Forest types have been delimited by altitudinal ranges, largely characterised by the presence or absence of species, as recorded by specimens made by collectors. Sampling by collectors of forest at the higher altitudes has been relatively low, and even the better surveyed lower altitudinal forest has not been exhaustively collected. Therefore the altitudinal ranges of the taxa used to characterise the forest types are liable to modification if, as is desirable, surveying continues. In short, some species cited as characterising one forest type are, with further collecting, very likely to be found in other forest types. Nonetheless, the forest types recognised here are thought to be robust enough to survive the test of additional data.

CONVENTIONS USED IN MAPPING UNIT TREATMENTS

Two sorts of species lists are included in the following vegetation treatments:

1. Lists based on the specimens collected in a plot by a team or by a single collector in one part of a vegetation type, giving a snapshot, or a series of snapshots of the species assemblages in one area on one day. Here species are listed in order of collection by the plot team or collector, and the list is prefixed by size data for the specimens, the collector, and the number of the first specimen upon which the sequence of specimens in based.

2. Lists extracted from the database of all records of species from a particular vegetation type, e.g. forest, 1300–1500 m alt. These lists are based on specimens collected throughout the history of botanical inventory at Bali Ngemba, by various collectors. Here species are listed in the order in which they appear in the checklist.

In both types of species lists, threatened taxa (those accepted by IUCN as 'Red Data' taxa, being either Vulnerable (VU), Endangered (EN) or Critically Endangered (CR)) are indicated by being annotated accordingly.

MAPPING UNIT 1. SUBMONTANE FOREST (1300 to 1900–2000 m alt.) (see Fig. 3)

The altitudinal definition of submontane forest used by Letouzey (1985: mapping unit 117) is adopted here, since it agrees with observations made at Bali Ngemba, and also with other studies of altitude-defined forest types (Cheek *et al.* 2000, Cheek *et al.* in press). The lower altitudinal limit of submontane forest in the works cited above is usually taken as the 800 m contour. However, this altitude is not attained in the Bali area, which occupies the 'low lava plateau' (900–1200 m alt.) which itself forms the perimeter of the Bamenda Highlands (Hawkins & Brunt 1965). The upper limit of submontane forest does occur within the forest reserve. Its demarcation with montane forest is less clear than that at e.g. Mt Oku and the Ijim Ridge (Cheek *et al.* 2000) in that several species that at Mt Oku and elsewhere in the Cameroon Highlands seem to characterise montane forest (1900–2000 m and above) have only been recorded at Bali Ngemba at 1400–1500 m altitude. These taxa are:

Xymalos monospora
Prunus africana
Pavetta hookeriana VU
Nuxia congesta

This phenomenon needs more investigation.

Sampling of submontane forest at Bali Ngemba is fairly satisfactory, especially at altitudes 1300–1700 m, since it has been our main focus for collection during surveys at the site. Furthermore, two plots were enumerated in this vegetation type, one at 1400 m, the other at 1700 m alt.

Species occurring more of less throughout this altitudinal range and therefore characterising submontane forest as a whole at Bali Ngemba are as follows. Altitudinal ranges are derived from the checklist and taxa are cited in the order that they appear here:

Salacia erecta
Garcinia smeathmannii
Strychnos tricalysioides
Memecylon dasyanthum VU
Ardisia staudtii
Chassalia laikomensis CR
Cremaspora triflora subsp. *triflora*
Ixora bauchiensis
Psychotria psychotrioides
Rothmannia urcelliformis
Deinbollia sp. 1
Synsepalum msolo
Quassia sanguinea VU
Leptonychia sp. 1 of Bali Ngemba
Boehmeria macrophylla
Cyphostemma rubrosetosum
Cyperus renschii var. *renschii*

Forest edge taxa

The interface between forest and grassland-savanna often harbours a different set of species to those away from the forest edge. Although the forest edge was not subjected to detailed scrutiny, the following list of taxa were not otherwise recorded elsewhere in the forest or in the adjoining grassland-savanna:

Dissotis irvingiana (1300 m)
Ctenium newtonii (1300 m)
Hibiscus noldeae (1400 m)
Guyonia ciliata (1400 m)
Sida javanensis (1500 m)
Ctenium ledermannii (1500 m)
Virectaria major (1560 m)
Cynoglossum coeruleum (1700 m)
Dombeya ledermannii CR (1700 m)
Crassocephalum bauchiense VU (1900 m)

Of those taxa that appear to characterise the submontane forest and its edge, five are threatened and other taxa, still unpublished, are also likely to prove to be threatened.

Threats to this forest type derive from human pressure to the west: removal of trees for timber on a small but steady scale, and clearance of the undergrowth for the farming of cocoyams (*Colocasia esculenta*), especially at lower altitude. Observations made by our survey teams over successive years in the period 2000–2004 inclusive have shown that this activity is slowly but steadily destroying the forest. The phytogeographical affinities of this vegetation type are with other submontane areas in the Cameroon Highlands, particularly with Mt Kupe and the Bakossi Mts: most of the species listed also occur there. About half the species listed are restricted to the Cameroon Highlands, with *Psychotria* sp. A and *Leptonychia* sp. 1 only known from Bali Ngemba. *Pterygota mildbraedii* itself, the single most conspicuous and representative species of this forest type, only occurs in the mountains of the Albertine Rift, east of the Congo Basin, apart from the Cameroon Highlands. Within the Cameroon Highlands it appears to have been restricted to the Bamenda Highlands, where its largest surviving tract is at Bali Ngemba.

The division of submontane forest

Occupying, as is generally accepted in this part of Africa, an altitudinal range of over 1000 m, it is only to be expected that, within this range, there are significant altitude-related changes in species composition which provide the impetus for further subdivision.

In a study of the Kupe, Mwanenguba and Bakossi Mts (Cheek *et al.* in press), which contains one of the largest intact areas of submontane forest surviving in West-Central Africa, submontane forest was subdivided into upper and lower types. This subdivision was based not on a change in physiognomy, but upon a discontinuity of common tree species that collection data showed occurred at about the 1400 m contour. However, at Bali Ngemba, the most obvious basis for a subdivision is *Pterygota mildbraedii*.

Submontane forest with *Pterygota mildbraedii* (1300–1700 m alt.)

This altitudinal zone is physiognomically distinctive because it is dominated by *Pterygota mildbraedii*. When mature these trees are canopy emergents, reaching c. 60 m tall and 4 m diam. at ground level. They occur in such numbers at Bali Ngemba that, when the forest is viewed in the dry season from the ridge above, to the E, or from the approach below from Bali, to the W, the dark green evergreen forest canopy of the lower part of the forest appears to be covered by a haze of grey, from the leafless crown branches of the *Pterygota*. Individual trees of this taxon appear to be fairly evenly spaced throughout this forest zone, and to be abruptly absent above the 1700 m contour.

The following species all have altitudinal ranges which correspond with the altitudinal range of this forest type (c. 1300–1700 m alt.) and therefore appear to characterise it. Altitudinal ranges are derived from the checklist and taxa are cited in the order that they appear there.

Alangium chinense
Impatiens filicornu
Kigelia africana
Drypetes sp. 3
Carapa grandiflora
Turraeanthus africanus
Strombosia scheffleri
Euclinia longiflora
Psychotria sp. A aff. *calva*
Psychotria sp. A of Bali Ngemba
Tricalysia sp. B aff. *ferorum*
Elatostema paivaeanum
Chlorophytum sparsiflorum
Amorphophallus staudtii

The *Pterygota* forest can be subdivided on altitude as follows:

a) Lower submontane forest with *Pterygota mildbraedii* (1300–1500 m alt.)

At Bali Ngemba a division between upper and lower submontane forest can be recognised on the basis of the same discontinuity of distribution of common tree species as at Kupe, Mwangenguba and Bakossi Mts. However, of the 17 characteristic and common tree taxa of lower submontane forest listed in the Kupe, Mwangenguba and Bakossi Mts area (Cheek *et al.*, in press), only seven occur at Bali Ngemba, and all three extend to the 1500 m contour rather than the 1400 m contour, two others extending as high as 1700 m. These are:

Allanblackia gabonensis (–1500 m) VU
Pseudagrostistachys africana (–1700 m) VU
Santiria trimera (–1430 m)
Chionanthus mildbraedii (–1500 m)
Polyscias fulva (–1500 m) Pioneer
Macaranga occidentalis (–1700 m) Pioneer
Cyathea spp. (–1400 m) Pioneers

On this evidence, the division between upper and lower submontane is raised from the 1400 m contour recognised in SW Province (Cheek *et al.* in press), to the 1500 m contour here, in the Bamenda Highlands. The following species have altitudinal ranges that are correlated to the examples above, and give support to adopting the 1500 m contour as the division:

Brillantaisia spp. (c. 1300–1560 m)
Justicia paxiana (1400–1560 m)
Achyranthes aspera (1310–1560 m)
Celosia pseudovirgata (1200–1500 m)
Trichoscypha acuminata (1300–1450 m)
Voacanga sp. 1 (1310–1500 m)
Ficus sur (1400–1500 m)
Vepris sp. A (1310–1500 m)
Psychotria sp. B of Bali Ngemba (1310–1560 m)
Strychnos tricalysioides (1310–1500 m)
Trichilia dregeana (1400–1500 m)
Stephania abyssinica var. *abyssinica* (1310–1500 m)

Common canopy and subcanopy trees in this forest type are seen in the list below, where all tree species in a plot at 1400 m were recorded.

Profile plot at 1400 m alt. on the Mantum-Kedeng path, at N5°49'17.5", E10°05'16.3". Enumeration began 5 Oct 2001, with Etuge, Njume, Rowe, Renggana, Nkeng, Ngolan. Canopy height 15–20 m. Species listed as recorded from *Cheek* 10730 *et seq.* and in fieldbook plot notes. Plot dimensions were 50 × 15 m. Herbaceous species were not recorded.

Common canopy and subcanopy trees:

Trichilia dregeana
Quassia sanguinea (understorey)
Turraeanthus africana
Garcinia smeathmannii
Drypetes sp. 3
Drypetes sp. aff. *leonensis* of Bali Ngemba
Strombosia scheffleri
Macaranga occidentalis (pioneer)
Euclinia longiflora (understorey)
Cola anomala
Alangium chinense (pioneer)
Isolona hexaloba
Synsepalum msolo
Polyscias fulva (pioneer)

Other species recorded only from forest within this altitudinal range, not in forest above 1500 m, are listed below. Most were recorded only once, and so a range for the species cannot be established. Several of these taxa are more typical of lowland evergreen forest than of submontane forest, and were rarely encountered possibly because they are at the edge of their altitudinal limits and so less frequent. Examples are *Funtumia elastica, Landolphia landolphioides, Bombax buonopozenze* and *Turraea vogelii.*

Dicliptera laxata
Artabotrys aurantiacus
Pseudospondias microcarpa
Trichoscypha acuminata
Funtumia elastica
Landolphia landolphioides
Marsdenia angolensis
Bombax buonopozenze
Canarium schweinfurthii
Rhipsalis baccifera
Ritchiea albersii
Stellaria mannii
Magnistipula sp. 1
Turraea vogellii
Tinaspora bakis
Ficus cyathistipula
Ficus glumosa
Ficus ingens
Ficus ovata
Ficus thonningii
Ficus vallis-choudae
Trilepisium madagascariense
Maesa lanceolata
Ochna membranacea
Fagopyrum snowdenii
Maesopsis eminii
Psychotria psychotrioides
Psychotria succulenta
Rytigynia laurentii
Zanthoxylum rubescens
Allophylus conraui EN
Eriocoelum macrocarpum
Boehmeria macrophylla
Laportea alatipes
Parietaria debilis
Pilea angolensis
Urera cf. *gravenreuthii*
Sanicula elata
Isachne buettneri
Oplismenus hirtellus
Setaria poiretiana
Hypselodelphys poggeana
Hypselodelphys scandens
Smilax anceps

Psydrax dunlapii (VU) characterises the 1400–1500 m belt at Bali Ngemba (six collections) as it does at Mt Cameroon. Strangely, it is not known from intervening areas of apparently suitable habitat, e.g. at Kupe-Bakossi.

In addition to the threatened and endemic taxa discussed earlier for the *Pterygota* forest as a whole, a further three threatened taxa occur in the 1300–1500 m band, with four further undescribed taxa which are likely to classed as threatened when published. Of these, *Psychotria* sp. B and *Magnistipula* sp. 1 appear to be entirely restricted (endemic) to this band.

Threats and phytogeographical links are as discussed above under submontane forest. The appearance of a substantial number of lowland forest species in this band has already been noted.

b) Upper submontane forest with *Pterygota mildbraedii* (1500–1700 m alt.)

This forest type lacks a characteristic set of common species such as is documented in the vegetation type above. Rather it appears to be distinguishable mainly by the absence of those taxa listed as specific to lower submontane forest, and the presence of those species that characterise the submontane and *Pterygota* forest types as a whole (see above), these being more frequent at this altitudinal range than below. This circumstance parallels exactly the situation seen in the submontane forest of Kupe, Mwanenguba and Bakossi Mts (Cheek *et al.*, in press).

It may be that common 'marker' species do exist which characterise this altitudinal range, but sadly the interval between 1500 m and 1700 m has been neglected by our survey. The only species that appear to be restricted to this range are as follows:

Acanthus montanus
Tabernaemontana sp. of Bali Ngemba
Pentarrhinum abyssinicum subsp. *ijimense* CR
Cordia millenii
Diospyros zenkeri

Other species recorded at 1700 m, but collected nowhere else, are listed below. They are as likely to represent the altitudinal band 1700–1900 m as 1500–1700 m. Certainly *Zenkerella citrina* is commonest at 1700–1900 m at Mt Kupe, and *Eugenia gilgii* at 1700–2000 m at Kilum-Ijim.

Justicia unyorensis
Tabernaemontana pachysiphon
Begonia fusialata var. *fusialata*
Pseudagrostistachys africana VU
Zenkerella citrina
Warneckea cinnamomoides
Eugenia gilgii CR
Thalictrum rhynchocarpum
Chrysophyllum gorungosum
Gouania longispicata
Oxyanthus formosus
Pauridiantha paucinervis
Rytigynia sp.

The list of species below can be taken as a snap-shot of the composition of this forest type.

25 m × 25 m plot, 1700 m alt. May 2002, led by Dr L. Zapfack.

List of species identified from the plot voucher series BALI-88 (naming incomplete), ordered by appearance in the series. The vouchers were not databased and so do not appear in the checklist that follows, unless they represent taxa not otherwise represented (namely the last four taxa in the list below):

Chrysobalanaceae sp. (probably *Magnistipula butayei* subsp. *balingembaensis*)
Strombosia scheffleri
Drypetes sp. 3
Garcinia sp. (probably *G. smeathmannii*)
Anchomanes difformis
Ardisia sp. (probably *A. staudtii*)
Drypetes sp. aff. *leonensis* of Bali Ngemba
Quassia sanguinea
Diospyros zenkeri
Amorphophallus sp. (probably *A. staudtii)*
Raphidiocystis phyllocalyx
Leptonychia sp. 1 of Bali Ngemba
Deinbollia sp. 1
Chrysophyllum sp. (probably *C. gorongosanum*)
Macaranga occidentalis
Psychotria peduncularis
Carapa (probably *C. grandiflora*)
Nephthytis poissonii
Memecylon dasyanthum
Piper capense
Elatostema paivaeanum
Tricalysia sp. B. aff. *ferorum*
Synsepalum msolo
Psychotria sp. A of Bali Ngemba
Peperomia retusa var. *mannii*
Newtonia camerunensis CR
Agelaea pentagyna
Bulbophyllum nigericum VU

Thus four threatened and two undescribed, but likely to prove threatened, taxa are known only from the 1500–1700 m band, in addition to those recorded above from the submontane and *Pterygota* forest types as a whole. No taxa strictly endemic to the 1500–1700 m band are known. Threats and phytogeographical links are discussed above under submontane forest. This band of forest being buffered, so to speak, from both Pinyin and Mantum villages by intervening bands of forest that are more accessible and so more easily exploited, is apparently slightly less damaged and threatened at least in the short term, than other forest bands documented here.

c) Extreme upper submontane forest (1700 to 1900–2000 m)

This division of upper submontane forest is physiognomically distinct from the forest type below due to the absence of *Pterygota* and its altitudinally correlated taxa (see above). The commonest tree species in this band include:

Xylopia africana VU
Magnistipula butayei subsp. *balingembaensis* CR
Drypetes sp. aff. *leonensis* of Bali Ngemba
Beilschmiedia sp. 1

These species occur at lower altitudes, but occur much more densely in this belt.

The tree species restricted to this band are as follows:

Oncoba lophocarpa VU
Oncoba sp. nov.
Cuviera longiflora
Entandrophragma angolense VU
Vepris sp. B
Lychnodiscus grandifolius
Manilkara obovata
Discopodium penninervium
Allophyllus bullatus VU
Ficus ardisioides subsp. *camptoneura*
Ficus exasperata

Climbers and herbs restricted to this band are:

Begonia oxyanthera VU
Raphidiocystis phyllocalyx
Raphiostylis beninensis
Adenia cf. *cissampeloides*
Adenia cf. *lobata*
Commelina cameroonensis VU

Regrettably, no plot was executed within this altitudinal range, partly because of the extent of destruction of the understorey of the areas surveyed.

Oncoba sp. nov. is restricted to this forest type and only known in the wild from Bali Ngemba. The *Drypetes, Beilschmiedia* and *Magnistipula* taxa listed above are also thought to be restricted to Bali Ngemba. Although they also occur at lower altitudes, collection evidence suggests that their main populations are in the 1700 to 1900–2000 m band.

Ternstroemia sp. nov?, if it survives in the wild at all (it is believed extinct at the only two known locations, one near Mt Oku, the other in the Bamboutos Mts) is most likely to be found in this altitudinal band at Bali Ngemba, given the altitudes at its former sites and that Bali Ngemba has the most substantial area of this forest type known to survive in the Bamenda-Bamboutos Highlands.

Vepris sp. B is also restricted to this forest type and is known only from Bali Ngemba. Known only from a single specimen collected in March 1951, it has not been seen since and may not be globally extinct. More efforts are needed to rediscover these taxa at Bali Ngemba.

A particularly notable record is that of *Apodytes dimidiata* (1920 m alt.). This tree is widespread and fairly common in montane East Africa, but very rare west of the Congo Basin. It is not mentioned either in Flore du Cameroun or in FWTA. One previous record has been found from West Province, Cameroon at Mt Bana, and three in adjoining Nigeria. This, it appears, is the first published record for Cameroon of the genus.

The fact that the 1700 to 1900–2000 m forest band is the main hope for the global survival of five tree taxa (*Magnistipula butayei* subsp. *balingembaensis, Vepris* sp. B, *Oncoba* sp. nov., *Beilschmiedia* sp. 1 and *Drypetes* sp. aff. *leonensis* of Bali Ngemba) is more than enough to make conservation of the entire Bali Ngemba forest an important priority. Threats to this band appear to be clearance of forest for agriculture especially for the cultivation of beans (*Phaseolus* spp.) by farmers from above the reserve, to the SE (*Pollard*, field notes 2001, 2002).

MAPPING UNIT 2. MONTANE FOREST, GRASSLAND AND SCRUB–FOREST EDGE (1900–2000 m to 2200 m alt.) (see Fig. 3)

Montane forest remains the least intensively investigated forest at Bali Ngemba, partly because it lay further from our base camp (at Mantum (1200 m alt.)) and partly because much has been destroyed by farming. *Podocarpus milanjianus*, a very distinctive gymnosperm of great phytogeographic interest, which is restricted to montane forest, was recorded here in 1951 by Ujor, but has not been refound by us. It may be locally extinct due to agricultural encroachment.

Canopy trees

Podocarpus milanjianus (possibly extinct locally)
Schefflera mannii VU
Agarista salicifolia
Bersama abyssinica
Ficus oreodryadum
Clausena anisata
Syzygium staudtii
Cassipourea malosana
Coffea liberica
Ixora foliosa VU
Gnidia glauca

The above list is comprehensive, no other tree taxa being known. Montane forest in the Cameroon Highlands is known for being taxonomically depauperate (Cheek *et al.* 2000), but this list is much smaller than is usual for the Cameroon Highlands. A curious phenomenon is that several typically montane species (*Nuxia congesta, Prunus africana, Xymalos monospora* and *Pavetta hookeriana* VU) do occur at Bali Ngemba, but are only recorded at 1400–1500 m alt. Several other trees typical of montane altitudes, *Schefflera abyssinica, Morella arborea* and *Ilex mitis*, have not been recorded at all at Bali Ngemba. While the last two taxa are often rare and easily overlooked, the first is usually common and conspicuous. More field investigation of these anomalies is needed: they may be purely a product of undersampling.

Montane forest climbers

Clematis simensis
Cyphostemma mannii

Montane forest herbs

Epistemma cf. *decurrens*
Impatiens sakerana VU
Girardinia diversifolia
Polystachya anthoceros (epiphyte) EN

Undersampling of montane forest is underscored by Darbyshire's discovery, in April 2004, of an *Epistemma.* This is probably *Epistemma decurrens*, a species known otherwise only from a single specimen at Banyo, in the N of the Bamenda Highlands, in Adamaoua Province. However, there are morphological differences; the Bali Ngemba material may therefore represent a new taxon.

The conservation value of the montane forest at Bali Ngemba seems relatively low, with only four Red Data species recorded. This partly reflects the low level of botanical surveying in this vegetation type.

Threats to montane forest

Threats appear to be clearance of forest for agriculture, especially for the cultivation of beans (*Phaseolus* spp.) by farmers from above the reserve, to the SE.

Montane grassland and scrub – forest edge community; 1900–2200 m alt.

Much of the intact montane area at Bali Ngemba consists of scrub – a mixture of shrubs, often fire-resistant, and grassland. Many of these taxa are restricted to the edge of forest in other parts of the Cameroon Highlands and it is likely that this diverse community is also forest-dependent at Bali Ngemba. Several species are rare and threatened, and their presence suggests strongly that this community is primary, at least in part, and is not derived solely from forest clearance, although its area may well have increased due to this. An explanation for the high proportion of scrub-forest edge vegetation to forest in the montane area may be the prevalence of thin soils over rock, providing too shallow a soil for the root-run of trees. Numerous species typical of thin soils over rock are present in this list, e.g. *Swertia* spp. and geophytes such as *Wurmbea tenuis* subsp. *tenuis, Hypoxis camerooniana* and *Moraea schimperi*.

The list of taxa below is derived by screening the checklist for 'grassland' taxa collected at 1900–2200 m alt. Taxa follow the checklist order:

Uebelinia abyssinica
Lobelia neumannii
Wahlenbergia ramosissima subsp. *ramosissima* VU
Wahlenbergia silenoides
Conyza attenuata
Conyza pyrrhopappa
Conyza stricta
Conyza subscaposa
Helichrysum forskahli
Clutia kamerunica EN
Euphorbia schimperiana var. *schimperiana*
Tragia benthamii
Geranium arabicum
Sebaea brachyphylla
Swertia mannii
Swertia quartiniana
Hypericum peplidifolium
Hypericum revolutum subsp. *revolutum*
Hypericum roeperianum
Psorospermum aurantiacum VU
Haumaniastrum sericeum
Isodon ramosissimus
Leucas oligocephala
Plectranthus kamerunensis
Satureja pseudosimensis
Satureja robusta
Adenocarpus mannii
Dalbergia oligophylla EN
Antopetitia abyssinica
Crotalaria incana subsp. *purpurascens*
Crotalaria ledermannii VU
Crotalaria orthoclada
Crotalaria recta
Eriosema erici-rosenii
Eriosema scioanum subsp. *lejeunei* var. *lejeunei*
Eriosema verdickii var. *schoutedenianum*
Indigofera atriceps subsp. *atriceps*
Lotus discolor
Trifolium baccarinii
Trifolium simense
Trifolium usambarense
Vigna gracilis var. *gracilis*
Vigna parkeri subsp. *maranguensis*
Vigna vexillata
Globimetula oreophila
Tapinanthus globiferus
Antherotoma naudinii
Dissotis bamendae VU
Dissotis longisetosa VU
Polygala tenuicaulis subsp. *tayloriana*
Clematis villosa subsp. *oliveri*
Delphinium dasycaulon
Alchemilla kiwuensis
Oldenlandia rosulata
Otomeria cameronica
Pentas ledermannii VU
Pentas schimperiana subsp. *occidentalis*
Alectra sessiliflora var. *senegalensis*
Rhabdotosperma densifolia
Veronica abyssinica
Triumfetta annua
Agrocharis melanantha
Peucedanum angustisectum
Wurmbea tenuis subsp. *tenuis*
Commelina africana var. *mannii*
Bulbostylis densa var. *cameroonensis* VU
Cyperus mannii
Digitaria abyssinica
Eragrostis atrovirens
Eragrostis volkensii
Hyparrhenia rufa
Panicum hochstetteri
Panicum nervatum
Hypoxis camerooniana
Moraea schimperi
Disa equestris
Disa welwitschii subsp. *occultans*
Satyrium volkensii

Eight threatened taxa are known from montane scrub at Bali Ngemba, several of which (*Clutia kamerunica, Dissotis bamendae, Dissotis longisetosa*) do not occur in any other protected area in the world, apparently making this vegetation type of particular conservation importance.

The main threat to montane scrub at Bali Ngemba is likely to be increasing agricultural encroachment. There is no evidence of grazing pressure, either from reports of collectors, or from presence of grazing-favoured species, such as *Sporobolus africanus*. Nor are many nitrophilous species, e.g. *Uebelinia abyssinica*, on the list above. The extent of the threat posed by arable farming needs further evaluation. It is possible that, if as is likely, the rarest species are confined to the shallowest soils, they are relatively secure, since farmers prefer deep soils for their crops.

Phytogeographic links of the montane vegetation types at Bali Ngemba are primarily with other areas of the Cameroon Highlands which are of sufficient altitude to bear montane grassland and forest edge communities, such as Bioko, Mt Cameroon, Mwanenguba, and, particularly, elsewhere in the Bamboutos-Bamenda Highlands.

Secondary links are with the mountains of East Africa, particularly the largest highland mass (above 2000 m alt.) in tropical Africa: Ethiopia.

MAPPING UNIT 3. SUBMONTANE INSELBERG GRASSLAND (1300–1650 m alt.) (see Fig. 3)

A pair of cone-like, herb-covered, inselbergs occur within the Bali Ngemba Forest reserve (see front cover). They are largely encircled by submontane forest. The largest ascends from c. 1350 m to 1650 m alt. (altimeter readings from Ben Pollard). The slopes vary from 30° to 60° to the horizontal, and achieving their summit can be precarious for collectors. The observations related here all derive from the field notes of Ben Pollard, who made several exploratory visits to the inselbergs in 2000, 2001 and 2002 (see species lists below). Apart from some trees of the savanna species *Entada abyssinica* at the base, the vegetation of these inselbergs consists of a sparse herb sward, sometimes with tussocks, on thin soils over basalt rock, with scattered surface rocks.

These areas coincide with the 'saxicolous communities' mapped by Letouzey (1985) from aerial photography. Sampling has been moderate, but far from comprehensive. No plots have been executed. Gramineae have been under-sampled. The main component of the vegetation are herbs of the Gramineae, Compositae, Labiatae and Scrophulariaceae, together with monocotyledonous geophytes, such as *Gladiolus*.

According to Pollard (pers. comm.), these slopes are not grazed by domestic animals but are subjected to annual burning, probably of both lightning and man-made origin.

Perhaps owing to the underlying rock, and resultant poor drainage, some peaty deposits can occur at the surface, which may result in the similarity with 'Basalt pavement grassland' reported from elsewhere in the Bamenda Highlands (Cheek *et al.* 2000: 18–19), with identical species or genera: *Drosera, Utricularia pubescens, Utricularia scandens, Scleria, Eriocaulon asteroides, Loudetia, Xyris.* These communities probably correspond with seasonal seepages.

Sampling of inselberg grassland has been moderate. Further studies should include plot work, and should attempt to cover the different faces of those inselbergs. Vegetation at the base of the inselbergs in particular would merit further examination.

The following species lists give a series of snapshots of the taxa evident at different seasons.

Pollard 519 *et seq.* 18 Nov. 2000. 1650 m alt.

'Summit of larger of two basalt inselbergs, half a mile SW of Mantum Earthwatch camp. Grassland with exposed basalt outcrops, some peaty deposits. Grasses forming dense tussocks.'

Aeollanthus angustifolius
Gynura pseudochina
Vernonia nestor
Physalis peruviana
Haumaniastrum caeruleum
Eriocaulon asteroides VU
Utricularia scandens
Utricularia striatula
Xyris sp. of Bali Ngemba
Scleria melanotricha var. *grata*
Drosera pilosa VU

Pollard 699 with Nucam & Wian *et seq.* 12 Oct. 2001. 1400–1500 m alt.

Inselberg behind Mantum Earthwatch Camp. 'Steep grassland and rocks with occasional pyrophytic shrubs.'

Aeollanthus angustifolius
Eriocaulon asteroides VU
Stachytarpheta cayennensis
Haumaniastrum caeruleum
Spermacoce pusilla
Eulophia odontoglossa
Sopubia mannii var. *mannii*
Utricularia scandens
Utricularia pubescens
Loudetia kagerensis
Panicum nervatum
Ctenium ledermannii
Digitaria diagonalis var. *diagonalis*
Hyparrhenia niariensis
Plectranthus decumbens
Protea madiensis subsp. *madiensis*

Pollard 972 *et seq.* 13 April 2002.

Bali Ngemba F.R. First granite inselberg behind base camp. 'Steep slopes – rocky grassland.'

Cycnium adonense subsp. *camporum*
Gladiolus unguiculatus
Gladiolus dalenii subsp. *andongensis*
Entada abyssinica

The following hill, lower than the inselbergs, occurs just outside the reserve, overlooking Mantum. It also has thin soils over rock and qualifies as a low-grade inselberg:

Pollard 933 *et seq.* 11 April 2002. 1450 m alt.

Tu Ndab hill just SE of Mantum compound. 'Grassland on shallow soil over rock.'

Aeschynomene baumii
Gladiolus dalenii subsp. *andongensis*
Echinops gracilis
Senecio lelyi
Gladiolus unguiculatus
Vernonia smithii
Senecio ruwenzoriensis
Pseudognaphalium undulatum
Clematis villosa subsp. *oliveri*
Pseudognaphalium oligandrum
Vernonia guineensis var. *cameroonica* VU
Helichrysum nudiflorum var. *nudiflorum*
Sopubia mannii var. *mannii*
Vernonia subaphylla
Ficus glumosa
Drimia altissima

While other of our collectors also surveyed submontane grassland areas at Bali Ngemba, their field notes do not distinguish between derived savanna and inselberg grassland, so their data are not included here.

None of the inselberg species thus far recorded from Bali Ngemba is endemic, while four (or five: see the introduction of the 'Red Data Plant Species of Bali Ngemba Forest Reserve' chapter) are threatened with extinction, usually on the basis of being known from ten or less sites, and being subjected to threats at one or more of these sites. However, no threats to inselberg grassland at Bali Ngemba were recorded, apart from some invasive alien plant species (see the chapter 'Invasive, Alien & Weedy Plants').

The phytogeographic links of this vegetation type are primarily with other areas of similar altitude in the Bamenda Highlands which have shallow soils over rocks. For example, three of the four threatened species are limited to this area. A substantial minority of taxa, however, also occur in other highland areas of Africa, or more usually, those to the E (e.g. *Senecio ruwenzoriensis*).

MAPPING UNIT 4. DERIVED SAVANNA AND FARM FALLOW (1300–1500 m alt.) (see Fig. 3)

a) Derived savanna

West of the area of forest that comprises Bali Ngemba is an area of savanna-grassland. This is representative, to a degree, of the grasslands that dominate the NW Province in Cameroon today, but which are thought to be largely secondary, that is, derived from clearance of the original forest that is now absent, apart from a few small scattered patches such as Bali Ngemba (Hawkins & Brunt 1965). Sampling of derived savanna remains highly incomplete since our focus for surveying was the forest. Although most woody taxa have been sampled, some herbaceous groups, such as Gramineae, are under-represented in the lists below.

The savanna trees that we have recorded at Bali Ngemba are as follows, in order of frequency:

Entada abyssinica
Combretum molle
Vitex doniana

Savanna shrubs:

Protea madiensis subsp. *madiensis*
Kotschya strigosa
Lippia rugosa
Psorospermum febrifugum

Elsewhere in the Bamenda Highlands a more diverse assembly of trees and shrubs is recorded, involving such taxa as: *Syzygium guineense, Annona chrysophylla, Lophira lanceolata, Daniellia oliveri, Hymenocardia acida, Terminalia glaucescens, Nauclea latifolia* and *Piliostigma thonningii* (Hawkins & Brunt, 1965). However savanna with mainly trees of *Entada abyssinica* is representative of lava-rich soils of the plateau according to Hawkins & Brunt (1965).

Parts of the derived savanna are farmed for food crops, larger areas still are fallow.

According to Hawkins and Brunt (1965) the savanna species of the Bamenda Highlands migrated to these higher altitudes, colonising areas denuded of forest, from lower altitude areas to which they are indigenous. However it is also likely that a contribution to these secondary areas was made by natural grassland areas such as the submontane inselberg vegetation discussed above (mapping unit 3). Both these habitats are diverse in the same families: Gramineae, Compositae, Labiatae and Leguminosae-Papilionoideae. The *Entada abyssinica* which Hawkins and Brunt (1965) contend has migrated up from lower altitudes, also occurs apparently naturally, at the base of the inselbergs (see mapping unit 3). In such cases it is difficult to establish the direction of migration. Certainly, some of the 'inselberg' species listed above are undoubted invasive weeds that have originated from the derived savanna-farmbush vegetation. Examples are *Stachytarpheta cayennensis* and *Physalis peruviana.*

In general, the deeper soils of the derived savanna, which once supported forest, now support a much denser, taller herbaceous vegetation than that which inhabits the shallow soils of the inselbergs. Diminutive, sun-loving herbs, such as *Eriocaulon asteroides* and other taxa listed above under inselberg grassland seepages, cannot compete in such an environment.

It is difficult to characterise representative herbaceous species of the 'derived savanna' at Bali Ngemba because it was not a major target of our inventory operation, and because those collectors who did collect in 'submontane grassland' did not distinguish between inselberg and derived savanna, apart from Ben Pollard who concentrated on inselberg vegetation. Nonetheless, the taxa listed below are probably representative of derived submontane savanna at Bali Ngemba (extracted from, and listed in order of appearance in, the checklist):

Sphaerocodon caffrum
Eriosema psoraleoides
Eriosema robustum
Dissotis brazzae
Microcharis longicalyx
Neonotonia wightii subsp. *psudojavanica*
Pseudarthria hookeri
Polygala persicariifolia
Buchnera welwitschii
Cyperus tenuiculmis var. *tenuiculmis*
Lipocarpa nana
Elymandra androphila
Hyparrhenia diplandra
Hyparrhenia umbrosa
Loudetia arundinacea

Species common to submontane derived grassland (1200–1500 m alt.) and to montane grassland and scrub (1900–2000 to 2200 m alt.)
The following taxa, extracted from the checklist, occur both in mapping units 2 and 4:

Echinops gracilis
Crotalaria subcapitata subsp. *oreadum*
Eriosema montanum var. *montanum*
Kotschya strigosa
Protea madiensis subsp. *madiensis*
Sopubia mannii var. *mannii*
Cyanotis barbata
Cyperus sesquiflorus subsp. *cylindricus*

No endemic and no threatened species are known from derived submontane savanna. No threats are known to this vegetation type. To the contrary, at Bali Ngemba, owing to human activity, derived savanna is set to expand at the expense of submontane forest, sadly.

The phytogeographic links of derived savanna, according to Hawkins and Brunt (1965) are mainly with lowland altitudes to which this community is indigenous. However, it also seems logical that indigenous submontane inselberg communities have also contributed taxa to this community.

b) Fallow in derived savanna

Farm fallow, and other ruderal habitats, were incompletely surveyed since our main focus was natural habitat. The following list gives a snapshot of some of the species involved.

Pollard 399 et seq. 3 Nov. 2000. 1340 m alt.

'Roadside and fallow. Mantum village.'

Oldenlandia herbacea
Eriosema robustum
Stachytarpheta cayennensis
Eriosema psoraleoides
Crotalaria subcapitata subsp. *oreadum*
Desmodium intortum

MAPPING UNIT 5. *EUCALYPTUS* PLANTATION (1300–1600 m alt.) (see Fig. 3)

A large area of the Bali Ngemba Forest Reserve has been planted up with *Eucalyptus*, presumably to provide quick-growing poles and firewood. The understorey consists of native herbaceous taxa. The plantation was not intensively surveyed by us since it was not seen as likely to be a priority for conservation.

Pollard 593 et seq. 4 Oct. 2001. 1560 m alt.

'Nchukob, track in *Eucalpytus* forest.'

Phyllanthus mannianus
Virectaria major var. *major*
Streptocarpus elongatus
Plectranthus tenuicaulis
Pilea angolensis var. *angolensis*
Parietaria debilis
Urena lobata
Margaritaria discoidea var. *discoidea*
Crassocephalum montuosum
Senna septemtrionalis
Asplenium dregeanum
Elephantopus mollis
Achyrospermum africanum
Dioscorea schimperiana
Cissus diffusiflora
Piper capense
Achyranthes aspera var. *sicula*
Aneilema umbrosum subsp. *ovato-oblongum*
Chlorophytum sp.
Justicia paxiana

None of the species listed above is threatened, nor are any endemics or undescribed taxa known from this vegetation type.

REFERENCES

Cheek, M., Onana, J.-M. & Pollard, B.J. (2000). The Plants of Mount Oku and the Ijim Ridge, Cameroon, A Conservation Checklist. RBG, Kew, UK. iv + 211 pp.

Cheek, M., Pollard, B.J., Darbyshire, I., Onana, J.-M. & Wild, C. (eds) (in press). The Plants of Kupe, Mwanenguba & the Bakossi Mts, Cameroon, A Conservation Checklist. RBG, Kew, UK.

Hawkins, P. & Brunt, M. (1965). The Soils and Ecology of West Cameroon. 2 Vols. FAO, Rome. 516 pp, numerous plates and maps.

Keay, R.W.J., ed. (1958). Flora of West Tropical Africa: Volume 1. Second edition. Crown Agents, London, U.K.

Lebrun, J. (1935). Les essences forestières des régions montagneuses du Congo oriental. Institut national pour l'étude agronomique du Congo Belge, Bruxelles, Belgium. 263pp.

Letouzey, R. (1968). Les Botanistes au Cameroun. Flore du Cameroun: 7. Museum National d'Histoire Naturelle, Paris. 110 pp.

Letouzey, R. (1985). Notice de la carte phytogéographique du Cameroun au 1: 500,000. IRA, Yaoundé, Cameroon.

GEOLOGY, GEOMORPHOLOGY, SOILS & CLIMATE

Martin Cheek
Herbarium, Royal Botanic Gardens, Kew, Surrey, TW9 3AE, U.K.

Bali Ngemba Forest Reserve is positioned facing west in a concave part of the steep scarp slope that connects the High Lava Plateau (1500–2100 m alt.) of the Bamenda Highlands with the Low Lava surface (900–1200 m alt.) on which the former appears to rest; see Fig. 1 (Hawkins & Brunt 1965). These basalts and trachytes are of ancient volcanic origin (Courade 1974). The Bamenda Highlands themselves are the most extensive of the mountainous areas, usually known as the Cameroon Highlands, that begin with the Atlantic Ocean islands of São Tomé, Príncipe, Annobon and Bioko and proceed NE, in a band 50–100 km wide. These mountains have their origin in a geological fault and are formed largely of igneous material. Three main periods of volcanic activity and one of plutonic uplift have been reviewed by Courade (1974). The Bamenda Highlands were largely formed in the Tertiary, in the second of the three main periods of volcanic activity 'the middle white series'. Courade classifies the soils in the area of Bali as being ancient, ferralitic, and of volcanic origin, being of medium fertility (3rd class out of 4 possible classes). Tye (1986) refers to soils in the Bamenda Highlands area as being 'clayey', resulting in more permanent streams than some of the mountain areas of more recent origin in the Cameroon Highlands, such as Mt Kupe and Mwanenguba.

Courade's rainfall map shows the area in which Bali Ngemba falls being somewhere between the 1 m and 2 m isohyets. However, rainfall data for a meteorological station at Bali, only 5 km W and at the same altitude as the lower part of Bali Ngemba (1300 m alt.) gives an annual total of c. 2.25 m (Hawkins & Brunt 1965). Most of this falls in the single long wet season in March–October inclusive, when 20 cm or more falls in each month, only 3–9 cm falling in each of the months from Nov. to Feb. when average max. monthly temperatures are 26.7–27°C, the coolest months being 21.8°C and 22.6°C. Average minimum monthly temperatures range from 11.1°C to 14.4°C. Temperatures at Bali, and at Bali Ngemba are partly depressed by temperature inversion, cold air draining off the high lava plateau (Hawkins & Brunt 1965: 27).

REFERENCES

Courade, G. (1974). Commentaire des cartes. Atlas Régional. Ouest I. ORSTOM, Yaoundé.

Hawkins, P. & Brunt, M. (1965). The Soils and Ecology of West Cameroon. 2 Vols. FAO, Rome. 516 pp, numerous plates and maps.

Tye, H. (1986). Geology and landforms in the Highlands of Western Cameroon: 15–17. *In* Stuart, S.N. (ed) (1986). Conservation of Cameroon Montane Forests. International Council for Bird Preservation, Cambridge, UK.

THREATS TO BALI NGEMBA F.R.

Martin Cheek
Herbarium, Royal Botanic Gardens, Kew, Surrey, TW9 3AE, U.K.

The following paragraph, taken from the BirdLife survey of forest sites in the Cameroon Highlands summons up the ambiguity of Bali Ngemba's position with regard to protection (Mackay 1994).

> 'Bali Ngemba (5°50'N, 10°7'E)
> This is the only site where there is continuous forest from low to high altitude (approximately 1400–2000 m), though most of this forest is degraded to some extent. Bannerman's Turaco was common above 1800 m, and Green Turaco below 1800 m. In one forest area at 1850 m Banded Wattle-eye and Scarlet-spectacled Wattle-eye occurred together. The forest is still a government forest reserve, but perhaps only in name. Most of the forest understorey has been cleared, and the fertile forest soils are used for growing coco-yams. Probably the only reason that the tree canopy remains is that coco-yams favour shady conditions, so the trees have not been felled for fuelwood. There are forest guards here, but they are powerless to stop the farmers. The conservation of this site may be facilitated by its forest reserve status, but its future looks fairly bleak at the moment, except for forests on the most inaccessible slopes.'

This statement seems as true today as when it was written. However, by comparison with other mountain forest patches assessed by Mackay (1994), Bali Ngemba is fortunate in having only suffered relatively slow deterioration. When these patches were re-surveyed by BirdLife 6 years after Mackay, some key areas of forest had almost entirely disappeared, such as Kejodsam. Satellite imagery from 1987 showed Kejodsam forest to be 30 km^2, but only c. 5 km^2 survived in 1994, according to Mackay, who believed then that it 'may have a long term future'. However, the latest BirdLife survey (Yana Njabo & Languy 2000) showed that only c. 5 Ha remained (a decline of 90% in forest area over 6 years), and that the outlook of this remnant was bleak.

The decline of Bali Ngemba forest, although minor in comparison with that suffered by Kejodsam, is nonetheless a cause of great concern, given its great biological importance as the most significant area of submontane forest known to survive in the Bamenda Highlands today.

Since the latest BirdLife survey (2000), the only objective observations on forest extent and health at Bali Ngemba are those gathered during our expeditions there. Ben Pollard, who spent 1–2 weeks at Bali Ngemba in 2000, 2001 and 2002, reported ever-increasing areas of cocoyam farming in the forest understorey (pers. comm.). Darbyshire (2004), leader of the most recent expedition to Bali Ngemba, reported active small-scale logging.

Pressures on Bali Ngemba appear to come from two centres of population, each in a separate tribal area (Fig. 4). To the west (at low altitude) is Bali, with a satellite village, Mantum, at the very foot of the forest. Mantum is probably the main base for small-scale logging. To the south, at high altitude, is Pinyin. This is believed to be the main base for those cultivating the understorey of the forest for cocoyams and clearing the upper reaches of the forest for agriculture. Since the forest is officially government, and not tribal (traditional) property, protection from the traditional authorities, such as exists at, and is so effective at preserving other areas of forest in the Bamenda Highlands, notably at Kilum-Ijim (Mt Oku and the Ijim Ridge) does not pertain. Instead, Bali Ngemba is a tribal no-man's-land, weakly guarded and so open to abuse. 'If we don't exploit the forest, others will anyway', seems to be the feeling on the ground. It is hoped that this book, by expounding the biological importance of the Bali Ngemba forest, will enhance its protection by both traditional and governmental authorities.

REFERENCES

Mackay, C.R. (1994). Survey of Important Bird Areas for Bannerman's Turaco *Tauraco bannermani* and Banded Wattle-eye *Platysteira laticincta* in North West Cameroon, 1994. Interim report. BirdLife Secretariat.

Yana Njabo, K. & Languy, M. (2000). Surveys of selected montane and submontane areas of the Bamenda Highlands in March 2000. Cameroon Ornithological Club, Yaoundé. Cyclostyled.

INVASIVE, ALIEN & WEEDY PLANTS

Martin Cheek
Herbarium, Royal Botanic Gardens, Kew, Surrey, TW9 3AE, U.K.

Invasive plants are usually non-native species that invade usually natural habitat. They are of concern because they can displace native species, and so threaten them.

Aliens are species that are not native. They are not necessarily invasive or weedy but are of interest because they may become so. Also termed exotics, many have been deliberately introduced for amenity horticulture, as beautiful flowering plants, while others as potential crop or fodder plants, and others inadvertently, often as seeds in the soil of intentionally introduced plants.

A major potential source of aliens in the Bamenda Highlands was a programme of introducing alien plant species to grasslands in order to improve their yield of fodder for domestic animals. This programme was executed at the Bambui Agricultural Station and at the Jakiri Livestock Improvement Centre. Details of the legume and grass species that were trialled are given in Hawkins & Brunt (1965). None of the species listed there appears to be alien at Bali Ngemba, however.

Weeds are any plants growing in what a person considers the 'wrong' place. This usually refers to plants growing in cultivated situations which have not been planted but are spontaneous, and may threaten the yield of, e.g., crop plants by competing for water, space, nutrient or light resources. Many weeds are pioneers in natural, disturbed situations, natives that have transferred to cultivated habitats. Other weeds are aliens.

It was not the main intention of our plant survey at Bali Ngemba to document invasive, alien or weedy species. Indeed, it is highly likely that some common weedy species have been omitted from this checklist. However, lately there has been much international interest in this subject, primarily because of the problems that invasive species can pose to natural habitats, and also because of the effect that weed species can have on the yields of crops. In both cases it is considered important to record the presence of invasive species and of weeds in order to document their spread, and also to record the existence of alien species, since these can become invasive or weedy in future years.

INVASIVE SPECIES

No invasive species were documented in forest habitats. However in inselberg grassland, the following two taxa were recorded growing in undisturbed natural habitat:

Stachytarpheta cayennensis (*Pollard* 701)

Physalis peruviana (*Pollard* 523)

Both are neotropical in origin, and already fairly common as weeds of cultivation in western Cameroon.

ALIEN TAXA

The following taxa are newly recorded here as aliens for Cameroon. Apparently they were growing in disturbed, not pristine, areas but more observations may show that they are also invasives or weeds, or both.

Rumex crispus (*Pollard* 461) appears to be previously unrecorded from the Gulf of Guinea. A European species, it is a significant arable weed in Australia and needs monitoring in Cameroon if it cannot be eliminated.

Caesalpinia decapetala (*Etuge* 5416) is not listed in the Flore du Cameroun account of 'Césalpiniïoidées', so is likely to have appeared in Cameroon only in the last few decades. A native of SE Asia (Keay 1958: 481), it is turning up throughout the tropics having been introduced as a hedge plant and then escaping (G. Lewis pers. comm.). The specimen referred to was recorded in 'secondary forest, near Mantum', so there is a particular concern that it may be beginning a career as an invasive species.

Oxalis latifolia var. *latifolia* (*Pollard* 956) a native of the neotropics, is a new record for Cameroon. Darbyshire observed this plant in Bali Ngemba in April 2004 when he found it to be very common in farmed areas in forest at 1650–1700 m alt. It may well be on the brink of joining the portfolio of existing weeds of the Cameroon Highlands.

WEEDY ALIENS

The following is a very incomplete treatment.

Biophytum umbraculum (*Onana* 1889) is a palaeotropical submontane weed.

Rubus rosifolius (*Ghogue* 1049, *Pollard* 415, 919a), introduced from Asia, presumably for its edible fruit, has become widely naturalised in the Cameroon Highlands, extending down to Mt Cameroon.

Desmodium uncinatum (*Biye* 109) a native of S. America.

Desmodium intortum (*Onana* 1801, 1802) also a native of S. America.

NATIVE WEEDS

The following list follows the order of appearance in the checklist.

Asystasia gangetica
Brillantaisia vogeliana
Achyranthes aspera
Impatiens burtoni
Cleome iberidella
Drymaria cordata
Ageratum conyzoides
Conyza attenuata
Emilia coccinia
Galinsoga parviflora
Ipomoea involucrata
Momordica foetida
Antopetitia abyssinica
Neonotonia wightii subsp. *pseudojavanica*
Pseudarthria hookeri
Urena lobata
Oldenlandia herbacea
Spermacoce pusilla
Veronica abyssinica
Triumfetta annua
Laportea ovalifolia
Parietaria debilis
Commelina benghalensis

Most of the weeds mentioned in the list above appear in the standard works on the subject: Ivens *et al.* (1978), Okezie Akobundu & Agyakwa (1987) and Terry (1983).

REFERENCES

Hawkins, P. & Brunt, M. (1965). The Soils and Ecology of West Cameroon. 2 Vols. FAO, Rome. 516 pp, numerous plates and maps.

Ivens, G.W., Moody, K. & Egunjobi, J.K. (1978). West African Weeds. Ibadan, Oxford Univ. Press. x + 225 pp.

Keay, R.W.J. (1958). Flora of West Tropical Africa: Volume 1. Second edition. Crown Agents, London, UK.

Okezie Akobundu, I. & Agyakwa, C.W. (1987). A Handbook of West African Weeds. International Institute of Tropical Agriculture, Ibadan. 521 pp.

Terry, P.J. (1983). Some Common Crop Weeds of West Africa and their control. USAID, Dakar. 132 pp.

THE EVOLUTION OF THIS CHECKLIST

Martin Cheek
Herbarium, Royal Botanic Gardens, Kew, Surrey, TW9 3AE, U.K.

As 'The Plants of Mt Oku & the Ijim Ridge' (Cheek *et al.* 2000) neared completion, the decision was taken to continue our botanical survey work in NW Province. Bali Ngemba Forest Reserve was selected as a focus for our efforts because it has the largest block of submontane forest known to survive in the Province's Bamenda Highlands, despite being less than 1000 Ha in extent. For this reason it is of great biological importance. Furthermore this forest was almost unknown botanically, and was, as it still is, under great threat from agriculturalists and loggers. It was hoped that a survey would enable, for the first time, the main vegetation types, and the species present in the forest, to be catalogued. This would enable the species to be assessed for threats, allowing a Red Data treatment to be produced, if needed.

Our first expedition to Bali Ngemba F.R. was made in October 2000, through the offices of the Bamenda Highlands Forest Project. From the outset the plan was to publish the results as a conservation checklist, as part of the series represented by Cable & Cheek (1998) and Cheek *et al.* (2000). This checklist for Bali Ngemba is one of three such outputs for the project 'Conservation of the Plant Diversity of Western Cameroon' (1999–2004) for which the main sponsor is the Darwin Initiative of the British Government. Further expeditions to Bali Ngemba were carried out in November 2001 and April 2002. Once duplicates from all three expeditions had reached RBG, Kew and were available for determination in the autumn of 2003, work began in earnest on the checklist, under the leadership of Yvette Harvey. By March 2004 the first draft of this checklist was completed. New and overlooked material then came to light, of which the analysis and incorporation into a new draft took further time. Completion of the first draft revealed probable gaps in coverage of the taxa, e.g. Orchidaceae, so a further expedition was executed in April 2004 with an orchid specialist in attendance. Finally, the last of the introductory chapters were completed in the autumn of 2004.

REFERENCES

Cable, S. & Cheek, M (1998). The Plants of Mount Cameroon, A Conservation Checklist. RBG, Kew, UK. lxxix + 198 pp.

Cheek, M., Onana, J.-M. & Pollard, B.J. (2000). The Plants of Mount Oku and the Ijim Ridge, Cameroon, A Conservation Checklist. RBG, Kew, UK. iv + 211 pp.

HISTORY OF BOTANICAL EXPLORATION OF BALI NGEMBA FOREST RESERVE

Martin Cheek, Benedict John Pollard, Iain Darbyshire & Yvette Harvey
Herbarium, Royal Botanic Gardens, Kew, Surrey, TW9 3AE, U.K.

Apart from our botanical surveys at Bali Ngemba (2000–2004), very little previous botanical exploration at this forest is known. Remarkably, other prolific specimen collectors, such as Letouzey, Thomas, Keay, and Hepper, although visiting the Bamenda Highlands, passed by Bali Ngemba in favour of larger and higher-altitude forest patches, such as Kilum-Ijim (Mt Oku) or the then larger and more accessible Bafut Ngemba F.R. (now almost bereft of natural forest).

Edwin Ujo Ujor (1918–?), Nigerian, is the earliest collector of plant specimens at Bali Ngemba, as far as we know. He was a roving collector for the Forest Herbarium Ibadan (FHI), Nigeria between 1945–1953, collecting in S. Nigeria, and, mainly Western Cameroon, becoming a Forest Guard in 1954. At that time Western Cameroon, the SW and NW Provinces of present day Cameroon, was administered by Nigeria, so forest inventory and management were co-ordinated and directed from Ibadan. For this reason the top set of specimens is at FHI, with other sets at FHO and K. The above biographical data is taken from Hepper & Neate (1971: 82), no other source being known. The current curator of FHI, Mrs Omokafe Ogbogu, has not been able to provide additional information on Ujor, nor have enquiries to FHI, nor searches of biographies held at the library of RBG, Kew, nor internet searches. Our specimen database shows that Ujor collected 49 numbers from Bali Ngemba in May and June of 1951, these being under the FHI number series 30322 to 30440. Some of these numbers contain more than one taxon, e.g. FHI 30324 consists of three orchids: *Brachycorythis ovata* var. *schweinfurthii*, *Satyrium volkensii* and *Disa welwitschii* subsp. *occultans*. Apart from this, his specimens mostly appear to be in chronological order and of good quality. Many of these are cited in FWTA Amongst his gatherings appear many of the Bali Ngemba taxa that we now know to be threatened, e.g. *Allanblackia gabonensis*, *Allophyllus conraui*, or to be new to science, e.g. *Deinbollia* sp. 1, *Beilschmiedia* sp. 1. His single most interesting specimen is that of a tree, *Vepris* sp. B, (Ujor FHI 30422), which appears to represent a new species to science known only from Bali Ngemba, but which has not been seen since he collected it over 50 years ago, despite extensive survey efforts in NW Province by our RBG Kew–National Herbarium of Cameroon teams in recent years. Apart from his collections at Bali Ngemba, he also collected specimens in other forest reserves, principally in today's NW Province.

D.A. Tiku (fl. 1951–1965) also listed as Tiko, probably Cameroonian, was a forest guard (Hepper & Neate 1971: 80) in the service of the Forest Department of Nigeria (which then administered the forests of W. Cameroon). He collected four specimens at Bali Ngemba in the first week of June 1951, coinciding with Ujor's last week of collecting, according to our specimen database. These specimens were:

FHI 30413 *Bersama abyssinica*
FHI 30417 *Kigelia acutifolia* (treated here as a synonym of *K. africana*)
FHI 30418 *Trichilia gilgiana*
FHI 30421 *Leptaulus daphnoides*

These were gathered on the 1st, 5th and 6th June. Two were recorded as being collected with Ujor (see above) and Dioh (see below), so it seems likely that these Forestry Department staff worked together, assisting Ujor in his survey of Bali Ngemba plants, Dioh as an officer rather than a guard, being presumably the senior, with overall management responsibility for this forest, together with Bafut Ngemba, Tiku probably being the guard for Bali Ngemba. These suppositions need verification in local archives.

Dioh (fl. 1951–1958) is recorded by Hepper & Neate (1971: 25) as being a Cameroonian forestry officer, based at Bafut Ngemba F.R. at which he assisted Hepper's botanical survey in 1958, collecting a few numbers in Hepper's series, these specimens being at K. He is included in this chapter because he collected at Bali Ngemba with Tiku and Ujor on 5th and 6th June (e.g. Tiku FHI 30421 'with Dioh and Ujor'). He is assumed to have been senior and to have acted as local host to Ujor (see note under Tiku).

Benjamin Obajide Daramola (1933–), Nigeria, is recorded by Hepper and Neate (1971: 23) as making numerous and extensive collections under FHI numbers in both Nigeria and Western Cameroon, his specimens being present at FHI, FHO, K, P and WAG. He is known to have collected three specimens from Bali Ngemba, 14th and 15th February 1959 as follows:

FHI 40476 *Impatiens hians* var. *hians*
FHI 40478 *Albizia gummifera* [*Albizia gummifera* var. *gummifera*]
FHI 40479 *Vernonia conferta* [*Vernonia doniana*]

None of these is particularly noteworthy. Since only three specimens over two days were gathered, the evidence suggests that these result from a quick, overnight reconnaissance visit by Daramola. Daramola has continued to collect plants in Nigeria during his retirement, usually targeting material of species requested by western scientists for laboratory research. He is believed to be active at the time of writing.

During our series of expeditions (2000–2004), but independent of them, a botanical expedition was also made to Bali Ngemba by Guy Chiron and Josiane Guiard of France, with the assistance of John DeMarco of the Bamenda Highlands Forest Project. The expedition was also based at Mantum, but was focused purely on orchids. The fieldwork was executed from 26 March until 6 April 2001, and was written up and published later that year (Chiron & Guiard 2001). 138 orchid plants were seen during this time, most of which were placed in cultivation by DeMarco and/or Guiard, and 82 of which were made into herbarium specimens, to be deposited at the National Herbarium of Cameroon, Université Claude Bernard, and K (Chiron & Guiard 2001). Fifty orchid taxa are listed, but about 20 incompletely identified taxa (e.g. *Bulbophyllum* spp. 1–4) were also found. Amongst these, 38 taxa were recorded for Bali Ngemba for the first time and were not found by our own expeditions.

BOTANICAL INVENTORIES OF THE BALI NGEMBA FOREST RESERVE, 2000–2004

Between 2000–2004, RBG, Kew, in collaboration with the Herbier National Camerounais, mounted four expeditions to Bali Ngemba Forest Reserve. These took place in 2000, 2001, 2002 and 2004. The following details are extracted from the relevant expedition reports. The names of the collectors who had their own series on these expeditions appear in bold.

Note: [Comments in square brackets have been added by YH, Sept. 2004] and Bali Ngemba F.R. has been abbreviated to BNFR in places.

Bali Ngemba Inventory 1: 8th–19th November 2000 (Martin Cheek)

Organised by RBG, Kew (funded by Earthwatch Institute) in collaboration with the Herbier National Camerounais (HNC-IRAD, Cameroon) (funded by the Darwin Initiative, DETR, British Govt. under the project for the conservation of the plant diversity of Western Cameroon) with the Bamenda Highlands Forest Project (administered by BirdLife International, funded by the National Lottery Commission, U.K.).

Fieldworkers:
ONADEF: Kpoumie Chouaibou (Provincial Chief of Bureau, Protected areas, NW Province), Chief of Post Bali, John Achu (Forest Guard)
Observer from HRH the Fon of Bali: Ba Doh Lawumma
Herbier National Camerounais, Yaoundé: **Jean-Paul Ghogue** (Technician, Leader HNC component), **Fulbert Tadjouteu** (Technician), Felicité Nana (Secretary), Boniface Tadadjeu (Driver).
Univ. Yaounde I, **Elvira Biye** (Asst. Lecturer, Botany).
RBG, Kew: **Martin Cheek** (Leader), **George Gosline** (Research Associate), **Ben Pollard** (Scientific Officer).
Earthwatch volunteers: Kate Johnston (Australia), Julia Garcia (Spain), Joe Carson, Hugh Guilbeau (USA); EW BAT fellows: Pettula Nabukeera (Uganda), Raza Zulfiqar (Pakistan), Christo van de Rheede (South Africa).
Earthwatch-funded Cameroonian staff, camp organisers: **Martin Etuge** (Botanist), Freddy Epie Emmanuel Ebong (Asst. botanist); Martin Mbong (administrator); Japhet Wain (driver).
Earthwatch-funded local workers from Mantum: Emmanuel Sama, Maurice Babila, Eric Nucam, Peter Nkankano (guides); Lilian Mankety, Patricia Lorga, Pamela Wolwonwo (cooks); Silvia Mbutigig Sakaah (cleaner, clothes washing); Yannick Njoya (water carrier); Laurence Tanketi (wood carrier).
Bamenda Highlands Forest Project: Innocent Wultoff, Ernest Keming, Clement Toh (Ecomonitors, KIFP); Grace Tima (Asst. Project Manager, BHFP); Clare Trinder (Env. Education Officer, BHFP); Vincent Fomba (Staff forester BHFP/MINEF); Ndamsa, driver BHFP; Rita Ngolan, Delphine Wanduku, Kenneth Tah (volunteers sent for training through BHFP); Francis Njie (Bird Survey team, BirdLife International); Clare Wirmum (Mbiame Forest).

ITINERARY

Weds. 8th Nov. 2000
Presentations on basic collecting principles and materials (Etuge & Cheek). To the forest, to 'first valley' c. 1400 m. Courtesy visit led by Gosline to Mantum quarterhead and Fon of Bali.

Thurs. 9th Nov. 2000
Divide into three teams: team 1 (low altitude trees): Etuge, Gosline, Maurice, v.d. Rheede, Zulfiquar, Epie; team 2 1500 m alt.: Ghogue, Johnston, Garcia, Chief of Post, Fons Representative and others; team 3 ('first valley' 1400 m): MC, BP, Emmanuel Sama, Peter Achu, Pettula Nabukeera, and others.

Fri. 10th Nov. 2000
To Bamenda [not reproduced here].

Sat. 11th Nov. 2000
One team to top (2000 m) of Bali Ngemba (Sama, Njie, Pollard, Ghogue, v.d. Rheede, Garcia, Nabukeera), other team data-basing and sorting specimens in am, pm to Bamenda to meet BHFP group.

Sun. 12th Nov. 2000
Rest day. Report writing. Data entry and specimen sorting.

Mon. 13th Nov. 2000
Team A to Nkom-Wum forest reserve [not reproduced here].
Team B continue in Bali Ngemba, investigating along first trail to summit ridge as far as 1900 m contour, where extensive farming occurs. Collection of possibly new *Beilschmiedia* [*Beilschmiedia* sp. 1 of Bali Ngemba, *Cheek* 10522], *Oncoba* sp. nov. [*Cheek* 10489] of Kilum-Ijim, thought extinct in the wild, discovered alive near cave at 1850 m.

Tues 14th Nov. 2000
Team A continue at Nkom-Wum.
Team B teaching session on woody plant identification at 1400 m. Collection of lower altitude tree species. Searching for additional plants of *Oncoba* (none found, apart from a dead plant at c. 1900 m alt. in recently cleared farm). v.d. Rheede, Garcia, Carson, Njie, Wanduku and Sama continue collecting at 1900–2000 m and stay night here.

Weds. 15th Nov. 2000
Team A continue at Nkom-Wum.
Team B reunite at 1900 m (Two Cups). Examine specimens collected from high altitude, including *Clutia kamerunensis* [*Clutia kamerunica*, *Cheek* 10523 (collected 14.xi.2000) & *Pollard* 512 (collected 14.xi.2000)], a Red Data candidate unknown from Mt Oku, but endemic to the Bamenda Highlands.

Thurs. 16th Nov. 2000
Team A continue at Bali Ngemba, investigating second trail to summit ridge; most forest along route found to be replaced by farms.
Team B to Oku-Elak [not reproduced here].

Fri. 17th Nov. 2000
Team A: part 1 spend day at public meeting in Mantum with Fon of Bali, explaining conservation work at Bali Ngemba. Earthwatch XI 2 vs Mantum XI 1 (Football match); part 2, reconnaissance of Mbei Royal Forest (via Santa) led by Ghogue; part 3, further exploration of lowland forest in Bali Ngemba by Etuge-led team.
Team B to Oku-Elak [not reproduced here].

Sat. 18th Nov. 2000
Team A; part 1 (Etuge) collect in forest along second trail, located fruiting population of ?*Gastrodia*; part 2 (Pollard, Johnston, Gosline) climb to top of inselberg, locate first Cameroonian population outside of Mt Oku area of *Eriocaulon asteroides* [*Pollard* 525], *Xyris* sp. nov. [*Xyris* sp. of Bali Ngemba, *Pollard* 528], interesting *Drosera* [*Drosera pilosa*, *Pollard* 531] and unusual *Protea* population [*Protea madiensis* subsp. *madiensis*]; part 3, complete specimen sorting and data entry (led by Guilbeau).
Team B. to Oku [not reproduced here].

Sun. 19th Nov. 2000
Rest day.

Summary of collections made in 2000:

Collector	Opening no.	Closing no.
Biye	75	137
Cheek	10431	10547
Etuge	4690	4820
Ghogue	1000	1153
Gosline	270	282
Pollard	386	532
Tadjouteu	395	432

Total number of collections from Bali Ngemba F.R.: 468

Bali Ngemba Inventory 2: 4th–13th October 2001 (Martin Cheek & Benedict Pollard)

Organised by RBG, Kew (funded by Earthwatch Institute) in collaboration with the Herbier National Camerounais (HNC-IRAD, Cameroon) (funded by the Darwin Initiative, DETR, British Govt. under the project for the conservation of the plant diversity of Western Cameroon) with the Bamenda Highlands Forest Project (administered by BirdLife International, funded by the National Lottery Commission, U.K.).

Abbreviations: HNC - Herbier National Camerounais; BHFP - Bamenda Highlands Forest Project; EW -Earthwatch Institute; KIFP- Kilum-Ijim Forest Project; LBG – Limbe Botanic Garden; ONADEF - National Office for the Development of Forestry; MINEF - Ministry of the Environment and Forests.

Fieldworkers:
MINEF: Patrick Fonji, Chief of Post Bali (two days); Waindah Nkemnya Mathias, observer for the Provincial Service of Forestry (duration of the expedition).
Herbier National Camerounais: Yaoundé: **Jean-Michel Onana** (Senior Researcher, Leader HNC component), Victor Nana (Technician).
RBG, Kew: **Martin Cheek** (Leader, UK), **Ben Pollard** (Scientific Officer, UK), Parmjit Bhandol (Technician, UK), Laszlo Csiba (ex RBG, Kew Diploma Student, Hungary).
Earthwatch volunteer: Michael DeVille (USA).
Earthwatch Rio Tinto Corporate Fellows: Agus Renggana (Indonesia), Ian Rowe (Australia), Mark Howson (UK).
Earthwatch Millennium Fellows: Carola Girling (UK), Elizabeth Merritt (UK).
Earthwatch-funded Cameroonian staff: camp organisers: **Martin Etuge** (Botanist), Edmondo Njume (Assistant botanist); Martin Mbong (camp manager), Robert Mesumbe (apprentice camp manager); Japhet Wain, Dansala Djire (drivers).
Earthwatch funded local workers from Mantum: Henry Nsonfon, Eric Nucam, Peter Nkankano (guides); Lilian Mankety Tanketi , Patricia Lorga Tanketi, Lovelyn Agem (cooks); Silvia Mbutigig Sakaah (cleaner, clothes washing); Vincent Tanketi (water carrier); Laurence Tanketi (wood carrier).
Local workers at Nkom-Wum F.R. at Bu in house rented from James Wakai: Belinda Mbong, Antoinette Mbong (cooks); Richard Walu Mbei, Dong Emmanuel Towah (guides); from Mbengkas: David H., Lah Mathew Loi (guides); Ngoh Jonathan Mbwai (observer for the Fon of Mbengkas).
Bamenda Highlands Forest Project: Walters Cheso (Environmental Education Officer); **John DeMarco** (co-Project Manager), Clare Trinder (Environmental Education Officer, BHFP: Bambili Lakes); Terence Suwinyi, Kenneth Tah (volunteers, BHFP).
Kilum Ijim Forest Project: Innocent Wultoff, Ernest Keming, Rita Ngolan, Clement Toh (Ecomonitors, KIFP).
Limbe Botanic Garden: **Philip Nkeng** (forester, Mount Cameroon Project), **Raphael Kongor** (curator, Limbe Herbarium).

ITINERARY

Thurs. 4th Oct. 2001
Presentations on basic plant specimen collecting principles and use and care of equipment of materials (ME & BP). To the Bali Ngemba F.R. forest, up main path towards Jerusalem Rock, at c. 1400 m. A potential Red Data species provisionally identified was *Allophyllus conraui*. This is a spiny, crenate-dentate *Allophyllus*, apparently known from only 4–5 collections, all in the vicinity of Bali Ngemba. Light rain in night; moderate rain in early pm.

Fri. 5th Oct. 2001
Team A led by JMO (with VN, CG, ME, MH, PF, EK, Raphel Kongor, LC guided by Henry Nsonfon) up path towards Jerusalem Rock, stopping below it.

Teams B & C (ME, BP, MC, PN, AR, IR, WC, RN, RK, guided by ENU and PNK) united to do 'forest profile' on path to Kedeng at c. 1400 m together with general collecting. Moderate rain at 2pm, heavy rain at 4 to 5 pm.

Sat 6th Oct. 2001
Teams A-C collect at Baba II [not reproduced here].

Sun. 7th Oct. 2001
Rest Day. One LR to Bamenda to deliver RN and EK, other to Bali for Church and swimming pool led by BP. BP takes group to visit Fon of Mantum Village. Libation and traditional dances performed. MH upholds Earthwatch honour with a couple of west-country songs for the elders. MC paperwork.

Mon. 8th Oct. 2001
Expedition divides into two groups until Friday pm. One party remains at Bali Ngemba, the other travels to Bu to investigate the Nkom Wum F.R. [not reproduced here].
Team A (JMO, VN & Henry Nsofon) to Jerusalem Rock.
Team B (BP, Mike DeVille, Carola Girling, Liz Merritt, & PN make courtesy visit to District Officer and Gendarmarie. Followed by further work on forest profile strip in Bali Ngemba.Training by BP.

Tues. 9th Oct. 2001
All to forest at Baba II [not reproduced here].

Weds. 10th Oct. 2001
Vehicle pressure greasing and oil change; shopping for supplies, at Bamenda.

Thurs. 11th Oct. 2001
Team A to grassland/forest patches to the West (JMO, VN, PN, Henry Nsofon).
Team B to the 1900 m ridge, searching for *Ternstroemia* (not located): BP, RK, EN, Peter Nkankano, CG.

Fri. 12th Oct. 2001
Team A surveying at Jerusalem Rock: JMO, VN, RK, Henry Nsonfon.
Team B surveying at top of grassy, rock-capped hill (difficult ascent): BP, Eric Nucam, JW conduct an hour-long thorough search for interesting *Drosera* found by him on 2000 expedition, BP suggests too early in season, large populations of *Eriocaulon asteroides* [*Pollard* 700] found. Two species of *Aeollanthus* (Labiatae) [*Aeollanthus trifidus*, *Pollard* 698 & *Aeollanthus angustifolius*, *Pollard* 699] found, one succulent species potentially new. CG, LM, PN, Simon collecting around base of rock.

Sat. 13th Oct. 2001
BP concludes work on forest profile and pre-expedition accounts. RK given training in plant description by MC. Remainder of group to Bambili Lakes [not reproduced here]. End-of-expedition social evening arranged by BP.

Sun. 14 th Oct. 2001
Rest day led by BP. Morning visit to the Paramount Chief, Fon of Bali, followed by football match arranged by BP and Maurice Babila. 60 minutes play in heavy rain - Mantum XI **1** vs Earthwatch XI **1**.
Raphael Kongor given further training in plant description by MC. Report writing. Packing ready for early departure next day.

Summary of collections made in 2001:

Collector	Opening no.	Closing no.
Cheek	10730	10906
De Marco	Not known	Not known
Etuge	4261r	4362r
Kongor	33	60
Nkeng	169	215
Onana	1800	1939
Pollard	575	715

Total number of collections from Bali Ngemba F.R.: 276

Bali Ngemba Inventory 3: 10th–21st April 2002 (Benedict Pollard)

Organised by RBG, Kew (funded by Earthwatch Institute) in collaboration with the Herbier National Camerounais (HNC-IRAD, Cameroon) (funded by the Darwin Initiative, DETR, British Govt. under the project for the conservation of the plant diversity of Western Cameroon) with the Bamenda Highlands Forest Project (administered by BirdLife International, funded by the National Lottery Commission, U.K.).

Abbreviations: AF - African Fellow; BEPA - Boyo Environmental Protection Agency; BERUDEP - Belo Rural Development Programme; BHFP - Bamenda Highlands Forest Project; CRES - Centre for the Reproduction of Endangered Species; EW - Earthwatch; HNC - Herbier National Camerounais; K - Kew herbarium; KIFP - Kilum-Ijim Forest Project; LBG - Limbe Botanic Garden; LR - LandRover; MINEF - Ministry of the Environment and Forests; NEPA - Natural Environment Protection Agency; NOWEBA - North-West Beekeeping Association; ONADEF - National Office for the Development of Forestry; ORDEP - Organisation for Rural Development and Environmental Protection; SATEC - Sustainable Agriculture Technicians; SPALE - Society for the Protection of Animal Life and the Environment; WWF - Worldwide Fund for Nature; YA - Yaoundé herbarium

Fieldworkers:
RBG, Kew: **Ben Pollard** (Project Leader, UK).
HNC: **Dr. Jean-Michel Onana** (Senior Researcher and Leader of HNC component); **Dr. Zapfack Louis** (Senior Lecturer, University of Yaoundé I and HNC Associate); Victor Nana and Nana Felicité (Technicians); Ela Ekeme Joseph (Accountant).
EW-funded African Fellows: Alex Asase (Teaching assistant, Department of Botany, University of GHANA); Bakari Salim Mohamed (Assistant Regional Catchment Forest Manager, East Usambara Conservation Area Management Programme, TANZANIA); Ignatius Malota, Senior Assistant Curator, National Herbarium & Botanic Gardens of Malawi, MALAWI); Lazzarus Oketch Ojiek (Faculty of Forestry & Nature Conservation, Makerere University, P.O.Box 7052, Kampala, UGANDA); O'Rorke Crentsil (Project Officer, Medicinal Plant Conservation Project, Aburi Botanic Gardens, GHANA).
MINEF: Nangmo Yves Nestor (Delegate for Momo Division); Kom Justin (Chief of Bureau, Bamenda Provincial Forestry Office); Patrick Fonji (Chief of Post, Bali).
EW-funded Cameroonian field staff: camp organisers: Martin Mbong (camp manager); Edmondo Njume (assistant botanist); Japhet Wain (driver).
EW- funded local workers from Mantum: Peter Nkankano (guide and ethnobotanical informant); Eric Nukam & Maurice Babila (guides & nightwatch); Lilian Mankety Tanketi, Patricia Lorga Tanketi & Sylvia Mbutigig Sakaah (cooks); Lovelyn Agem (camp cleaner, clothes washer); Regina (assistant to Lovelyn Agem); Vincent & Simon Tanketi (water carriers); Laurence Tanketi (wood carrier); Henry Ngupi (camp assistant).
BERUDEP: Cha Kenneth Yuh (senior bee-keeping trainer) (Week 1).
NEPA: Florence Nkenya Azamah (senior forest technician) (Week 1 & 2).
SATEC: Cletus Monju (field technician) (Week 2).
ORDEP: Marcellus Che (project execution manager) (Week 2).

ITINERARY

Weds. 10th April 2002
0800 plant specimen and collecting equipment demonstrations by ZL, JMO & EN.
0900 whole group except EN to forest. Collecting. 2 interesting *Allophyllus* spp. [*Allophylus* sp. A, *Onana* 1984 & *Deinbollia* sp. 2, *Onana* 1981] found, including spiny species collected in 2001. Other interesting taxa include possible new species of *Deinbollia* (Sapindaceae) [*Deinbollia* sp. 1, *Pollard* 923], collected in mature fruiting stage, to complement flowering material of 2001. Team returns to camp at 1400. HNC staff get mission orders signed at Gendarmerie, and visit Senior Divisional Officer (SDO). Formal visit of whole team to Paramount Chief of Bali and the Fon of Mantum village pre-arranged for Sunday 14th April. Quarterhead met and given lift back to Mantum. He donates a flagon of mimbo (Palm Wine). Quarterhead arrives for mimbo and *Cola* nuts. Stays for dinner.

Thurs. 11th April 2002
0700 BP gives 1 hr demonstration of how to check drying specimens. BP remains behind at camp to finalise team itinerary and start expedition report. Heavy morning rain.
Teams A (JMO) & B (ZL) to forest. Interesting collections (JMO) include *Tabernaemontana* sp. nov.? (ridged fruits) [*Voacanga* sp. 1, *Onana* 1992] and *Memecylon dasyanthum* (Melastomataceae) [*Onana* 1989], a Cameroon endemic known from only c. 10 earlier collections. ZL's team collected several flowering orchid species, including *Stolzia repens* [*Zapfack* 2034] and *Bulbophyllum tentaculigerum* [*Bulbophyllum sandersonii* subsp. *sandersonii*, *Zapfack* 2035] (both Orchidaceae); a fruiting member of the Sapotaceae, perhaps *Englerophytum* sp. [*Synsepalum msolo*, *Zapfack* 2040] and a fruiting Anacardiaceae [*Trichoscypha lucens*, *Zapfack* 2037].

1600 BP completes camp work and decides to do some late afternoon collecting with JW at Tu Ndab hill, just behind basecamp. Clear evidence of severe burning in recent times, but now that the rains have been falling for about six weeks, much lush vegetation present. Species encountered include *Gladiolus dalenii* subsp. *dalenii* [*Gladiolus dalenii* subsp. *andongensis*, *Pollard* 934] (perhaps distinct from *G. sp. nov.*, *sensu* Pollard in Cheek *et al.*, 2000, in having up to 7 or more flowers per inflorescence, outer tepals splayed, flowering after dry season); another as yet unidentified *Gladiolus* [*Gladiolus unguiculatus*, *Pollard* 937]; *Urginea altissima* (Hyacinthaceae) [*Drimia altissima*, *Pollard* 949]; and a stout terrestrial *Clematis* (Ranunculaceae), formerly included in *Clematopsis*, recently transferred by R.K.Brummitt to *Clematis* as *C. villosa* subsp. *oliveri* (Kew Bull. 55: 104 (2000) [*Pollard* 942].

Fri. 12th April 2002
Team A to Baba II (JMO) [not reproduced here].
Team B to Jerusalem Rock (ZL) LO, AA, VN, COR, PF, KJ, MB (guide). Some interesting forest encountered, with *Diospyros* sp. (Ebenaceae) [no collection appears to have been made] and again the two dominant orchid species, *Bulbophyllum tentaculigerum* [*Bulbophyllum sandersonii* subsp. *sandersonii*, *Zapfack* 2059] & *B. scaberulum var.?* (Orchidaceae) [only one orchid collected by Zapfack on this day].
Team C to top of ridge towards Pinyin (BP) ENU (Nukam), EN, FA, LO, KC.
0830 depart camp (1400 m alt.) for grassland ridge-top (2200 m alt.). Stop en route to collect 30 m tree, *Englerophytum* sp.? (Sapotaceae) [*Synsepalum msolo*, *Pollard* 950], in flower and fruit (EN successfully climbs branchless fluted bole to c. 10 m.). The *Oncoba* sp.nov. (Flacourtiaceae) [*Pollard* 952] is re-located at the cave at 1800 m, and about 30 flower-buds counted.
1300 reach valley just below grassland peak. Collect *Cassipourea* sp. (Rhizophoraceae) [*Cassipourea malosana*, *Pollard* 957]. Ascend grassy hill and encounter 2 terrestrial orchids, one a *Habenaria*, the other indet. [apart from the *Brachycorythis* (see later), 3 orchids were collected by Pollard on this day: *Bulbophyllum oreonastes* (951); *Bulbophyllum sandersonii* subsp. *sandersonii* (955); and *Disa equestris* (963)]; also encounter many interesting geophytic monocotyledons: *Hypoxis camerooniana* (Hypoxidaceae) [*Pollard* 961], a blue-flowered *Iris*-like herb, *Moraea schimperi* (Iridaceae) [*Pollard* 958], swathes of the diminutive *Wurmbea tenuis* (Colchicaceae) [*Wurmbea tenuis* subsp. *tenuis*, *Pollard* 959] and small populations of the dicotyledonous *Swertia quartiniana* (Gentianaceae) [*Pollard* 965]. On the descent collect spectacular 50 cm tall epiphytic *Brachycorythis* (Orchidaceae) [*Brachycorythis kalbreyeri*, *Pollard* 971] on fallen branch. Return to camp at 1800. Forest habitat degradation is considerable in Bali Ngemba Forest Reserve, BP noting huge patches cleared in only the past six months (since his ascent of the same path in Oct 2001). One method removes large trees by hollowing out the base between the buttresses, and lighting a fire in the centre, ultimately leading to felling of trees without the need of access to a chainsaw. The understorey is cleared manually and generally planted with beans, potatoes and cocoyams. It seems there are no significant barriers to agricultural expansion in much of the reserve, particularly at altitudes between 1600 and 1900 m.

Sat. 13th April 2002
Team A (JMO) to hills west of camp JMO, NF, FA, ELA, NV. *Diospyros* sp. [*Isolona hexaloba*, *Onana* 2030] again encountered; unusual *Clerodendrum* sp. (Verbenaceae) [*Clerodendrum violaceum*, *Onana* 2051], *Striga* sp. (Scrophulariaceae) [*Cycnium adonense* subsp. *camporum*, *Onana* 2048].
Team B (ZL) to Bamenda in LR [not reproduced here].
Team C (BP) to summit of granite Inselberg No. 1 near camp BP, KC, ENU, LT New route discovered up the inselberg. Recently burned soil has left much lush growth, much of it too weak to use as a handhold. At summit many specimens of *Gladiolus dalenii* subsp. *dalenii* [*Gladiolus dalenii* subsp. *andongensis*, *Pollard* 975] found as both orange and yellow (1 specimen) forms. Date palm seen on sides of mountain. Very strong southerly wind and spectacular views over many miles past Bali town and towards the Nigerian border.

Sun. 14th April 2002
Morning: COR, AA to Presbyterian Church, Bali. LR returns to camp to collect BP, HN, ENU, DT to waterfall near Bali. Car returns to Bamenda to collect ZL and drop him and waterfall group in central Bali, then returns to camp to collect rest of group to see HRH Paramount Fon of Bali, Dr Ganyonga III.
1530 group well-received and granted 1.5 hour reception in Palace, photo opportunities, discussion of conservation options of Bali Ngemba Forest Reserve. BP presents group to Fon, and hands over copies of 2000 and 2001 expedition reports, conservation poster of *Eriocaulon asteroides* and Sylvia Phillips' paper describing it, signed with compliments of the author.
1800 group visits Fon of Mantum. Heavy rain.

Mon. 15th April 2002: Plot Day
0830 whole group departs for forest, taking path towards Jerusalem Rock to do 25 × 25 square metres plot following Hall & Swaine method. GPS reading: 05 49 27N, 10 06 04E, 1700 m. Submontane forest plot completed, 89 vouchers made.
1630 BP gives 2 hour introduction to plant identification skills, principles and family characters.

Tues. 16th April 2002
0830 team A, Rural Training Centre (RTC), Fonta lowland forest [not reproduced here].
0900 BP starts scientific talk day.

Weds. 17th April 2002
0730 team C to Sabga [not reproduced here].
Small collecting trip to Bali Ngemba Forest Reserve with ZL.
Evening, most of team go to Bali to watch Cameroon vs Austria soccer match on TV.

Thurs. 18th April 2002
Team A to Lake Aweng, Bafut Ngemba Forest Reserve [not reproduced here].
Team B (all AFs, NF, FA, PN) remain in camp and enter forest with traditional herbalist to learn about local medicinal plants. 11 ethnobotanical specimens collected and vouchers made.

Fri. 19th April 2002
Team A to RTC Fonta [not reproduced here].
Team C (BP, AA, COR, MC, CM, EN, PN, MB) ascend path towards Jerusalem Rock and then branch to the right towards Lale I. Ascend to c. 1800 m, stop and collect. Many taxa not previously seen collected. *Drypetes* sp. (Euphorbiaceae) [*Drypetes* sp. aff. *leonensis* of Bali Ngemba, *Pollard* 1041] in flower and fruit. Several other small to medium size trees collected in flower. Unusual terrestrial orchid encountered, *Nervilia bicarinata* [*Pollard* 1043]. Team splits at 1330, AA, COR return to camp with MB and help EN with specimen sorting. BP leads rest up to 2000 m, in search of *Ternstroemia polypetala* (Theaceae). No luck in finding it, but more collections made of other taxa. Descend towards Lale II and arrive at camp c. 1515.
1600 BP calls Cheek in London. Ascends hill behind camp and collects *Drosera* sp. (Droseraceae) [*Drosera pilosa*, Pollard 1044]. Appears not to have aerial stem, leaves borne at ground level, with burnt remnants of previous year's growth visible.

Sat. 20th April 2002
am PN, EN sent out to look for 'Jangang', *Newtonia cameroonensis*, as informed by Cheek yesterday on satellite phone that Barbara Mackinder is almost certain an Etuge collection from 2000 in the reserve represents this species.
AF training day.

Sun. 21st April 2002
Labelling finalised, no collecting.

Summary of collections made in 2002:

Collector	Opening no.	Closing no.
Onana	1980	2053
Pollard	917	1048
Zapfack	2000	2125

Total number of collections from Bali Ngemba F.R.: 209

Bali Ngemba Inventory 4: 9th–20th April 2004 (Iain Darbyshire)

Executed by RBG, Kew (funded in part by the Earthwatch Institute) in collaboration with the Herbier National du Cameroun (HNC-IRAD, Cameroon) (funded in part through a grant from the Overseas Fieldwork Committee, RBG, Kew) with the aid and co-operation of local conservation organisations: North West Province (BHFP and ANCO) and South West Province (ERuDeF).

Abbreviations: ANCO - Apicultural and Nature Conservation Organisation (formerly North-West Beekeeping Association - NOWEBA); BHFP - Bamenda Highlands Forest Project; BNFR - Bali Ngemba Forest Reserve; CRES - Center for the Reproduction of Endangered Species; ERuDeF - Environment and Rural Development Foundation; EW - Earthwatch; HNC - Herbier National du Cameroun; K - Herbarium of the Royal Botanic Gardens, Kew; LBG

- Limbe Botanic Garden; MINEF - Ministry of the Environment and Forests; SCA - Limbe Botanic Garden herbarium; YA - Herbarium of HNC.

Fieldworkers:
RBG, Kew: **Iain Darbyshire** (ID - EW Principal Investigator); **Dave Roberts** (DR - Orchidaceae specialist); **Nina Rønsted** (NR - PhD. Student, working on the genus *Ficus* (Moraceae), sponsored by Carlsberg).
HNC: **Nana Felicité** (NF - technician), **Barthelemy Tchiengue** (BT - researcher), Nicole Guedje (NG - researcher).
University of Yaoundé I: **Simo Placide** (SP - PhD. student under Dr L. Zapfack, joining the expedition to collect orchid taxa, particularly *Polystachya* and to collaborate with DR).
LBG: Enow Kenneth (EK - joining the expedition to gain fieldwork experience).
ANCO: Terence Suwinyi (TS - joining the expedition to gain fieldwork experience), **Kenneth Tah** (KT - botanist).
ERUDEF: Terence Atem (TA - leader of excursion to Fossimondi, Bamboutos and joining expedition to gain experience of botanical inventory work).
RBG, Kew-sponsored field assistant: Walters Cheso (CW - formerly of BHFP).
Earthwatch volunteers and corporate sponsor (where appropriate):
Zeynep Barlas (ZB - Istanbul, Turkey, HSBC); Armelle de Martrin (AM - Paris, France, HSBC); Abdullah Shah (ASh - Islamabad, Pakistan, British & American Tobacco); Adam Surgenor (ASu - Epsom, UK, private volunteer).
Earthwatch-funded Cameroonian field staff:
Ngolle Ngolle Hoffmann (NH - camp manager); Edmondo Njume (EN - equipment organiser and botanical assistant); **Martin Etuge** (ME - botanist and research assistant); Gilbert Tanyi (GT - driver of EW vehicle).
Earthwatch-funded local Cameroonian camp assistants:
Maurice Babila (MB), Eric Nukam (ENu) and Peter Nkankano (PN) (guides and ethnobotanical informants). Lilian Mankety Tanketi and Sylvia Mbutigig Sakaah (cooks); Mdme Pamela (cleaner); Mdme Anna (water carrier); Laurence Mbenbwo (firewood carrier).

ITINERARY

Fri. 9th April 2004
ID gives opening talk to volunteers on the project, the fieldwork plan and on Health & Safety. ME, EN, DR and PN begin collection in BNFR on trail to Pinyin; ME collects *Crinum zeylanicum* [*Etuge* 5308], the *Scadoxus multiflorus* [*Etuge* 5318] seen by ID and NR on 6th (both new to the checklist); *Drypetes* sp., *Leptonychia* sp. and several new orchid taxa for the checklist. They report that a large bushfire has cleared much of the undergrowth in the lower reserve and that good forest is scarce.

Sat. 10th April 2004
Am: EN and ID demonstrate specimen drying technique to volunteers.
Group 1: ID, EN, NF, ENu, ASh, ASu, ZB, AM to BNFR, where ID demonstrates collecting and pressing techniques with EN using specimen of *Drypetes* sp. Notable collections: *Deinbollia* sp. 1 (fr.), *Memecylon dasyanthum* (fl.), *Ardisia* sp. (fl.) and *Urera trinervis* [*Darbyshire* 352] (new for checklist). On return, one plant of *Psorospermum* cf. *tenuifolium* [*Darbyshire* 353] was found in fruit in a recently burned area (taxon previously known only from fl.). NF collections include *Rinorea preussii* and *Allophylus* sp.
Group 2: ME, TA, CW, EK, MB along Ntanyan River towards Pinyin; notable finds: *Tabernaemontana* sp. (fr.), *Prunus africana* [*Apodytes dimidiata*, *Etuge* 5334], 2 Sapotaceae spp.
Group 3: DR, SP, NR, PN collecting orchids in BNFR.

Sun. 11th April 2004
Group 1: ID, NR, TA, ENu on Ntanyan trail up to ridge at top of reserve (2100 m alt.). Notable finds: *Oncoba* sp. nov. (fr.), with material collected in silica gel for study by S. Zmartzy (K), an unsual Sapotaceae sp.[*Epistemma* cf. *decurrens* (*Asclepiadaceae*), *Darbyshire* 378] at the tree line, *Psorospermum aurantiacum* [*Darbyshire* 363], *Impatiens sakerana*, an epiphytic *Begonia* sp. and *Bridelia* sp. [*Darbyshire* 358] (new for checklist). Widespread cultivation and *Eucalyptus* plantation is noted throughout the reserve, with burning noted on the ridge.
Group 2: ME, EK, EN, ASh, MB continue on route of previous day and head into high forest. Collect *Entandrophragma angolense* [*Etuge* 5369], *Maytenus* sp. [*Maytenus gracilipes* subsp. *gracilipes*, *Etuge* 5361], *Trichoscypha acuminata* [*Etuge* 5360], all new for the checklist.
Group 3: DR, SP, ZB, PN continue orchid collection, recording further new taxa.

Mon. 12th April 2004
Group 1: ID, NF, ZB, AM, TA, ENu to forest along main trail into BNFR, climbing to 1800 m alt. Notable finds: *Vepris* sp. A (fl.), *Tricalysia* sp. and *Ochna membranacea* (fr.) [*Darbyshire* 400] (all ID) and *Homalium* sp. [*Magnisptipula butayei* subsp. *balingembaensis*, *Nana, F.* 105].
Group 2: ME, EK, ASu, NR, MB collect in forest patches within the reserve E of Mantum. Finds include: *Beilschmiedia* sp. nov.? (fr.), *Sorindeia* sp. (fr.) [*Trichoscypha* sp., *Etuge* 5390, collected 13.iv.2004]. NR collects *Trilepisium madagascariense* [*Rønsted* 230] for the first time in BNFR.
Group 3: DR, SP, PN collect orchids around coffee plantations next to BNFR; several additional taxa recorded; now up to 22 new orchid taxa for the checklist, with a further 18 non-orchid taxa.

Tues. 13th April 2004
Group 1: to Baba II [not reproduced here].
Group 2: NF, SP, EK, ENu collect in lower forest in BNFR.
Group 3: DR, PN, ASu, EN collect orchids in BNFR.

Weds. 14th April 2004
Group 1: to Finge [not reproduced here].
Group 2: ID, NF, TA, ASu, ASh, ENu collect in Kupp Pinyin forest and then onto lower grassland (1600 m alt.) and forest patches on western side of BNFR. Collections include *Cordia* sp., *Psorospermum febrifugum* [*Darbyshire* 407] (new for checklist), *Nervilia bicarinata* [*Darbyshire* 405], *Disa* aff. *nigerica* [*Darbyshire* 409] (numerous on terraced grassland, not matched clearly in Fl. Cameroun by DR) and *Ficus vallis-choudae* [*Darbyshire* 418]. Several other taxa new to the checklist recorded.
Group 3: ME, ZB, NG, SP, EK, EN, PN collect in forest patches and farmbush on western side of reserve. Collections include an interesting Annonaceae, a *Costus* sp. [*Etuge* 5432], the first for the reserve, and 2 further new taxa for BNFR (*Cissus producta* [*Etuge* 5430], *Plukenetia conophora* [*Etuge* 5424]).

Thurs. 15th April 2004
Group 1: to Mujung [not reproduced here].
Group 2: ID, NF, ASh, EK, ENu to small forest patches beyond reserve boundary east of Mantum. Few new taxa recorded, but include *Urera cordifolia* [*Darbyshire* 424], *Cissus oreophila* [*Darbyshire* 421] and *Viscum congolense* [*Darbyshire* 426], all new for the checklist. NF collects an interesting flowering and fruiting Apocynaceae, and a flowering *Vitex* sp.
Group 3: ME, ZB, SP, EN, MB, CW collect in forest patches on the BNFR boundary, recording several interesting taxa including *Dichapetalum* sp. [*Etuge* 5443], *Jasminum* sp. [*Jasminum pauciflorum*, *Etuge* 5447] and *Neoboutonia mannii* [*Neoboutonia mannii* var. *mannii*, *Etuge* 5435] together with a second *Costus* sp. [*Etuge* 5449] (sterile).

Fri. 16th April 2004
Group 1: to Fossimondi [not reproduced here].
Group 2: ME, CW, AM, ASh, EN, SP, MB collect in farmbush in and along edge of BNFR.
Group 3: DR, ZB, PN to western slopes of reserve to see first-hand and collect further specimens of the *Nervilia* and *Disa* collected by ID on 14th.

Sat. 17th April 2004
Group 1: to Fossimondi [not reproduced here].
Group 2: ME, SP, EN, MB, AS, AM, CW to BNFR for further collections in the main reserve, exploring different routes; few extra taxa recorded, return mid-afternoon.

Sun. 18th April 2004
Rest day. ID, ME, DR, ZB, ASh to Bali to meet His Royal Highness, the Fon of Bali.

Mon. 19th April 2004
Group 1: to Baba II [not reproduced here].
Group 2: BT, SP, ASh, ENu collect in BNFR, taking different routes to those previously collected on. However, no new taxa come to light.
Group 3: KT, ASu, PN trek to high ridge of BNFR, instructed by ID to search for *Ternstroemia polypetala.* No plants are found, but several interesting collections made, including *Chassalia laikomensis.*
ME and DR to Bali to meet MINEF District Officer who is happy that we are working in BNFR.

Tues. 20th April 2004
Expedition team process specimens, complete databasing and label production, separate out CITES material and package up completed bundles. ME agrees to continue the drying of remaining specimens in Nyasoso and to separate out all remaining CITES material.
ID produces (a) a provisional checklist of plants for Fossimondi, (b) a list of all collections made during the expedition to aid Dr Achoundong in processing export and CITES permits, (c) a summary of collections made during teams I–III [teams I & II not reproduced here].

Summary of collections made in 2004

Collector	Opening no.	Closing no.
Darbyshire	340	448
Etuge	5308	5467
Nana (Felicite)	78	134
Rønsted	222	237
Simo	140	273
Tah	286	328
Tchiengue	1846	1972

Total number of collections from Bali Ngemba F.R.: 371

REFERENCES

Cheek, M. (2001). Report on an expedition for botanical inventory and conservation management to the Bali Ngemba Forest Reserve and other forest sites in the Bamenda Highlands, North West Province, Cameroon 6-20 Nov. 2000. R.B.G., Kew. Cyclostyled. 6 pp.

Cheek, M. & Pollard, B. J. (2002). Report on an expedition for botanical inventory and conservation management to the Bali-Ngemba Forest Reserve, Nkom-Wum Forest Reserve and other forest sites in the Bamenda Highlands, North West Province, Cameroon 3-15th Oct. 2001. R.B.G., Kew. Cyclostyled. 7 pp.

Chiron, G.R. & Guiard, J. (2001). Etude et conservation des orchidées de Bali Ngemba (Cameroun). Richardiana 1(4): 153–186.

Darbyshire, I., Cheek, M.,and Onana, J.-M. (2004). Report on the RBG Kew & Herbier National du Cameroun expedition for botanical inventory and conservation management in Mefou Proposed National Park and Bali Ngemba Forest Reserve, March-April 2004 R.B.G., Kew. Cyclostyled. 26 pp.

Hepper, F.N. & Neate, F. (1971). Plant collectors in West Africa; a biographical index of those who have collected herbarium material of flowering plants and ferns between Cape Verne and Lake Chad, and from the coast to the Sahara at 18° N. International Bureau for Plant Taxonomy and Nomenclature - Utrecht, Oosthoek. [(Regnum vegetabile, v. 74) xvi, 89 pp. 8 p. of photos.]

Pollard, B. J. (2002). Report on the RBG Kew March-May 2002 expedition for botanical inventory and conservation management in the Bamenda Highlands, North West Province, Cameroon. R.B.G., Kew. Cyclostyled. 20 pp.

ETHNOBOTANY

Benedict John Pollard
Herbarium, Royal Botanic Gardens Kew, Surrey, UK, TW9 3AB.

There are two main groups of plants in our checklist area that are significant to the local people: crop plants (mainly introduced) and mostly-native wild species. Expansion of cropland and non-sustainable harvesting of wild species contribute threats to natural vegetation, particularly the expansion of illegal small-scale agricultural holdings within the reserve.

In compiling this checklist of plants from the Bali Ngemba Forest Reserve, it became apparent that a number of our plant specimen collectors (particularly Cheek, Etuge and Pollard) invested time investigating the ethnobotanical knowledge held by the local people. Although this chapter represents very much a preliminary survey, it is of great interest, with 61 different plant uses recorded. Several of these are derived from a day's 'Herbal Plants Workshop' held at Mantum under the tutelage of a traditional practioner, Pa Foncham Foseleng, in April 2002. This was organised in response to the request (among others) of Earthwatch African Fellow Crentsil O'Rorke, manager of the Medicinal Botanic Garden in Aburi, Ghana.

The local names have been incorporated in the checklist, near the end of each taxon account, and are all Bali names, excepting a few in Pidgin English and occasionally local English names. The data regarding plant uses that were reported to us and recorded during our fieldwork are summarised with reference to standardised terms proposed by Cook (1995).

Wild plants are used in a multitude of ways in the Bali tribal area, and are listed below according to LEVEL 1 states. The dashed lines indicate separate plant groups: Dicotyledonae are listed first, followed after the dashed lines by Monocotyledonae, and thirdly Gymnospermae and Pteridophyta.

Caveat: be advised that the indigenous traditional knowledge contained in this volume has been given in good faith as a contribution to the common good and the furtherance of mutual understanding and the preservation of all life on our planet. It is not be used for personal or commercial gain and must be treated with respect and used only for the purpose for which it was gifted. Anyone who reads this volume assumes the moral and ethical obligations implied by this statement.

CROP PLANTS

There are a number of crop plants that are utilised for food, in agroforestry, for timber or for other inherent properties such as their medicinal powers. Within the reserve, the most significant threats to natural vegetation are removal of timber for MATERIALS, and transformation of forest to agricultural holdings. The main crops grown are cocoyams and beans.

WILD PLANT USES

FOOD

Mangifera indica (Anacardiaceae)
Pseudospondias microcarpa var. *microcarpa* (Anacardiaceae)
Trichoscyphya acuminata (Anacardiaceae)
Begonia oxyanthera (Begoniaceae)
Echinops gracilis (Compositae)
Vernonia hymenolepis (Compositae)
Beilschmiedia sp. 1 of Bali Ngemba (Lauraceae)
Kotschya strigosa (Leguminosae: Papilionoideae.)
Passiflora edulis (Passifloraceae)
Rubus rosifolius (Rosaceae)
Cuviera longiflora (Rubiaceae)
Deinbollia sp. 1 (Sapindaceae)
Synsepalum msolo (Sapotaceae)
Physalis peruviana (Solanaceae)
Solanum betaceum (Solanaceae)
Aframomum sp. 1 of Bali Ngemba (Zingiberaceae)

FOOD ADDITIVES

Echinops giganteus (Compositae)
Ocimum gratissimum subsp. *gratissimum* var. *gratissimum* (Labiatae)
Aframomum cf. *zambesiacum* (Zingiberaceae)

ANIMAL FOOD
Acmella caulirhiza (Compositae)
Synsepalum brevipes (Sapotaceae)
Synsepalum msolo (Sapotaceae)

MATERIALS
Trichoscypha lucens (Anacardiaceae)
Echinops gracilis (Compositae)
Garcinia cf. *smeathmannii* (Guttiferae)
Vitex doniana (Labiatae)
Beilschmiedia sp. 1 of Bali Ngemba (Lauraceae)
Sida cf. *acuta* subsp. *carpinifolia* (Malvaceae)
Memecylon dasyanthum (Melastomataceae)
Carapa grandiflora (Meliaceae)
Entandrophragma angolense (Meliaceae)
Turraeanthus africanus (Meliaceae)
Ficus exasperate (Moraceae)
Maesopsis eminii (Rhamnaceae)
Pouteria altissima (Sapotaceae)
Synsepalum brevipes (Sapotaceae)
Aframomum sp. 1 of Bali Ngemba (Zingiberaceae)

FUELS
Nuxia congesta (Buddlejaceae)

SOCIAL USES
Ageratum conyzoides (Compositae)
Piper capense (Piperaceae)
Piper umbellatum (Piperaceae)

NON-VERTEBRATE POISONS
Plectranthus glandulosus (Labiatae)

MEDICINES
Mangifera indica (Anacardiaceae)
Rauvolfia vomitoria (Apocynaceae)
Begonia oxyanthera (Begoniaceae)
Acmella caulirhiza (Compositae)
Ageratum conyzoides (Compositae)
Allanblackia gabonensis (Guttiferae)
Garcinia smeathmannii (Guttiferae)
Leucas deflexa (Labiatae)
Plectranthus occidentalis (Labiatae)
Satureja robusta (Labiatae)
Desmodium repandum (Legumiosae: Papilionoideae)
Hibiscus noldeae (Malvaceae)
Piper capense (Piperaceae)
Piper umbellatum (Piperaceae)
Paullinia pinnata (Sapindaceae)
Solanum aculeastrum var. *albifolium* (Solanaceae)
Anchomanes difformis (Araceae)
Aframomum sp. 1 of Bali Ngemba (Zingiberaceae)

ENVIRONMENTAL USES

Cupressus lusitanicus (Cupressaceae)

REFERENCES

Cook, F.M. (1995). Economic Botany Data Collection Standard. RBG, Kew, UK. 146 pp.

THE BIRDS OF BALI NGEMBA FOREST RESERVE

Iain Darbyshire
Herbarium, Royal Botanic Gardens, Kew, Surrey, TW9 3AE, U.K.

INTRODUCTION

The highlands of western Cameroon, south-eastern Nigeria and Bioko are recognised as a priority area for bird conservation in west Africa, together comprising the 'Cameroon mountains' Endemic Bird Area (EBA)[1] (Stattersfield *et al.* 1998). This area contains 30 restricted range species, 10 of which are recognised as Vulnerable or Endangered under IUCN (2001) criteria, and two of which represent endemic monotypic genera (BirdLife International 2003).

The montane and submontane forests of the Bamenda Highlands of Cameroon are recognised as an area of conservation priority within this EBA, being an area containing two near-endemic bird species, the Bannerman's Turaco (*Tauraco bannermani*) and Banded Wattle-eye (*Platysteira laticincta*), and known to be under considerable threat from anthropogenic pressures. This region has therefore been a focus of bird-driven conservation activity in the recent past. Both the Kilum-Ijim Forest Project and the Bamenda Highlands Forest Project, which have sadly now been terminated, were established in the region by BirdLife International, the former focusing on wildlife monitoring and conservation management in the protected Kilum-Ijim forest on Mt Oku and the latter studying the biodiversity of forest patches in the remainder of the much-denuded Bamenda Highlands, and promoting community-based forest conservation. As a result, the bird populations of many of these forest areas are well-documented.

Although a government-run Forest Reserve, and thus officially outside the remit of the Bamenda Highlands Forest Project, Bali Ngemba was recognised by BirdLife International as a significant area of remnant submontane forest in the Bamenda Highlands and its bird populations were therefore surveyed up to the year 2000. The result is that it is now recognised as an Important Bird Area (IBA), qualifying on two of the three grounds for designation (BirdLife International 2003), namely that it holds populations of more than one globally threatened bird species and that it is home to a suite of restricted range species.

During botanical inventory work at Bali Ngemba in the period 6th to 21st April 2004, the author made records of the bird species seen and their apparent abundance, together with observations on the threats to their habitats. As the focus of the 2004 expedition was not on birds, the time available for observation was highly limited and the records are biased towards those species found in the lower elevation primary forest and the adjacent farmland and disturbed forest patches. In addition, due to the author's unfamiliarity with the calls of West African birds, records were largely reliant upon sightings alone. It is therefore doubtless that many of the forest taxa, particularly those at higher elevations, were under-recorded. This is reflected in the low total species count during 2004 of only 74 species, only a small proportion of the total 185 so far recorded at the site (BirdLife International 2003). However, as several of the threatened or restricted range species were recorded, together with several uncommon species of more widespread distribution, this paper presents an updated summary of our knowledge of the birds of the reserve and their conservation significance.

KEY BIRD SPECIES

Of the 185 species recorded to date at Bali Ngemba, five are currently listed by BirdLife International on the IUCN redlist (IUCN 2003) as being globally threatened or near threatened:

Bannerman's Turaco (*Tauraco bannermani*)
Endangered (EN B1+2abcde)

This striking species, readily distinguished by its red crown feathers, is Endangered due to the high anthropogenic pressure upon its submontane and montane forest habitat in the Bamenda Highlands, where it was believed endemic until recent discoveries of populations at both Fossimondi-Alou and Fomenji-Magha in the Lebialem Highlands, western Bamboutos Mts (Nkembi & Atem 2003). Borrow & Demey (2001) record it as a common resident in the remaining forest patches but postulate that it is only likely to survive if the Kilum-Ijim forest is preserved. At Bali

[1] Endemic Bird Areas are defined by BirdLife International as those in which two or more landbird species of restricted breeding range (defined as less than 50,000 km^2) are recorded (Stattersfield *et al.* 1998).

Ngemba it is recorded as common above 1800 m but extending down to 1600 m (BirdLife International 2003). Observations in 2004 largely reflect this finding, it being seen at 1600 m in disturbed forest, and heard regularly in the steep forested slopes at 1800–2000 m. However, it is very retiring here and BirdLife International (2003) believe that the high densities may be due to immigration of birds displaced by forest clearance elsewhere. It is apparently more numerous and approachable at the nearby Baba II community forest (Walters Cheso, pers. comm.), where it was heard frequently by the author in April 2004. The ease of distinguishing this species by its call, taking care to separate it from the deeper and more raucous Green Turaco (common in the reserve), should allow a population census to be carried out at Bali Ngemba without difficulty.

The crown feathers of this species are revered as symbols of status in local communities, thus it is recognised as a species key to promoting community-based forest protection in the Bamenda Highlands, through the argument that without the forest, the Bannerman's Turaco will be lost, and with it these local traditions and customs (Cheso Walters, pers. comm.).

Banded Wattle-eye (*Platysteira laticincta*)
Endangered (EN B1+2abcde)

Previously treated as a subspecies of the Black-throated Wattle-eye (*P. peltata*) of E, E central and SE Africa, the Banded Wattle-eye is now widely recognised as a true species, believed endemic to the montane forest of the Bamenda Highlands (Borrow & Demey 2001). However, it too has recently been recorded at Fossimondi (Nkembi & Atem 2003). It is recognised as endangered for the same reasons as Bannerman's Turaco. It is recorded as not uncommon in forest patches above 1800 m alt (IUCN 2003). At Bali Ngemba it was found in 2004 on the upper limits of the forest, where an adult male and immature bird were recorded, suggesting breeding had occurred. However, the birds were seen within an area which had recently been burned, a common occurrence within the Reserve, which may have an adverse effect upon the population here.

Green-breasted Bush-shrike (*Malaconotus gladiator*)
Vulnerable (VU B1+2abcde, C2a)

This species, endemic to the Cameroon mountains EBA, is considered Vulnerable as it has a highly restricted and fragmented distribution, with significant losses of its submontane forest habitat in parts of its range. BirdLife International (2003) postulate that, as much of the forest at Bali Ngemba is between 1400 m and 1800 m, this could be an important site for this species. However, it is at this altitudinal range that forest clearance has been extensive at this site, thus this species is clearly threatened here. It was not observed in 2004, though it is secretive and the author is not familiar with its call, thus it could easily be overlooked.

Cameroon Greenbul (*Andropadus montanus*)
Near Threatened (NT)

This species, restricted to the Cameroon mountains EBA, almost qualifies as Vulnerable under IUCN criterion B (IUCN 2003), being local in montane forest. It was not recorded in 2004, but is unobtrusive, thus easily overlooked. The continued presence of forest on the steeper upland slopes at Bali Ngemba may aid its survival here.

Bangwa Forest Warbler (*Bradypterus bangwaensis*)
Near Threatened (NT)

Another species restricted to the Cameroon montains EBA, it also almost qualifies under IUCN criterion B, being restricted to dense undergrowth on the edges of montane forest, in the northern mountains of the EBA. Its habitat is threatened at Bali Ngemba by regular burning to clear the undergrowth for cocoyam farming, and it was not recorded here in 2004. It was, however, seen in the nearby Baba II community forest.

In addition a further 28 resident species at Bali Ngemba are recorded as key taxa under the IBA designation, 11 of which are restricted range species of the Cameroon mountain EBA (table 1).

During the April 2004 botanical expedition, several other bird species worthy of note were recorded by the author. Species recorded within the closed forest and forest patches included Petit's Cuckoo-shrike (*Campephaga petiti*, recorded on two occasions), Yellow-spotted (*Buccanodon duchaillui*) and Double-toothed (*Lybius bidentatus*) Barbets and Black-throated Apalis (*Apalis jacksoni*). At least one pair of Red-necked Buzzards (*Buteo auguralis*) were holding territory in the western section of the reserve.

SPECIES	HABITAT	OBSERVATIONS IN APRIL 2004
Cameroon Olive-pigeon* *Columba sjostedti*	Montane forest and forested gullies	Not seen
Bar-tailed Trogon *Apaloderma vittatum*	Moist montane forest	Not seen: this species is highly unobtrusive, thus was likely overlooked
Western Green-tinkerbird *Pogoniulus coryphaeus*	Montane forest and forest patches	Not seen: many Tinkerbirds were heard, but those seen belonged to the much commoner *P. bilineatus*
Grey Cuckoo-shrike *Coracina caesia*	Montane forest	One adult seen at c. 1500 m alt.
Grey-throated Greenbul* *Andropadus tephrolaemus*	Montane forest and forest patches	Seen near the lower edge of the reserve in disturbed forest, and in remnant gulley forest immediately southwest of the reserve
Cameroon Olive Greenbul* *Phyllastrephus poensis*	Montane forest and forested ravines	Not seen
Yellow-breasted Boubou* *Laniarius atroflavus*	Dense undergrowth in forest clearings and forest patches	Two birds seen in disturbed forest at 1700–1800 m alt., several heard here and at the nearby Baba II community forest, probably locally not uncommon
Mountain Boubou* *Laniarius poensis*	Dense undergrowth of montane forest margins and clearings	Not seen: this species is secretive and easily overlooked
Mountain Robin-chat* *Cossypha isabellae*	Montane and submontane forest	Not seen
Abyssinian Hill-babbler *Pseudoalcippe abyssinica*	Dense undergrowth of montane forest and clearings	Not seen: this species is shy and unobtrusive, thus easily overlooked
Grey-chested Illadopsis *Kakamega poliothorax*	Ground level of montane forest	Not seen: again shy and unobtrusive, thus easily overlooked
Brown-backed Cisticola* *Cisticola (chubbi) discolor*	Dense undergrowth of forest edges, clearings and abandoned cultivation	One seen on the edge of cultivation near Mantum village. *C. discolor* is often considered a subspecies of Chubb's Cisticola *C. chubbi*
Green Longtail* *Urolais epichlora*	Montane and submontane forest	Several seen in closed-canopy forest at c. 1450–1550 m alt., where it is possibly locally numerous
Black-collared Apalis *Apalis pulchra*	Dense undergrowth of montane forest and forest edges	Not seen
Bamenda Apalis* *Apalis bamendae*	Forest canopy	One seen in disturbed forest at c. 1450 m alt. near the lower limit of the reserve
Black-capped Woodland Warbler* *Phylloscopus herberti*	Wet montane and submontane forest, understorey to canopy	Not seen; this species has only recently (2000) been recorded for the first time in the reserve
White-bellied Crested Flycatcher *Trochocercus albiventris*	Montane forest, usually in undergrowth	One recorded in open forest at c. 1700 m alt.
Cameroon Sunbird* *Nectarinia oritis*	Undergrowth of mid-elevation and montane forest	Not seen; this is unusual in view of the fact that sunbird species are active and conspicuous and that it was recorded during a brief stay in nearby Bamenda
Northern Double-collared Sunbird *Nectarinia preussii*	Open montane forest, clearings and secondary bushland	Frequently recorded within the reserve and around Mantum village, particularly at forest / woodland edges
Thick-billed Seedeater *Serinus burtoni*	Edges of montane forest and in patches of scrub at higher elevations	Not seen at Bali Ngemba but recorded in the nearby Baba II village at the lower edge of the community forest
Oriole Finch *Linurgus olivaceus*	Montane forest edges and abandoned farmland	Not seen
Fernando Po Oliveback* *Nesocharis shelleyi*	Montane forest edges and clearings, plantations	Not seen
Red-faced Crimson-wing *Cryptospiza reichenovii*	Dense undergrowth and edges of montane forest	Not seen: this species is again secretive and easily overlooked
Baglafecht Weaver *Ploceus baglafecht*	Edges of montane forest	Fairly common in cultivation at forest edges and clearings in the reserve. Note: this species is common in E Africa but is much more local in W Africa, where it is confined to the Cameroon mountains EBA
Black-billed Weaver *Ploceus melanogaster*	Dense undergrowth of montane forest edges and clearings	Not seen
Brown-capped Weaver *Ploceus insignis*	Canopy and mid-levels of montane and mid-elevation forest	One female recorded in open woodland on the edge of Mantum village near the lower limits of the reserve
Waller's Starling *Onychognathus walleri*	Canopy of montane forest	Small flocks, believed to be of this species, seen flying into the reserve from the west in the early mornings and evenings

Table 1. The key species of bird highlighted in the designation of Bali Ngemba Forest Reserve as an Important Bird Area (IUCN 2003) with records of their occurrence and abundance noted during the April 2004 botanical expedition. * designates restricted range species of the Cameroon mountains EBA. Habitat information is derived from Borrow & Demey (2001).

BALI NGEMBA AS A PRIORITY SITE FOR BIRD CONSERVATION IN CAMEROON

The designation of Important Bird Area (IBA) status to the Bali Ngemba Forest Reserve by the IUCN highlights its importance as a site for bird diversity within the Cameroon highlands. IBAs are used by both international and local environmental organisations as a tool to prioritise sites for conservation. Bali Ngemba is one of only 33 such sites in Cameroon and, holds populations of 3 of Cameroon's 15 globally threatened species, 2 near-threatened species and 16 of the total 30 restricted range species recognised within the Cameroon mountains Endemic Bird Area. The lattermost figure is comparable to that of Mt Kupe, with 17 restricted range species, and the Bakossi Mts, with 18 such species (Wild *et al.*, in press), sites long renowned for their high bird diversity. In fact, the only restricted range species of submontane habitats which has not been recorded at Bali Ngemba is the globally threatened (VU) Bannerman's Weaver (*Ploceus bannermani*), which may well be recorded during future surveys as its favoured habitats of forest edges, clearances and edges of cultivation are common within the reserve.

It is therefore clear that Bali Ngemba is of high significance for conservation of the Cameroon mountains EBA. However, although the 2004 botanical expedition recorded several of the key species, it did highlight the continued decline in habitat quality throughout large sections of the reserve, with losses of both natural forest cover, through conversion to farmland or *Eucalyptus* plantation, and understorey cover through clearance for cocoyam cultivation, these two habitats being of prime importance for the key bird species. It is clear that the status of Forest Reserve is insufficient to afford the site adequate protection from excessive disturbance and encroachment of anthropic habitats.

Recognition of the site as an IBA is a positive first step towards halting of the decline in forest habitat at Bali Ngemba but active management and enforcement of the reserve boundaries against illegal farming and logging activities are required in order to preserve the remaining natural habitats of this important site. BirdLife recognise the importance of immediate conservation efforts, though with the recent closure of their projects in the Bamenda Highlands it is unclear to what extent they will maintain an active interest. Recent moves by the Cameroon government's agency for forestry development to encourage conservation of parts of the reserve and to promote reforestation of native trees along the watercourse (BirdLife International 2003) are to be encouraged, though there was little evidence of such activity during the 2004 botanical expedition. Local environmental education and community-led conservation would appear key to the future survival of the rare fauna and flora here. Finally, in view of the high number of localised or endangered bird species to be found in the reserve, the promotion of ecotourism may provide an economic incentive for long-term conservation measures.

REFERENCES

BirdLife International (2003). BirdLife's online World Bird Database: the site for bird conservation. Version 2.0. Cambridge. Available at http://www.birdlife.org (accessed Sept. 2004).

Borrow, N. & Demey, R. (2001). Birds of western Africa. Christopher Helm, London.

IUCN (2003). The IUCN Red List of Threatened Species. Available at http://www.redlist.org (accessed Sept. 2004).

Nkembi, L.& Atem, T. (2003). A report of the biological and socio-economic activities conducted by the Lebialem Highlands Forest Project, South West Province, Cameroon. Environment and Rural Development Foundation (ERuDeF), Cameroon.

Stattersfield, A.J., Crosby, M.J., Long, A.J. & Wege, D.C. (1998). Endemic Bird Areas of the world: priorities for biodiversity conservation. BirdLife International, Cambridge.

Wild, C., Morgan, B. & Fotso, R. (in press). The Vertebrate Fauna. *In* Cheek, M., Pollard, B.J., Darbyshire, I., Onana, J.-M. & Wild, C. (eds). The Plants of Kupe, Mwanenguba & the Bakossi Mountains, Cameroon, A Conservation Checklist. RBG, Kew, UK.

ENDEMIC, NEAR-ENDEMIC & NEW TAXA AT BALI NGEMBA

Martin Cheek
Herbarium, Royal Botanic Gardens, Kew, Surrey, TW9 3AE, U.K.

Of the 23 endemic or near-endemic taxa listed below, 12 are considered as strict endemics, i.e. only known from Bali Ngemba, and 11 as near-endemics (known at one other site).

Monanthotaxis sp. of Bali Ngemba (more material needed to confirm and describe)
Tabernaemontana sp. of Bali Ngemba (also at Tabenken)
Voacanga sp. 1 (also at Kupe-Bakossi)
Salacia sp. nr. *erecta*
Magnistipula butayei subsp. *balingembaensis* (see Fig. 5)
Magnistipula sp. 1 of Bali Ngemba (more material needed to confirm and describe)
Drypetes sp. 3 (also at Kupe-Bakossi)
Drypetes sp. aff. *leonensis* of Bali Ngemba (see Plate 4(A))
Erythrococca sp. cf. *hispida*
Oncoba sp. nov.
Psorospermum cf. *tenuifolium* (also Bamenda)
Beilschmiedia sp. 1 of Bali Ngemba (see Plate 5(A))
Psychotria sp. A *aff. calva* (also Kupe-Bakossi)
Psychotria sp. A of Bali Ngemba (also elsewhere in Bamenda Highlands)
Psychotria sp. B of Bali Ngemba
Rytigynia sp. (more material needed to confirm)
Tricalysia sp. B aff. *ferorum* (also Kupe-Bakossi)
Vepris sp. A (also elsewhere in Bamenda Highlands)
Vepris sp. B (see Fig. 7)
Allophylus sp. A (more material needed to confirm)
Leptonychia sp. 1 of Bali Ngemba (see Plate 6(D))
Bulbophyllum sp. nov.
Polystachya anthoceros (also SE Nigeria)

NEW TAXA

In addition to the new taxa contained in the list of endemics the following three new taxa are also known from Bali Ngemba, but are not endemic or nearly endemic, being more widespread and known at at least two other locations in addition to Bali Ngemba.

Strombosia sp. 1 (also Kupe-Bakossi, Bamboutos & Mambilla) (see Fig. 6)
Deinbollia sp. 1 (also known from Mt Cameroon and Kupe-Bakossi)
Commelina sp. B of FWTA (also known from SE Nigeria and elsewhere in the Bamenda Highlands)
Scleria afroreflexa (also Kupe-Bakossi and Ijim Ridge) (see Fig. 8)

The number of taxa new to science described or due to be described, solely or partly from material derived from Bali Ngemba, thus numbers 25.

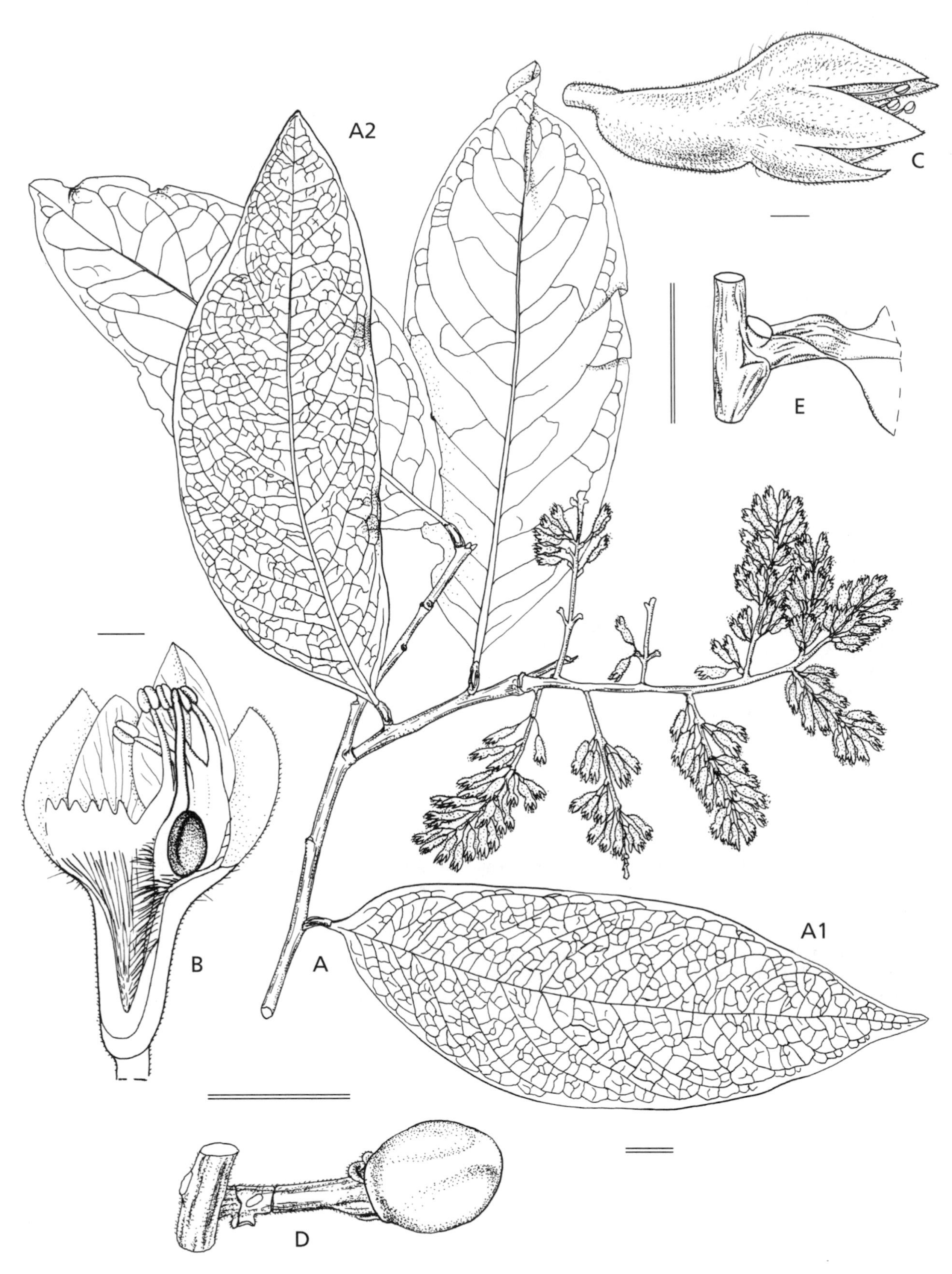

Fig. 5. *Magnistipula butayei* subsp. *balingembaensis*. **A** habit; **A1** abaxial laminal surface; **A2** adaxial surface; **B** flower (longitudinal section); **C** flower (external side view); **D** immature fruit and pedicel; **E** base of lamina with petiole. **A, D & E** from *Pollard* 1028 (K); **A1, A2, B & C** from *Zapfack* 2038 (K). Drawn by Linda Gurr. Double line scale bar represents 1 cm and single line scale bar represents 1 mm.

VALIDATION OF NEW NAME

A NEW SUBSPECIES OF *MAGNISTIPULA* ENGL. (CHRYSOBALANACEAE) FROM THE BALI NGEMBA FOREST RESERVE

Benedict John Pollard, Cynthia A. Sothers & Ghillean T. Prance
Herbarium, Royal Botanic Gardens, Kew, Richmond, Surrey, UK, TW9 3AE

Summary. *Magnistipula butayei* De Wild. subsp. *balingembaensis* Sothers, Prance & B.J.Pollard is described and illustrated. It is the eleventh subspecies to be described in *M. butayei*, and represents the first of several new taxa to have been described from collections made during the recent expeditions to the Bali Ngemba F.R. (2000–2002). Notes are provided on its habitat, ecology, conservation (for which an IUCN Red-Listing assessment is made), and its taxonomic relationships.

INTRODUCTION

Magnistipula butayei De Wild. occurs both in rainforests of the Guineo-Congolian phytochorion, and in woodland and wooded grassland of the Sudanian and Zambezian phytochoria (Prance & Sothers 2003). White (1976) recognised 9 subspecies and Champluvier (1990) described a tenth, subsp. *ituriensis* Champl., from the Democratic Republic of the Congo (formerly Zaïre). She also modified White's subsp. *transitoria*, which he created to accommodate those individuals occurring in the geographical area separating subsp. *bangweolensis* (R.E.Fr.) F.White from subspp. *butayei* and *montana* (Hauman) F.White. Champluvier (op. cit.) found that subsp. *transitoria* was a mixture of several subspecies and separated them accordingly, and reduced subsp. *transitoria* to synonymy. Currently 10 subspecies are accepted. This new subspecies brings to 8 the number of taxa in *Magnistipula* Engl. known to occur in Cameroon, including *M. butayei* De Wild. subsp. *sargosii* (Pellegr.) F.White from S Province (see Letouzey & White 1978: 79). Champluvier (pers. comm. 2004) suspects that that taxon may need to be raised to specific rank, based on the absence of endocarpic golden hairs, but further examination of material held at K, P and several other herbaria is necessary to resolve this issue. Herbaria acronyms follow Holmgren *et al.*, excepting BHFP, for which an explanatory note is given.

THE NEW TAXON

Magnistipula butayei *De Wild.* subsp. **balingembaensis** *Sothers, Prance & B.J.Pollard* **subsp. nov.**, *M. butayei* De Wild. subsp. *glabriusculae* (Hauman) Champl. similis, sed ramulis petiolis et costis nervisque secundariis foliorum glabris (non indumentosis), foliis anguste ovatis ellipticis vel elliptico-oblongis (non obovatis), receptaculo ventraliter paullo gibboso (non valde gibboso), floribus 7–8 mm longo (non 6–7 mm) differt; et ab aliis subspeciebus exocarpis fructuum glabris (non tomentulosum) differt. Typus: Cameroon, NW Province, Bali Ngemba Forest Reserve, Lalé 1, 05°49'27"N, 10°06'04"E, 1700 m, fl. 19 April 2002, *Pollard* 1028 (holotypus K!, isotypi BHFP!, BR!, MO!, P!, SCA!, WAG!, YA!).

Tree to 15 m tall. *Branches* glabrous, lenticellate. *Leaves* narrowly-ovate, elliptic or elliptic-oblong, 9.9–13.4(–18.5) × 3.4–5.4(–6.8) cm; lamina glabrous throughout, drying pinkish above and on the veins beneath; apex acute to (long-) acuminate; base rounded to cuneate (rarely subcordate), occasionally almost shortly-attenuate; glands on lower leaf surface usually in 1–4(–several) basal pairs either side of midrib, rarely with a basal pair on leaf-margin, with some also scattered randomly in apical half of blade; secondary veins 7–10(–16) pairs, glabrous; petiole glabrous, woody, articulate, 5–6(–10) mm long, pulvinate. *Stipules* caducous (not seen). *Inflorescences* terminal or subterminal panicles; branches horizontal or patent, to 9 cm, with up to 25 cymules each (usually many fewer); rachis and branches fulvous; bracteoles often caducous, narrowly lanceolate, c. 4 mm long; pedicels c. 2 mm long. *Flowers* 7–8 mm long. *Receptacle tube* 3.5–4.0 mm long, ventrally slightly gibbous, puberulous outside, occasional glands seen on exterior, usually nearer the base. *Sepals* 3–4 mm long. *Petals* white, ciliolate at apex. *Stamens* 7, on one side, connate for a third of their length. *Staminodes* connate for most of their length, forming a comb-like ligule, extending c. 1 mm above the receptacle face. *Style* to 4 mm, glabrous. *Ovary* glabrous. *Fruit* ellipsoid, drying black; exocarp glabrous, smooth, to 9 × 8 mm (immature, as in *Pollard* 1082), or c. 3 × 2.5 cm (mature, as in *Ghogue* 1081); endocarp with many golden hairs embedded, each c. 1–3 mm long. *Seed* 1. Fig. 5
SPECIMENS. NW Province. Bali Ngemba Forest Reserve: road from Mantum to Pigwin, 05°48'23"N, 10°05'55"E, 1950 m, fr. 11 Nov. 2000, *Ghogue* 1081 (BHFP, K!, MO!, SCA, YA) & 05°49'32"N, 10°05'42"E, 1310 m, fl. 11 April 2002, *Zapfack* 2038 (BHFP, K!, SCA, YA); Lalé I, 05°49'27"N, 10°06'04"E, 1700 m, fl. & fr.

19 April 2002, *Pollard* 1028 (holotype K!, isotypes BHFP!, BR!, MO!, P!, SCA!, WAG!, YA!); 05°50'00"N, 10°04'39"E, fl. 12 April 2004, *Nana* (*Felicité*) 105 (BR, K!, MO, P, SCA, WAG, YA).

NOTES. *Ghogue* 1081 includes leafy branchlets and 2 mature fruits collected from the ground, the latter having fallen from the parent plant. We cannot be sure that these fruits belong to the new taxon, as they are not attached to vegetative parts in our material, and are poorly preserved, but we have tentatively included them in the description. They do, however, possess the densely arranged endocarpic golden-hairs, characteristic of the genus, and also appear to be unilocular, characteristic of all the species of *Magnistipula* Engl., excepting *M. tessmannii* (Engl.) Prance, which is bilocular. It seems that in *Ghogue* 1081, the exocarp may have decomposed or been eaten (it may be thin and crustaceous), but it is difficult to interpret the external morphology based on these two gatherings. Further complete carpological collections, preserved in spirit, would help us fully understand the mature fruiting morphology.

Zapfack 2038 was originally sorted at K as a Legume, and Champluvier (1990) also found specimens of *M. butayei* subsp. *ituriensis* Champl. originally curated under Leguminosae. The conspicuously zygomorphic flower with a gibbous receptacular base, and the presence of pulvini, make them easily confusable with Legumes (when in flower). In Chrysobalanaceae the gynobasic style, excentric ovary, receptacle tube, and connate stamens and staminodes render the apparent similarities superficial.

HABITAT & ECOLOGY. Submontane forest understorey and riverine forest; (?1310–)1700–1950 m altitude. The altitudinal reading of *Zapfack* 2038, 1310 m, is probably inaccurate and best treated with caution; the lowest point of the actual reserve is c. 1400 m.

CONSERVATION. Within the boundaries of the reserve, there is widespread felling of canopy trees (e.g. *Synsepalum msolo* (Engl.) T.D.Penn. (Sapotaceae)), large emergent species (e.g. *Pterygota mildbraedii* Engl. (Sterculiaceae)), and many smaller tree species, as well as extensive clearance of shrubby and herbaceous undergrowth vegetation (Pollard pers. obs. 2000–2002). This is the work of local farmers, particularly those from the far side of the higher altitude grassland ridge towards Pinyin (c. 2000 m), who are increasing the number and scale of their agricultural small-holdings to increase productivity of various crops, notably coco-yams and potatoes. *Ghogue* 1081 was collected from 'disturbed plantation area in grassland', which alludes to the fact that the natural habitat of this taxon is being altered by farming practices at the higher elevations of the reserve, towards the grassland ridge. All of the reserve's native vegetation is thus under threat, and there are currently no efficient protective measures in place to monitor or prevent such activities. Without conservation-directed intervention, it is likely that this decline in habitat quality will continue unchecked and could contribute towards the extinction of several threatened plant taxa known to occur within the reserve. Much of the conservation initiatives and actions in the NW Province are co-ordinated at the nearby Bamenda Highlands Forest Project, which has overseen promising community-based forest protection throughout the Province. They also hold a reference collection of an estimated 1500 plant specimens from NW Province, collected during the three RBG, Kew - Herbier National Camerounais expeditions since 2000. This 'herbarium' is referred to here by the acronym BHFP. The new taxon should become a priority target for seed-collection and cultivation for introduction at other suitable sites nearby, and could be used to highlight the urgent need for *in-situ* conservation measures. It here becomes a 'flagship' taxon for the conservation of the plants of the Bali Ngemba Forest Reserve, which may be the last remaining patch of submontane forest in the whole of NW Province, 96.5% of the forest having been lost in the Bamenda Highlands region (Cheek *et al.* 2000). *Magnistipula butayei* De Wild. subsp. *balingembaensis* Sothers, Prance & B.J.Pollard is only known from four collections inside the reserve (Area of Occupancy, and Extent of Occurrence are c. 8 km^2). Our estimate of generation time for this subspecies is c. 20 years, so that within three generations (60 years) from now, we suspect c. 80% loss of forest cover, based on observations of current habitat loss (Cheek, Etuge, Onana, Pollard & Zapfack pers. obs. 2000–2002). It is therefore here assessed as Critically Endangered, following the guidelines of IUCN (2001).

ETYMOLOGY. This species is named *balingembaensis* to draw attention to the critical importance of the Reserve of the same name, both in regional and global terms.

IUCN RATING. CR A3c; B1ab(iii)+2ab(iii).

TAXONOMIC RELATIONSHIPS. The most striking diagnostic character of subsp. *balingembaensis* is that the whole plant (excepting the inflorescence) is completely glabrous, most notably the exocarp of the fruit, which is tomentulose in all the other subspecies. In Prance & Sothers (2003), the new subspecies keys out to subsp. *glabriuscula* (Hauman) Champl., but differs in having glabrous branchlets (not pubescent); glabrous midribs and secondary nerves (not sparsely white-hairy beneath and sometimes above); glabrous petioles (not brown-hairy, with hispid hairs to 1 mm long); leaves narrowly-ovate, elliptic or elliptic-oblong (not usually obovate), with: bases usually rounded to cuneate, rarely subcordate (not subcordate to rounded, rarely cuneate), apices (long-)acuminate (not sometimes obtuse); receptacles only slightly gibbous ventrally (not with pronounced ventral gibbosity); flowers 7–8 mm (not 6–7 mm); staminodes extending ≤ 1 mm beyond the receptacle face (not 1–1.5 mm).

See Table 2 (below) for an interpretation of the varying degrees of gibbosity between all the subspecies, most of which are illustrated in Prance & White (1988), of which only subsp. *montana* is less gibbous (<) than the new subspecies.

Subspecific epithet	gibbosity	ventral / dorsal
balingembaensis	=	v
bangweolensis	>>	v, d
butayei	>	v
glabriuscula	>	v
greenwayi	>	v
ituriensis	>	v
montana	<	-
sargosii	>	v
tisserantii	>>	v, d
youngii	>	v

Table 2. Relationships of gibbosity of the receptacule tube for all of the accepted subspecies in *Magnistipula butayei* De Wild., as compared to subsp. *balingembaensis*. (Explanation of terms: < less than; > more than; >> much more than; - / = same as; v ventral, d dorsal)

The ratio of the length of the dorsal sepal to the receptacular tube is ± stable at c. 1 : 1 in our newly described taxon, but in the other subspecies is never this ratio, e.g. in subsp. *montana* (Hauman) F.White, the receptacle is shorter than the dorsal sepal. The angle of the throat of the receptacle tube is variable, from flat to quite oblique, more often flat, even in flowers on the same inflorescence, and is not thought by us to be a useful diagnostic character for our new subspecies. The new taxon displays a strikingly angled dorsal side, which is usually cornute (horn-shaped).

ACKNOWLEDGEMENTS

We thank Linda Gurr (K) for the illustration, Mark Coode (K) for the Latin diagnosis and Dominique Champluvier (BRLU) for helpful comments on the taxonomic relationships. The fieldwork which resulted in the discovery of this subspecies was partly funded through sponsorship received from the Earthwatch Institute (Europe) with the support of DGVIII of the European Commission and by a Darwin Initiative Grant from the Department of the Environment, Transport and the Regions, UK. We are especially grateful for the co-operation of Dr Gaston Achoundong, chief of HNC, and his staff at Yaoundé for facilitating fieldwork in Cameroon and for assistance with obtaining research permits. We are indebted to The Paramount Fon of Bali (Dr Ganyonga III); The Fon, Quarterhead and people of Mantum for assistance and hospitality during our time in the lands of the Bali people; staff of the Ministry of Environment and Forests (MINEF), NW Province, for enabling us to enter and conduct research within the reserve; Michael Vabi, John DeMarco, Anne Gardener and all of the BirdLife International staff at BHFP, for introducing us to the reserve and supporting our work therein.

REFERENCES

Champluvier, D. (1990). Une nouvelle sous-espèce de *Magnistipula butayei* De Wild. (*Chrysobalanaceae*) de l'Ituri (Zaïre). Bull. Jard. Bot. Natl. Belg. 60: 393–403.

Cheek, M., Onana, J.-M. & Pollard, B.J. (2000). The Plants of Mount Oku and the Ijim Ridge, Cameroon, A Conservation Checklist. RBG, Kew, UK. iv + 211 pp.

IUCN (2001). IUCN Red List Categories and Criteria. Version 3.1. IUCN, Gland, Switzerland. 30 pp.

Holmgren, P.K., Holmgren, N.H. & Barnett, L.C. (1990). Index Herbariorum. 8th ed. New York Botanical Garden. 693 pp.

Letouzey, R. & White, F. (1978). *Chrysobalanaceae*. Flore du Cameroun 20: 3–138. Museum National d'Histoire Naturelle, Paris.

Prance, G.T. & White, F. (1988). The genera of Chrysobalanaceae. Philos. Trans., Ser. B, v. 320, no. 1197: 161.

Prance, G.T. & Sothers, C.A. (2003). Chrysobalanaceae 2: *Acioa* to *Magnistipula*. Species Plantarum: Flora of the World, Part 10: 1–268.

White, F. (1976). The taxonomy, ecology and chorology of African Chrysobalanaceae (excluding *Acioa*). Bull. Jard. Bot. Natl. Belg. 46: 265–350.

Fig. 6. *Strombosia* sp. 1 (Olacaceae). Drawn by Hazel Wilks.

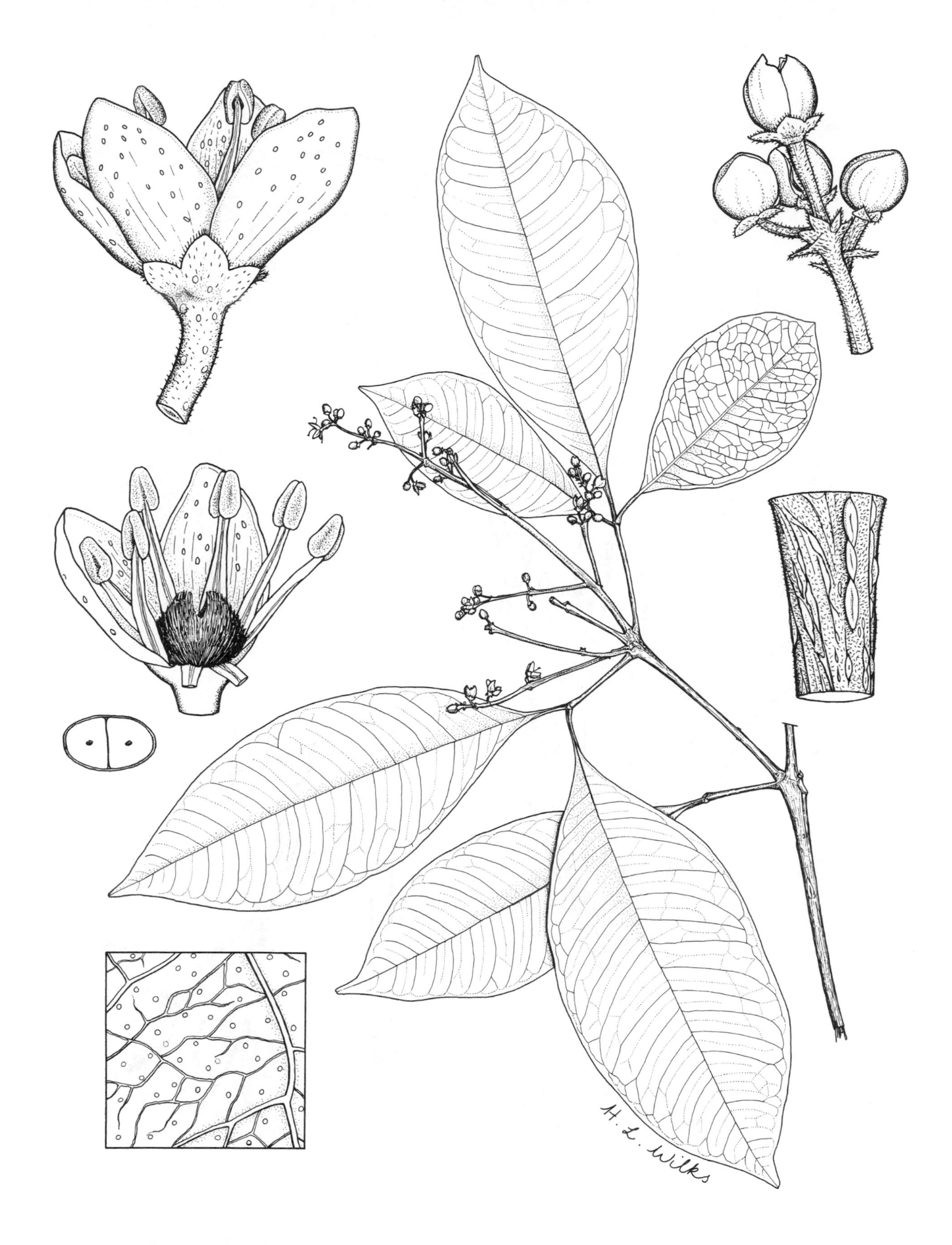

Fig. 7. *Vepris* sp. B (Rutaceae). Drawn by Hazel Wilks.

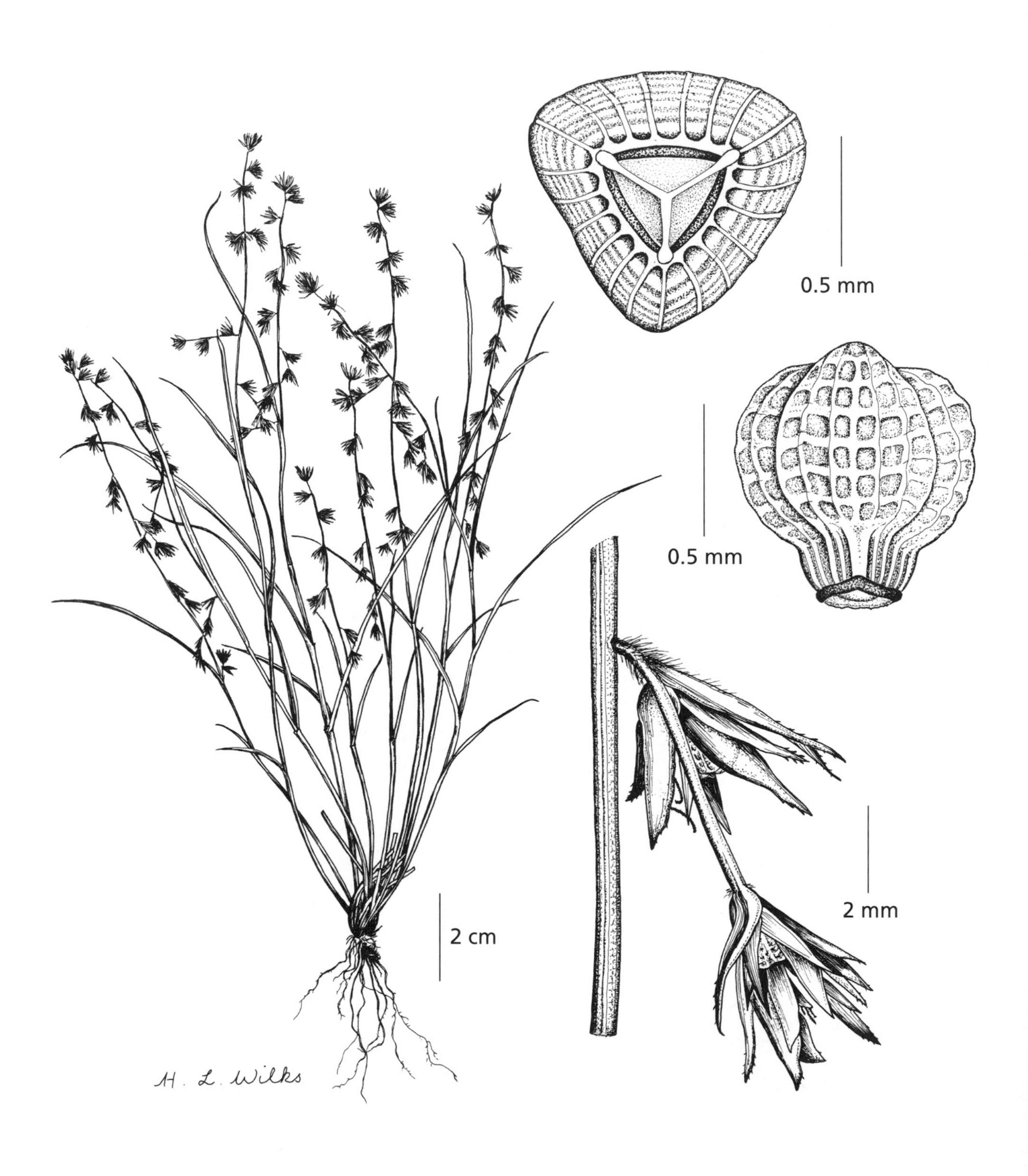

Fig. 8. *Scleria afroreflexa* (Cyperaceae). Drawn by Hazel Wilks.

RED DATA TAXA OF BALI NGEMBA FOREST RESERVE

Martin Cheek & Yvette Harvey
Herbarium, Royal Botanic Gardens, Kew, Surrey, TW9 3AE, U.K.

INTRODUCTION

As in the previous conservation checklist in this series (The Plants of Kupe, Mwanenguba and the Bakossi Mts, Cheek *et al.* in press), **all** flowering plant taxa recorded in the checklist have been assessed on a family by family basis for their level of threat, i.e. as threatened (CR – critically endangered; EN – endangered; VU – vulnerable), near threatened (NT) or of least concern (LC).

The main part of this chapter consists of taxon treatments, giving detailed information on the 39 Red Data species known to be present in the checklist area. Several of these taxa were assessed for the first time in the process of writing this book while others had been assessed as threatened in other publications and are reassessed here. All of the treatments have been reviewed by Craig Hilton-Taylor, the IUCN Red Data officer. All have been accepted. IUCN rules do not allow acceptance of taxa unless they are either published or on the brink of publication. Consequently, the estimated 24 new species to science that are not yet at this point, many of which are only known from Bali Ngemba, have not been assessed and are not mentioned in this chapter.

This, the introductory part of the chapter, details the methodology used in making the assessments, followed by a series of lists in which Red Data taxa are detailed by vegetation type.

ASSESSMENTS – METHODOLOGY

Taxa of Least Concern

In the first place all taxa that were found to be fairly widespread e.g. Extent of Occurrence (IUCN 2001) greater than 20,000 km^2 (and/or from 20 or more localities) were listed as LC. These facts were established principally using FWTA as an indication of range and number of collections sites, supplemented by other published sources, such as Flore du Cameroun, or by research into specimens at Kew, in cases of doubt. Taxa which, by these measures, are widespread and common do not qualify as threatened under Criterion B of IUCN (2001), the main criterion used here for assessing threatened species. Under Criterion A, widespread and common taxa may, in contrast, still be assessed as threatened, but only if their habitat, or some other indicator of their population size has been, or is projected to be, reduced by at least 30% in the space of three generations, so long as this does not exceed 100 years.

Threatened and Near Threatened Taxa

Those taxa in the checklist that were not assessed as LC, were then checked for level of threat using IUCN (2001). Most of these taxa were at least fairly narrowly endemic (restricted) in their distributions, e.g. endemic to Cameroon, or to NW Province, Cameroon and Nigeria. Criterion C, which demands a knowledge of the number of individuals of a species, was not used, since this data were not available. Similarly, neither was Criterion D, except in the special case of VU D2 (see IUCN 2001). Criterion E, which depends on quantitative analysis to calculate the probability of extinction over time, was also not used.

i) Using Criterion B

About one-third of the assessments were made using Criterion B (usually B2ab), since the nature of the data available to us for our taxa lends itself to this criterion. Knowledge of the populations and distributions of most tropical plant species is mainly dependent on the existence of herbarium specimens. This is because there are so many taxa, most of which are poorly known and have never been illustrated. For this reason observations based only on sight-records are particularly unreliable and so undesirable in plant surveys of diverse tropical forest. Exceptions can be made when a family or genus specialist, working with a monograph at hand, or a proficient tree spotter identifying timber tree species. In contrast, surveys of birds and primates are not specimen dependent, since species diversity in these groups is comparatively low, and comprehensive, well-illustrated identification guides are available.

For the purpose of Criterion B we have taken herbarium specimens to represent 'locations'. Deciding whether two specimens from one general area represent one or two locations is open to interpretation, unless they are from the same individual plant. Generally in the case of several specimens labelled as being from one town e.g. 'Bipinde', or one forest reserve e.g. 'Bafut Ngemba F.R.', these have been interpreted as one 'location'. Where a protected area has been divided into several geographical subunits, as at Mt Cameroon (see Cable & Cheek, 1998), and it is known that, say six specimens of a taxon occurring at Mt Cameroon fall into two such subunits, then this is treated as two locations. 'Area of occupancy' (AOO) and 'extent of occurrence' (EOO) have been measured by extrapolating from the number of locations at which a species is known. The grid cell size used for calculating area of occupancy has depended on the taxon concerned - figures of 1 or 2 km^2 have often been employed. Information on declines in Criterion B has been obtained from personal observations, sometimes supplemented by local observers.

ii) Using Criterion A

The remaining assessments, about one half of the total, were made using Criterion A. This criterion was used more than any other because montane and submontane taxa of the Cameroon Highlands that extend to the Bamenda Highlands were relatively easy to assess thanks to a recent GIS study of forest loss over 8 years in one part of the Bamenda Highlands, in the Kilum-Ijim area (Moat in Cheek *et al.* 2000).

These data have been used to estimate that during the last century, over 30% of the forest habitat of montane (above 2000 m alt.) and submontane (800–2000 m alt.) species has been lost in the Cameroon Highlands, so qualifying species with this range as Vulnerable. *Schefflera mannii* is an example of such a species. Formerly such taxa were often treated as unthreatened (e.g. Cable & Cheek 1998) because they are secure on Mt Cameroon.

iii) Using Criterion D

Assessments using Criterion D generally depend on a knowledge of the global numbers of mature individuals (D1). This knowledge is generally not available for plant taxa in Cameroon. However VU D2 is a special case which allows threatened status for taxa on the basis simply of a low number of sites (five or less) or a restricted area of occupancy (20 km^2 or less), in the absence of other factors, such as declines or direct threats, e.g. *Drosera pilosa*. VU D2 has therefore been used in a small number of cases in this Red Data chapter.

Changes in IUCN criteria

It is to be hoped that there will be a moratorium on the almost annual changes in IUCN criteria which have been made in recent years. These frequent changes have made the work of assessors more difficult and have also reduced comparability of assessments made in different years. Assessments made in earlier Cameroonian checklists (Cable & Cheek 1998, Cheek *et al.* 2000) under IUCN (1994) criteria have been updated according to IUCN (2001) criteria where the taxa occur in the present checklist.

Red Data species by vegetation type

The Red Data species of Bali Ngemba forest reserve are presented below under each vegetation type in which they occur. The vegetation types follow the classification used in the vegetation chapter. No species occurs in more than one vegetation type. More detailed information on the placement of the Red Data species within the subdivisions of each vegetation type is given in the vegetation chapter.

Species are listed below as they appear within the vegetation chapter, so that for example, within montane vegetation (mapping unit 2) forest species are listed first, and grassland-scrub-forest edge species last. In the Red Data taxon treatments that conclude this chapter the species are presented in the same order as in the main checklist, i.e. alphabetically by family and within family first by genus, then by species. The author(s) of the family assessment is given.

By far the most important vegetation type for threatened taxa is the submontane forest (22 Red Data taxa), with significant numbers also occurring in montane vegetation (13 Red Data taxa) and submontane inselberg grassland (4 or 5 Red Data taxa). The remaining vegetation types have none.

MAPPING UNIT 1. SUBMONTANE FOREST (1300 to 1900–2000 m alt.)

Pavetta hookeriana VU
Memecylon dasyanthum VU
Quassia sanguinea VU
Chassalia laikomensis CR
Crassocephalum bauchiense VU
Dombeya ledermannii CR
Allanblackia gabonensis VU
Pseudagrostistachys africana VU

Allophylus conraui EN
Psydrax dunlapii VU
Pentarrhinum abyssinicum subsp. *ijimense* CR
Eugenia gilgii CR
Newtonia camerunensis CR
Bulbophyllum nigericum VU
Xylopia africana VU
Magnistipula butayei subsp. *balingembaensis* CR
Oncoba lophocarpa VU
Entandrophragma angolense VU
Allophylus bullatus VU
Begonia oxyanthera VU
Commelina cameroonensis VU
Vittaria guineensis var. *camerooniana* VU

MAPPING UNIT 2. MONTANE FOREST, GRASSLAND AND SCRUB–FOREST EDGE (1900–2000 to 2200 m alt.)

Schefflera mannii VU
Ixora foliosa VU
Impatiens sakerana VU
Wahlenbergia ramosissima subsp. *ramosissima* VU
Clutia kamerunica EN
Psorospermum aurantiacum VU
Dalbergia oligophylla EN
Crotalaria ledermannii VU
*Dissotis bamendae*VU
Dissotis longisetosa VU
Pentas ledermannii VU
Bulbostylis densa var. *cameroonensis* VU
Polystachya anthoceros EN

MAPPING UNIT 3. SUBMONTANE INSELBERG GRASSLAND (1300–1650 m alt.)

Vernonia guineensis var. *cameroonica* VU
Drosera pilosa VU
Aeollanthus trifidus VU
Scleria afroreflexa EN. Habitat needs clarification.
Eriocaulon asteroides VU

MAPPING UNIT 4. DERIVED SAVANNA AND FALLOW (1300–1500 m alt.)

No Red Data taxa are known.

MAPPING UNIT 5. *EUCALYPTUS* PLANTATION

No Red Data taxa are known.

RED DATA TAXON TREATMENTS

DICOTYLEDONAE

ANNONACEAE (M. Cheek) (see Plate 3(A))

Xylopia africana (Benth.) Oliv.
VU A2c
Range: São Tomé (1 collection); SE Nigeria (Obudu (1 collection)); Cameroon (Mt Cameroon (4 collections), Bakossi (10 collections), Bali Ngemba (numerous collections)).
Of all the forests that we have surveyed, *Xylopia africana* has been found in greatest density at the Bali Ngemba Forest Reserve in North West Province. This, now the largest remnant of the forest that cloaked the Bamenda Highlands at the 1300–1900 m range, is less than 1000 Ha in extent and shrinking fast due to illegal clearance for farming.
Presumably this species was once common in the Bamenda Highlands where it is now all but extinct. While there are no figures for rates of forest loss in the Bamenda Highlands as a whole, in one area which has been studied, the Kilum-Ijim area, forest loss of 25% over 8 years in the 1980s–1990s has been recorded (Moat in Cheek *et al.* 2000). Past forest loss in the Bamenda Highlands is therefore the main basis for the threat to *Xylopia africana*. On Mt Cameroon it appears rare, being found only twice in the surveys of 1992–1994. Elsewhere in the mountains of the Cameroon line it is also known from the extension into Nigeria: the Obudu Plateau where it is also threatened due to forest clearance, if indeed it is still extant there. It is also known from São Tomé in the Gulf of Guinea. Strangely, it is not known from Bioko or from the Rumpi Hills or the Bamboutos Mts. Mt Kupe and the Bakossi Mts probably now support the largest single subpopulation of *Xylopia africana*. This taxon has also been assessed for Cheek *et al.* (in press).
Habitat: submontane and lower montane forest; 800–2000 m alt.
Threats: clearance of forest for timber and agricultural land.

Management suggestions: if proposals to protect forest above 1000 m alt. in much of Bakossi are enacted and respected, this subpopulation seems secure.

ARALIACEAE (M. Cheek)

Schefflera mannii (Hook.f.) Harms
VU A2c
Range: Annobon (1 collection); São Tomé (1 collection); Bioko (10 collections at 4 sites); SE Nigeria (Obudu, 1 collection); Cameroon (Mt Cameroon (7 collections), Mt Oku (7 collections), Kupe and Mwanenguba Mts (6 collections at 4 sites), Bamboutos, Bafut Ngemba & Bali Ngemba (all 1 collection)).
One of the very few montane (above 2000 m alt.) trees that are endemic to the Cameroon uplands (another is *Morella arborea*), this evergreen canopy tree begins life as an epiphytic shrub. This taxon has also been assessed for Cheek *et al.* (in press). The only change in the assessment has been the addition of Bali Ngemba F.R. to the range.
Habitat: montane forest; (1400–)2000–2400 m alt.
Threats: forest clearance for agriculture and wood has reduced the habitat of this species by an estimated 30% or more over its whole range due principally to loss in the Bamenda Highlands, which, having the largest area above 2000 m in the Cameroon uplands, was probably once the stronghold for this species. Between 1987 and 1995, 25% of forest was lost in one area of the Bamenda Highlands (Moat in Cheek *et al.* 2000). Extensive losses of habitat have also occurred at Mwanenguba, Obudu, Bamboutos and Bafut Ngemba.
Management suggestions: the status of *S. mannii* on the Gulf of Guinea islands requires elucidation, perhaps by survey. At Mts Kupe, Cameroon and Oku (Kilum-Ijim) the species seems currently secure.

ASCEPIADACEAE (Y.B. Harvey)

Pentarrhinum abyssinicum Decne subsp. *ijimense* Goyder
CR A1c
Range: Cameroon (Mt Oku and the Ijim Ridge (5 collections), Bali Ngemba F.R. (1 collection)).
This endemic forest climber was discovered to be a potential novelty by Dr Goyder in 1998 while identifying specimens of Asclepiadaceae from the 1996 expedition to the Ijim Ridge. At that time only a single specimen was known. *Etuge* 3565 was collected near Tum at Ijim on 21 November 1996. A second specimen from Kilum, collected slightly earlier (*Buzgo* 798) lacks collecting data and so had been passed over. A new site was discovered for this plant by Pollard at the Akwamofu Medicinal Forest at Laikom in December 1998 (*Pollard* 368). In November 1999 Etuge led a large party of botanists back to the Tum forest patch which had yielded his specimen in 1996. Here about a dozen plants were counted flowering and fruiting profusely on both sides of the stream that bisects the forest. On the west side of the stream this climber is locally abundant over several square metres of *Solanecio mannii*, to the extent that it is difficult to distinguish one individual from another. On the east side of the stream, three plants were seen, each separated by 10–15 m. New material, including flowers in spirit, was collected (*Cheek* 9943) so that the taxon could be formally described. Etuge (pers. comm.) reported that this forest patch had been reduced by about two-thirds in the three years since he was there last. A third site for this rare plant was discovered by the Ghanaian botanist Amponsah at 'Back valley', Mbingo on 10 November 1999 (*Cheek* 10063). In October 2001, Etuge (*Etuge* 4267) found this taxon at a fourth site, Bali Ngemba F.R. This taxon is completely unprotected in Mt Oku and the Ijim ridge. This taxon was assessed as above in Cheek *et al.* (2000). No new data have come to light since then, apart from the recent collection at Bali Ngemba F.R., so the original conservation assessment is maintained here.
Habitat: lower montane forest with *Garcinia smeathmannii, Solanecio mannii, Cuviera longiflora,* and *Pouteria altissima*; 1750–1900 m alt.
Threats: clearance of forest for cultivation of crops.
Management suggestions: effort should be made to protect the surviving forest patch at Tum (site of the largest known subpopulation of this taxon but outside the present protected area boundary) and to establish whether other sites for this variety exist apart from the three listed.

A. Mantum village, Earthwatch basecamp, 2000–2004, Oct. 2001, Parmjit Bhandol
B. Rita Ngolan leading group on path from Mantum to Kedeng: forest at 1400 m alt., Oct. 2001, Parmjit Bhandol
C. Martin Cheek collecting data: forest canopy intact, but understorey planted with cocoyams, Oct. 2001, Parmjit Bhandol
D. Stream near western boundary of the reserve, Apr. 2004, Dave Roberts
E. Forest with inselbergs beyond, Apr. 2004, Dave Roberts
F. Effects of felling trees in the reserve for building materials, Oct. 2001, Benedict Pollard
G. Effects of 'slash and burn' agriculture in the reserve, Oct. 2001, Benedict Pollard

Plate 1

Reserve viewed from the grassland ridge, c. 2000 m alt., April 2002, Benedict Pollard

Reserve viewed from farmland around Mantum, c. 1400 m alt., April 2002, Benedict Pollard

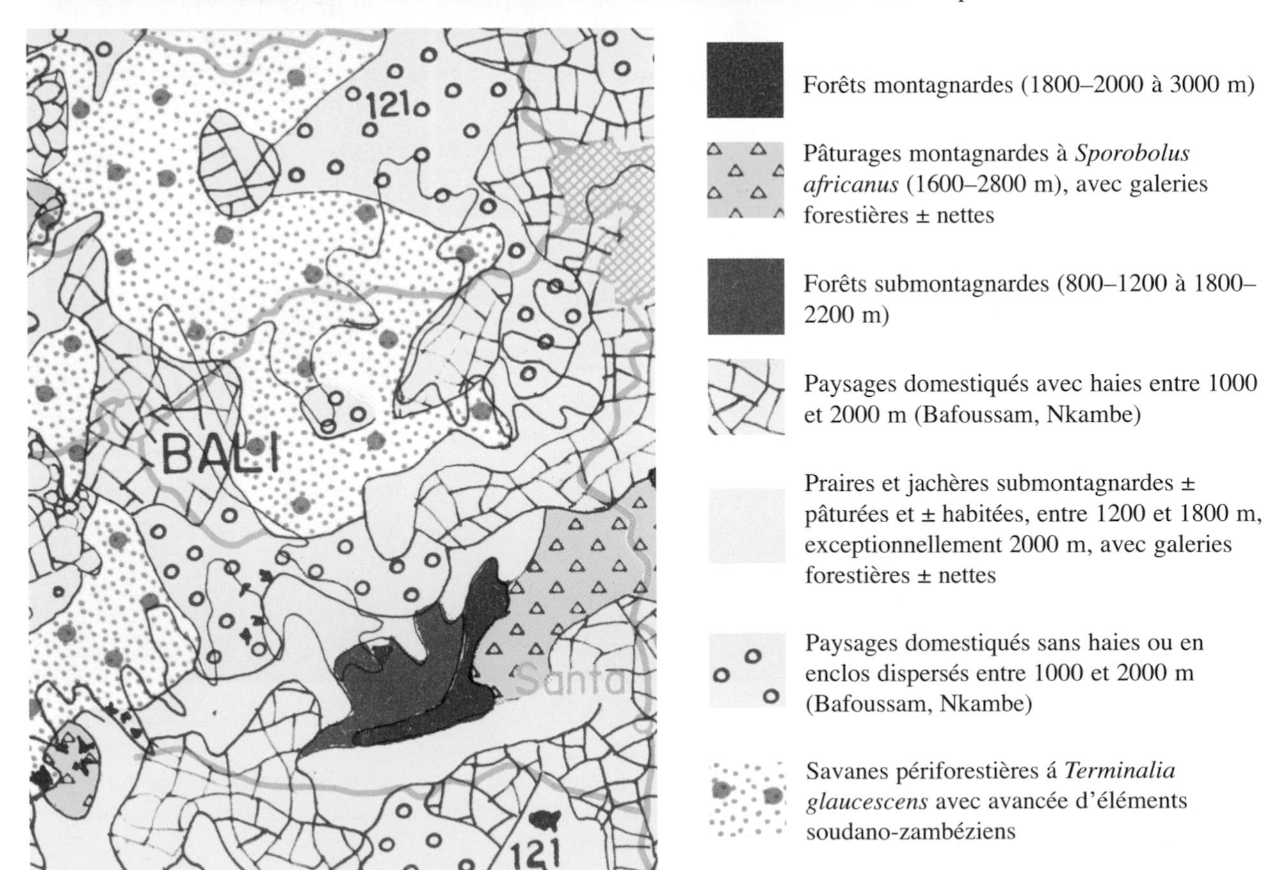

C. Phytogeographic map of the reserve and surrounding area. Reproduced with the permission of IRAD, Yaoundé, Cameroon from Letouzey (1985).

Plate 2

A. *Xylopia africana* (Annonaceae), Nov. 2000, *Cheek* 10527, Martin Cheek
B. *Impatiens sakerana* (Balsaminaceae), Apr. 2004, Dave Roberts
C. *Echinops gracilis* (Compositae), Apr. 2004, Adam Surgenor
D. *Gynura miniata* (Compositae), Apr. 2004, Adam Surgenor
E. *Rhaphidiocystis phyllocalyx* (Cucurbitaceae), Apr. 2004, Dave Roberts
F. *Clutia kamerunica* (Euphorbiaceae), Nov. 2000, Christo van de Rheede

Plate 3

A. *Drypetes* sp. aff. *leonensis* of Bali Ngemba (Euphorbiaceae), Apr. 2002, *Pollard* 1034, Benedict Pollard
B. *Oncoba lophocarpa* (Flacourtiaceae) (fruiting), Nov. 2000, Martin Cheek
C. *Allanblackia gabonensis* (Guttiferae), Apr. 2004, Iain Darbyshire
D. *Psorospermum febrifugum* (Guttiferae), Apr. 2004, Adam Surgenor
E. *Aeollanthus trifidus* (Labiatae), Oct. 2001, *Pollard* 698, Benedict Pollard

Plate 4

A. *Beilschmiedia* sp. 1 of Bali Ngemba (Lauraceae), Nov. 2000, Martin Cheek
B. *Zenkerella citrina* (Leguminosae – Caesalpinioideae), Nov. 2000, Martin Cheek
C. *Warneckea cinnamomoides* (Melastomataceae), Nov. 2000, *Cheek* 10529, Martin Cheek
D. *Ficus thonningii* (Moraceae), Apr. 2004, *Rønsted* 223, Nina Rønsted
E. *Pittosporum viridiflorum 'ripicolum'* (Pittosporaceae), Apr. 2004, Adam Surgenor
F. *Mussaenda erythrophylla* (Rubiaceae), Apr. 2004, Adam Surgenor

Plate 5

A. *Synsepalum brevipes* (Sapotaceae), Nov. 2000, *Cheek* 10519, Martin Cheek
B. *Psychotria camptopus* (Rubiaceae), Apr. 2004, Iain Darbyshire
C. *Cycnium adonense* subsp. *camporum* (Scrophulariaceae), Apr. 2002, *Pollard* 972, Benedict Pollard
D. *Leptonychia* sp. 1 of Bali Ngemba (Sterculiaceae), Nov. 2000, *Cheek* 10521, Martin Cheek

Plate 6

A. *Crinum zeylanicum* (Amaryllidaceae), Apr. 2004, Dave Roberts
B. *Scadoxus multiflorus* (Amaryllidaceae), Apr. 2004, Iain Darbyshire
C. *Commelina africana* var. *krebsiana* (Commelinaceae), Oct. 2001, Parmjit Bhandol
D. *Dracaena arborea* (Dracaenaceae), Apr. 2004, Iain Darbyshire
E. *Eriocaulon asteroides* (Eriocaulaceae), Oct. 2001, *Pollard* 700, Benedict Pollard
F. *Angraecopsis* sp. vel. *tenerrima* (Orchidaceae), Oct. 2001, Benedict Pollard

Plate 7

A. *Brachycorythis kalbreyeri* (Orchidaceae), Apr. 2004, Dave Roberts
B. *Disa* aff. *nigerica* (Orchidaceae), Apr. 2004, Dave Roberts
C. *Bulbophyllum sandersonii* (Orchidaceae), Apr. 2004, Dave Roberts
D. *Nervilia bicarinata* (Orchidaceae), Apr. 2004, Dave Roberts
E. *Polystachya odorata* var. *odorata* (Orchidaceae), Apr. 2004, Dave Roberts
F. *Cyathea camerooniana* var. *camerooniana* (Cyatheaceae), Nov. 2000, *Cheek* 10507, Christo van de Rheede

Plate 8

BALSAMINACEAE (M. Cheek)

Impatiens sakerana Hook.f. (see Plate 3(B))
VU A2c
Range: Bioko; Cameroon (Mt Cameroon (6 collections), Mwanenguba (2 collections), Bali Ngemba F.R. (2 collections), Bamenda Highlands (1 collection) to Mt Oku (12 collections), W Bamboutos Mts (1 collection)).
This often robust, locally common terrestrial herb is found at the highest altitude of all Cameroonian *Impatiens*. Secure on Mt Cameroon, and probably Bioko, forest at Mwanenguba, the Bamenda Highlands and the Bamboutos Mts has been under pressure from grassland fires set by graziers and for clearance for agriculture; forest in the Kilum-Ijim area outside the protected area having seen a reduction in cover of c. 25% in eight years of the 1980s and 1990s (Moat in Cheek *et al.* 2000). Assuming a generation time of ten years, it is estimated that about 30% habitat loss may have occurred in the last 30 years. This taxon has also been assessed for Cheek *et al.* (in press). The only change in the assessment has been the addition of Bali Ngemba F.R. to the range.
Habitat: understorey of montane forest; 2000 m alt.
Threats: see above.
Management suggestions: enforcement of protected area boundaries. Demographic studies are needed to elucidate generation time and ecological requirements of this taxon.

BEGONIACEAE (M. Cheek)

Begonia oxyanthera Warb.
VU A2c
Range: Nigeria (1 collection); Bioko (6 collections from 2 sites), Cameroon (Mt Cameroon (4 collections), Mt Kupe and the Bakossi Mts (5 collections), Rumpi Hills (1 collection), Bamboutos Mts (2 collections), Mwanenguba (1 collection), Bamenda Highlands (17 collections), Bali Ngemba F.R. (1 collection)).
The range data above are taken from the account of the species in the recently published revision of *Begonia* sect. *Tetraphila* (de Wilde 2002) (with the additional Bali Ngemba F.R. collection). In previous Red Data assessments (Cable & Cheek 1998 and Cheek *et al.* 2000) this taxon was listed as LR nt. Here it is reassessed as Vulnerable because of habitat destruction in what appears to be its main subpopulation in the Bamenda Highlands from which most of the c. 30 specimens listed by de Wilde derive. Moat (in Cheek *et al.* 2000) records forest loss in the Kilum-Ijim protected area itself. Since the 'generation' duration of this taxon might easily be five years, it is estimated that habitat loss for the species over its area of occupancy as a whole is likely to have been over 30% in the last 15 years. The increasing density of this species, moving northwards (Mt Cameroon and Bioko only c. 2 sites each; Bamenda Highlands numerous sites and collections) is perhaps a reflection of a longer dry season requirement than its southern congeners. This taxon has also been assessed for Cheek *et al.* (in press). The only change in the assessment has been the addition of Bali Ngemba F.R. to the range.
Habitat: submontane and montane forest; 1200–2200(–2400)m alt.
Threats: forest clearance for wood and agriculture (mainly in the Bamenda Highlands).
Management suggestions: subpopulations in SW Province, and probably also in Bioko, are fairly secure, lacking threats. In NW Province the substantial subpopulation in the well protected Kilum-Ijim site (Mt Oku and the Ijim Ridge) may be the only locality where the taxon will survive, unfortunately, thus protection here is important.

CAMPANULACEAE (M. Cheek)

Wahlenbergia ramosissima (Hemsl.) Thulin subsp. *ramosissima*
VU B2ab(iii)
Range: Nigeria (Mambilla Plateau (2 collections)); Cameroon (Mt Cameroon (4 collections), Mwanenguba, Chappal Waddi (1 collection each), Bamenda Highlands (2 collections), Bali Ngemba F.R. (1 collection)).
This small (4–15 cm tall) erect, blue-flowered annual herb is found between grass tussocks and is known from only 11 specimens at six mountain sites along the Cameroon line. This taxon has previously been assessed in Cable & Cheek 1998, Cheek *et al.* 2000 and Cheek *et al.* in press, as Vulnerable. No new data have come to light since then, apart from the recent collection at Bali Ngemba F.R.
Habitat: between *Sporobolus* grass tussocks and on shady banks; 1500–2600 m alt.
Threats: unknown, but possibly trampling by cattle during the wet (growing) season.
Management suggestions: assessing the size of the subpopulations, ideally when the plants are in flower and most conspicuous, is advisable. Monitoring variation in numbers of individuals present from year to year would also help better assess the threat to this taxon.

CHRYSOBALANACEAE (B.J. Pollard)

Magnistipula butayi De Wild. subsp. *balingembaensis* Sothers, Prance & B.J.Pollard (see Fig. 5)
(see chapter 'Validation of New Name' for the assessment).

COMPOSITAE (M. Cheek)

Crassocephalum bauchiense (Hutch.) Milne-Redh.
VU B2ab(iii)
Range: Nigeria (Bauchi Plateau, Jos Plateau, Pawpaw Mt, Amedzefe (1 collection each)); Bioko (3 collections); Cameroon (Bakossi (3 collections at 2 sites), Mt Cameroon (11 collections), Mt Oku (6 collections), Bali Ngemba F.R. (1 collection)).
This erect blue-flowered herb is known from nine montane sites in Lower Guinea. If new sites are discovered it may be downgraded to NT, especially in view of the fact that little data are available regarding loss of its habitat. This taxon has also been assessed for Cheek *et al.* (in press). The only change in the assessment has been the addition of Bali Ngemba F.R. to the range.
Habitat: open woodland, savanna, forest edge; 900–2000 m alt.
Threats: clearance of trees for agriculture and wood.
Management suggestions: the effect of forest loss on this, and other rare, sun-loving Compositae, many of which share the same range of habitats, is poorly understood. Research is needed to redress this. Limited 'habitat destruction' at intervals of years, in the form of fires or limited long-fallow small-holder agriculture or small scale tree felling may be helpful to the survival of such taxa by keeping their habitat open and allowing establishment of probably short-lived perennials such as these.

Vernonia guineensis Benth. var. *cameroonica* C.D.Adams
VU B2ab(iii)
Range: Nigeria (Gashaka, 1 collection); Cameroon (NW Province: Ndop; Bamenda-Banso; Bamenda; Bambuluwe; Nchan; Bali Ngemba; Bambili (1 collection each). W Province: Ndinimeki; Bangwa (1 collection each)).
This pyrophyte is restricted to the Bamenda Highlands of NW Province, Cameroon, apart from a few outlying areas.
Habitat: grassland and savanna; 1200–2000 m alt.
Threats: probably conversion of grassland to agricultural land, whether for crops or as pasture for cattle.
Management suggestions: threats to this and other endemic species of grassland in the Bamenda Highlands are poorly known and need researching. Enclosing areas of grassland containing similar (but ideally less threatened) species of *Vernonia* and subjecting them to varying regimes of (e.g.) fire and grazing over several years, and monitoring the response would provide data that would aid management of such taxa.

DROSERACEAE (M. Cheek)

Drosera pilosa Exell & Laundon
VU D2
Range: Cameroon (Bamenda Highlands (3 sites)); Tanzania (Mt Rungwe (3 collections)); Kenya (Mau Forest, Cherangani Hill (1 collection each)); Rwanda (Prov. Bururi (1 collection)).
Distinct from the more widespread *D. burkeana* in the pilose, non-glandular peduncle, this sundew is known from one site each in Tanzania, Kenya and Burundi. In Cameroon it was only known from two collections, both in the Bamenda Highlands (Nchan and Bum) but appears not to have been recollected since 1931 until Nov. 2000, at a new site, Bali Ngemba (*Pollard* 531). *Pollard* 1044 from the same locality, but sterile, probably represents the same entity. The area of occupancy of *D. pilosa* is estimated at 8 km^2, 1 km^2 per site.
Pollard (pers comm.) notes that this species is rare, just one or two individuals being found after extensive searching. By contrast other species of the genus are often gregarious being found in dense colonies. Other apparently similar areas of habitat searched yielded no individuals of this plant.
Habitat: peaty soil between grass tussocks on steep slopes; 1450 m.
Threats: extremely rare, otherwise unknown.
Management suggestions: a survey of the site at Bali Ngemba F.R. should be made to assess the size and demography of the subpopulation. This species may be an annual or short-lived perennial, flowering at the end of the wet season. Regular burning of grassland (reported by *Pollard* 531) may favour this plant by removing the shade inimicable to most *Drosera* (pers. obs.)

EUPHORBIACEAE (M. Cheek)

Clutia kamerunica Pax (see Plate 3(F))
EN A2c
Range: Nigeria (Chappal Waddi (1 collection)); Cameroon (Bamenda to Bafut Ngemba (7 collections); Bali Ngemba F.R. and Baba II (3 collections); Bamboutos-Dschang (1 collection), Markt Singwa (1 collection)).
First collected in the Bamenda Highlands in the early 20th Century (*Ledermann* 1658, Markt Singwa) this white, downy shrub is notably absent from the Kilum Ijim area (Mt Oku and the Ijim Ridge) that has been surveyed in some detail recently (Cheek *et al.* 2000). Over half the known collections were made on various sites near Bamenda in or near the Bafut Ngemba reserve (Mba Kokeka, Bambuluwe etc.). Once this was the largest area of natural forest in the Bamenda Highlands but has since been decimated.
Habitat: submontane and montane forest, forest edge or riverine forest where, usually, single isolated plants are found; 1300–2300 m.
Threats: forest clearance for wood and agriculture. Bafut Ngemba F.R., by 2000, had been 99% converted to *Eucalyptus* plantation or agriculture, forest near Dschang is under enormous pressure from agriculture. More than 50% of *Clutia kamerunica* appears to have been lost over recent decades, largely through the demise of Bafut Ngemba as a natural forest.
Management suggestions: a survey for *C. kamerunica* should be executed at Bafut Ngemba F.R. and the adjoining area to establish whether the species is indeed extinct there, or not. Surveys of this sort at other known locations for the species would also be valuable. Conservation efforts to protect the species are probably best focussed on the sites at Bali Ngemba F.R., and at the village forest of Baba II, since plants have been seen here in the last few years. Education of members of the public likely to affect the survival of the plants could be carried out by a poster campaign, ideally by such organisations as the Bamenda Highlands Forest Project who are already carrying out conservation work in the area.

Pseudagrostistachys africana (Müll.Arg.) Pax & K.Hoffm. subsp. *africana*
VU A2c; B2ab(iii)
Range: Ghana; São Tomé; Bioko; SE Nigeria (Obudu Plateau (1 site)); Cameroon (Mt Etinde, Mt Kupe and Bakossi Mts (4 sites), Bali Ngemba F.R. (1 collection)).
Listed as VU A1c; B1+2c by Hawthorne in 1997 ((IUCN 2003) www.redlist.org), this monotypic genus, a tree restricted to submontane forest (apart from at one lowland site in Ghana), probably now has its largest subpopulation in Bakossi, where it is fairly secure. It is reassessed here on the basis of more extensive disturbance data in Cameroon and according to the modified IUCN criteria of 2001. In the field it is readily recognisable by the large Irvingiaceae-like sheathing apical stipule and the long-scalariform tertiary leaf venation. This taxon has also been assessed for the Kupe-Bakossi checklist'.
Habitat: submontane, or rarely lowland, forest; 500–1500 m alt.
Threats: forest clearance for wood and agriculture (Obudu Plateau and Bali Ngemba F.R.).
Management suggestions: the status of this taxon in São Tomé and Bioko needs more investigation. Bali Ngemba F.R. represents the most easily accessible and dense population of the taxon, followed by Mt Kupe; these are the more promising sites for demographic studies of the taxon.

FLACOURTIACEAE (M. Cheek)

Oncoba lophocarpa Oliv. (see Plate 4(B))
(syn.: *Calancoba lophocarpa* (Oliv) Gilg)
VU A2c
Range: Cameroon (Mt Cameroon (c. 6 collections), Kupe-Bakossi (13 collections at 7 sites), Bali Ngemba F.R. (1 collection).
This canopy tree of submontane forest is notable for the rope-like inflorescences that can bear fried-egg-like flowers several metres from the trunk, where they emerge from the leaf litter like parasites. Previously this species was listed as LR nt (Cable & Cheek 1998) on the basis that it was restricted to Mt Cameroon where its habitat was considered unthreatened. Independently it was assessed as VU D2 under the name *Calancoba lophocarpa* on the basis of its occupancy at Bangem, Mamfe, Bakossi Mts and Mt Cameroon (WCMC, 1997 based on forms by Peguy, T (IUCN 2003) www.redlist.org). We now know that its range and number of sites are larger than previously thought (see above) and that it no longer qualifies as VU D2, but as VU A2c since forest losses in the Bamenda Highlands part of this extended range, equate to over 30% habitat loss over the last 100 years over the species range as a whole. This taxon has also been assessed for Cheek *et al.* (in press).
Habitat: submontane evergreen forest, (400–)800–1950 m alt.

Threats: forest clearance for wood and agriculture. This is most significant at the lower parts of the altitudinal range (below 1000 m) at all three areas. In the Bamenda Highlands the tree is probably now restricted to the Bali Ngemba Forest Reserve. These extensive highlands were probably once home to the main subpopulation of the species given the density of the species at the Bali Ngemba remnant. Elsewhere in the Bamenda Highlands, 25% loss of forest in the eight years between 1987–1995 is recorded by Moat in Cheek *et al.* (2000). With 13 collections at six sites, Kupe-Bakossi appears to be the stronghold for this species, where it is relatively secure and unthreatened.
Management suggestions: a population census, focusing upon the known sites at Kupe-Bakossi and Bali Ngemba, should be carried out and specific threats recorded. A poster campaign focussing on this striking species may serve to promote forest conservation to the local communities.

GUTTIFERAE (M. Cheek)

Allanblackia gabonensis (Pellegr.) Bamps (see Plate 4(C))
VU A2c
Range: Cameroon (Kupe-Bakossi (11 collections at 5 sites), Bali Ngemba, Bamenda Highlands (several collections), Mt Bana, Batcham, Yaoundé, Ebolowa, Sangmelima (1 collection each)); Gabon (Moubighou, Moucongo and Tcyengue (1 collection each)).
This tree is conspicuous for carpeting the forest floor with its pale lemon-coloured or red fallen flowers, each about 6 cm across. The largest part of its domain was probably once the Bamenda Highlands, where submontane forest is now confined to a few small parcels, the largest of which is at Bali Ngemba. Forest in these highlands is still being lost: 25% of forest in one area disappeared in an 8-year period ending in 1995 (Moat in Cheek *et al.* 2000). Overall, more than 30% of the habitat of this species has probably disappeared over the last 100 years. Mt Kupe and the Bakossi Mts are now probably the stronghold for *A. gabonensis*. Elsewhere the species occurs on several of the small hills dotted through the forest belt in South and Central Province, finally extending into the Crystal Mts of Gabon. This taxon has also be assessed for Cheek *et al.* (in press).
Habitat: submontane forest; 700–1500 m alt.
Threats: forest clearance for agriculture and wood particularly in the Bamenda Highlands and Yaoundé area.
Management suggestions: this species is fairly secure in the upper part of its altitudinal range in Kupe-Bakossi, but enforcement of the forest reserve boundary is needed if it is to survive at Bali Ngemba F.R. Surveys should be conducted on the forested hills from which it has been collected elsewhere in Cameroon and also in the Crystal Mts, to determine whether it survives there and whether it can be protected at any of these sites.

Psorospermum aurantiacum Engl.
VU B2ab(iii)
Range: Nigeria (Obudu Plateau (5 collections)); Cameroon (NW Province: Bambui; Bamenda, Kumbo-Oku; Bafut Ngemba; Bali Ngemba (12 collections). W Province: Kounden (1 collection)).
This tree or shrub is distinctive for the dense orange-brown hairs on the lower surface of the leaf, the upper surface a contrasting glossy-black when dried. It appears confined to the Bamenda Highlands, with outliers in the adjoining Obudu Plateau and Bamboutos Mts (Kounden).
Habitat: edge of gallery forest; 1500–1800 m alt.
Threats: dry season grassland fires, usually set by man, burn into the montane and submontane forest in the Cameroon Highlands, reducing its area. It is possible, even likely, that *P. aurantiacum* by the nature of its habitat, has some resistance to fire and may even benefit from occasional fires. However, the current frequent and intense fires may affect individuals adversely. Conversion of forest to farmland, by contrast, is an undoubted threat. Over 25% of forest in one sample area of the Bamenda Highlands was lost in the 1980s–1990s (Moat in Cheek *et al.* 2000).
Management suggestions: research to explore the effect of different fire regimes on this species is advised. In the short term, the highest priority is to re-find individuals at the known sites and to seek means to protect these. While almost all natural forest at Bafut Ngemba F.R. has disappeared already, Bali Ngemba F.R. still remains fairly intact and may be the best prospect for the conservation of *P. aurantiacum*.

LABIATAE (B.J. Pollard)

Aeollanthus trifidus Ryding (see Plate 4(E))
VU B1ab(i,ii,iii,iv,v)+B2ab(i,ii,iii,iv,v)
Range: Nigeria (Mambilla plateau (2 collections)); Cameroon (Bamboutos Mts, Bali Ngemba F.R., Bamenda, Bui (1 collection each)).

The holotype (*Saxer* 64) was collected at Dschang, Bamboutos Mts, W Province, Cameroon on 29 June 1955. Two further specimens were collected on 10 August 1973 across the border in Nigeria: *Chapman* 103 from Ngurogi, and *Medler* 941, near Chappal Wadi, probably on the road from Takum. J.F.F.E. de Wilde collected a specimen from Jakiri, Bamenda, NW Province, Cameroon, which remained unidentified until now, and represented the most recent collection until those of *Etuge* 4736 from Mbiami and *Pollard* 699 from the larger granite inselberg at Mantum by the Bali Ngemba Forest Reserve, both in November 2000.
Habitat: rocks or shallow soil, subject to burning during the dry season (Dec.–Mar.); 1500–2200 m alt.
Threats: change of land use, although this taxon is known mainly from rocky areas unsuitable for agriculture.
Management suggestions: introduce this species into cultivation and distribute living specimens to local and regional botanic gardens for *ex-situ* conservation.

LEGUMINOSAE – MIMOSOIDEAE (B.J. Pollard)

Newtonia camerunensis Villiers
CR A2ac+4ac
Range: endemic to the Bamenda Highlands and Bamboutos Mts of Cameroon.
This species was assessed as CR A1c in Cheek *et al.* (2000), when it was known from only five collections. A reassessment is necessary here, both to follow the 2001 IUCN guidelines, and to present additional information recorded since 2000. Three further collections were made during the April-May 2002 expedition to NW Province, a focus of which was to look for known Red Data species, monitor their population sizes and search for additional locations. The first of these collections was a sterile sapling collected at Bali Ngemba F.R. during the 25 m × 25 m plot survey (*BAL*52), and this collection was not identified until 2004 at Kew. The return to Laikom in May 2002 heralded the rediscovery of this species, feared to be extinct in the wild, and also the first opportunity to carry out detailed survey work, which resulted in two fruiting collections: *Ghogue* 1401 & *Pollard* 1097. Despite exhaustive searching, the flowers of this species remain unknown to science, probably appearing earlier in the season, around March. On our return to the Akwamofu sacred forest a large fruiting tree was seen, to 30 m and dbh c. 2.5 m, and on the way down the path as far as the bridge across the stream 6 very large trees were seen, with dbh of up to 3 m or more. Many seedlings were found, including c. 20 within 5 m radius of the tree collected under *Pollard* 1097. The largest tree observed was being smothered and strangled by *Schefflera abyssinica* (Araliaceae), and the second largest tree with two juvenile *Schefflera* growing on the upper part of trunk, resulting in considerable die-back in the crown. The canopy in the Akwamofu sacred forest consists mainly of *Albizia gummifera* and *Newtonia*; the latter having a darker appearance when viewed from the ground, partly due to the darker green colour of the leaflets, but perhaps also the closer arrangement of the leaflets that form a denser covering, allowing less light through to the understorey. Three mature fruiting specimens were seen on the way to the waterfall, and at least four other fruiting trees seen on the way back across to the main path. Seedlings were seen to occur in huge numbers in the understorey, including c. 50 in a 5 m^2 patch, but only perhaps c. 30 plants with dbh between 1 and 5 cm were seen in the whole forest, suggesting poor recruitment to maturity, with even fewer specimens of dbh between 5 and 100 cm. Fruit is set abundantly. *Ghogue* 1401 was collected from a further subpopulation recorded in a forest pocket up along the Ijim ridge, which suggests that these fragments are important for the survival of this species. A further subpopulation has been located by Kenneth Tah at Finge (Tah pers. comm. 2002).
Habitat: submontane to montane forest with *Albizia gummifera, Carapa grandiflora, Syzygium staudtii, Prunus africana*; 1600–2030 m alt. (this represents an increase in the known upper altitudinal limit from 1800 m). Known to occur in forest patches in grassland on the Ijim Ridge.
Threats: the sites at Kilum-Ijim fall outside the boundary of protection, except at Akwamofu where a low level of protection is afforded by the Kom, who do not allow clearance of forest for traditional reasons. Unfortunately this is only enforced on one side of the path, and most of the specimens of *N. camerunensis* appear to occur on the non-protected side. It seems that the proliferation of *Schefflera* poses a threat to mature individuals, from which most of the seed-rain originates. It is likely that these substantial trees are still used as timber and it may well be exploited elsewhere in its known range. At Bali Ngemba F.R. the threat from forest clearance for timber, firewood and small-scale agriculture is increasing every year (Pollard pers. obs.).
Management suggestions: further to those suggestions mentioned in Cheek *et al.* (2000), a project has been inititiated to introduce this species into cultivation. At Laikom, BP collected about 100 seeds from fallen pods which were given to Kenneth Tah of the Bamenda Highlands Forest Project (Pollard 2002) to plant in their nursery. The project has unfortunately since closed down, and it is not certain what fate has befallen the seedlings which had grown to c. 1 m by 2004 (Darbyshire pers. comm.). Efforts are being made to secure them and ensure that they are kept in a good state of health to allow possible reintroductions into the wild at suitable selected sites across the province. There is scope to investigate the biology of this species more closely, particularly at Laikom, to investigate: the causes of mature specimen mortality, including the effects of parasitic and strangler plants; the causes of sapling mortality, and the dynamics of recruitment to maturity; its ecological relationships with *Albizia*

gummifera; the morphology of the inflorescences and flowers; identification of pollinators and seed-dispersal agents; cultivation regimes. This would greatly assist our conservation efforts.

LEGUMINOSAE – PAPILIONOIDEAE (M. Cheek)

Crotalaria ledermannii Baker f.
VU D2
Range: Nigeria (Mambilla Plateau (1 collection)); Cameroon (Mwanenguba Mts (1 collection), Bamenda Highlands (7 collections)).
This is a rare species, having been gathered only once in the course of our three inventories at Mt Oku and Ijim, once at Bali Ngemba F.R., and Mwanenguba Mts, and twice at Bafut Ngemba. The collection from the Mambilla Plateau is taken from a *Eucalyptus* plantation, so *Crotalaria ledermannii* may be able to tolerate some disturbance to its habitat. This is an annual or short-lived perennial species, so a deleterious change in the habitat or poor seed set in one year could drastically reduce the population within 12 months. This herb was assessed as above in Cheek *et al.* (2000). No new data have come to light since then, apart from the recent collections at Bali Ngemba F.R. and Mwanenguba Mts , so the original conservation assessment is maintained here. This taxon has also been assessed for Cheek *et al.* (in press).
Habitat: grassland and forest edge; 1200–2200 m alt.
Threats: unknown but possibly conversion of land to cultivation and grazing, trampling or fires.
Management suggestions: the natural habitat of *Crotalaria ledermannii* is not well characterised . More study is needed to rectify this. A survey to find the range of this species and levels of regeneration is advisable.

Dalbergia oligophylla Baker ex Hutch. & Dalziel
EN A2c+3c
Range: Cameroon (Mt Cameroon (4 collections), Bafut Ngemba (1 collection), Bali Ngemba F.R. (1 collection), Mt Kupe (5 collections), Mwanenguba (1 collection)).
Known only from six sites, this shrub-liana dries a distinctive black, helping to distinguish it from similar species from Sierra Leone and Gabon, with which it has been confused. This taxon has also been assessed for Cheek *et al.* (in press).
Habitat: submontane and montane forest edge; (900–)1500–2000 m alt.
Threats: forest clearance for agriculture and wood: the subpopulations at Bafut Ngemba, Obudu Plateau and Mwanenguba have quite possibly been lost due to this already. The Bamenda Highlands may once have been the main range of this species (judging by its existence at Bali Ngemba and Bafut Ngemba) but forest loss here has been as high as 25% between 1987–1995 in one area studied (Moat in Cheek *et al.* 2000). Overall, over 50% habitat loss is postulated over the last 100 years, and at least 50% of that which remains could be lost in the next century.
Management suggestions: unless Bali Ngemba F.R. is protected from incursion more vigorously, the only hope for the survival of this species is the summit area of Mt Kupe and the upper tree line of Mt Cameroon both of which are reasonably secure from threat.

MELASTOMATACEAE (M. Cheek)

Dissotis bamendae Brenan & Keay
VU B2ab(iii)
Range: Nigeria (Mambilla Plateau, Nguroje F.R. (1 collection)); Cameroon (NW Province: Bamenda (2 collections); Nchan (1 collection); Bafut Ngemba F.R. & Bali Ngemba F.R. (3 collections). W Province: Bamboutos Mts (1 collection)).
This many-stemmed, shapely shrub was sunk into *D. princeps*, a widespread species (South Africa to Ethiopia) by Jacques-Felix, but since it is abundantly distinct (e.g. 3 leaves per node, not 2) we follow FWTA in maintaining it as different. Only 7 sites are known.
Habitat: grassland, often near streams; 1500–2200 m alt.
Threats: poorly known, but including conversion of natural habitat to farmland.
Management suggestions: see comments for *D. longisetosa* (below) and *Vernonia guineensis* var. *cameroonica* (Compositae).

Dissotis longisetosa Gilg & Ledermann ex Engl.
VU B2ab(iii)
Range: Nigeria (Obudu (2 collections), Mambilla Plateau (Ngurdje F.R. to Gangirwal (4 collections)); Cameroon (NW Province: Bafut Ngemba F.R. (2 collections); Bali Ngemba F.R. (2 collections); Tabenken. Adamowa Prov.: Ngaoundere (4 collections). W Province: Bamboutos Mts (2 collections)).
This striking, long-setose *Dissotis* is not easily confused with anything else. *Lock* 84/56, from Togo, 740 m, referred to this taxon at K, 'appears to belong to another taxon', it has soft hairs and a subsessile inflorescence.
Habitat: heavily grazed and sometimes burnt grassland; 1600–2100 m.
Threats: conversion of grassland to tilled land.
Management suggestions: ecologically this appears such an undemanding species that it may be expanding its range since, on the whole, burnt, heavily grazed grassland appears to be increasing within its range! Known from only 7 sites, the factors accounting for its rarity and a detailed study of its ecology are desirable. Further surveys may reveal more sites, so downgrading the threat level to this taxon.

Memecylon dasyanthum Gilg & Ledermann ex Engl.
VU A2c
Range: Cameroon (Mt Kupe (1 collection at one site) to Bamenda Highlands and Bamoun (numerous collections at several sites)).
This tree of upper altitude forest appears often to have been confused (e.g. in Cable & Cheek 1998) with the lower altitude *M. myrianthum*. In the Bamenda Highlands it is the commonest Memecylonoid in e.g. Bali Ngemba, but this forest is diminishing rapidly. There is no doubt that over 30% of its habitat will be lost in the next century. It is likely that if better data were available for loss in the Bamoun area the assessment would be as EN. This taxon has also been assessed for Cheek *et al.* (in press).
Habitat: upper submontane forest; 1550–1900 m alt.
Threats: clearance of forest for agriculture and wood, especially in the Bamenda Highlands where between 1987–1995 25% of forest cover was lost in one sample area (Moat in Cheek *et al.* 2000). The forest of the Bamoun highlands are also afflicted by forest loss, particularly due to dry season fires (Ndam pers. comm.).
Management suggestions: although the species is reasonably secure at Mt Kupe, this seems to be an outlying subpopulation. The Bali Ngemba F.R. has the densest known subpopulation of *M. dasyanthum*, and should be the priority for protection of the species. Enforcement of existing boundaries would probably be adequate to secure the future of the species there. More data are needed on the survival of forest habitat in the Bamoun area and the continued existence of this species there.

MELIACEAE (M. Cheek)

Entandrophragma angolense (Welw.) C.DC.
VU A1cd
Entandophragma angolense is one of the five African internationally-traded timber species of the mahogany family that were listed as Vulnerable by Hawthorne (Hawthorne 1997, www.redlist.org, in IUCN 2003) using the 1994 criteria of IUCN. They all have a wide range in Africa and were they reassessed in this book, without reference to their use as timber, they would probably be downlisted. Hawthorne cites over-exploitation, poor levels of regeneration, fire damage, and slow growth to support his assessments of these species.

MYRTACEAE (M. Cheek)

Eugenia gilgii Engl. & Brehm.
CR A1c
Range: Nigeria (Mambilla Plateau (13 collections)); Cameroon (Bamboutos Mts (3 collections), Bamenda Highlands (7 pre-1996 collections, 4 recent collections), Ngaoundere (1 collection)).
The assessment above was made in Cheek *et al.* (2000) in IUCN (2003) (www.redlist.org). That assessment is maintained here.
Described in 1917 from *Ledermann* 2131 (NW Cameroon, Tapare to Riban, gallery forest, 1300 m, Jan. 1909). Although many herbarium collections of *Eugenia gilgii* exist, it seems highly threatened by the fact that:

a) its natural habitat has been almost completely destroyed
b) what is left is disappearing rapidly and
c) by the fact that, apart from a few trees on Laikom Ridge at Ijim, no individuals are in protected areas.

Nonetheless, this was probably once a relatively common species in the Bamenda Highlands. Apart from two widely separated trees in the Anyajua area and small populations at Laikom and Bali Ngemba, the only trees of

Eugenia gilgii that I have seen are in an extensive subpopulation of at least 50 trees discovered in 1999 on the path to Mbingo 'Back Valley'. This may be the main refuge for the species in the Bamenda Highlands, although it has no formal protection. The relatively numerous collections from the Mambilla Plateau were all made over 23 years ago. It seems reasonable to infer that the situation with regard to habitat destruction in Nigeria over this time has been no better than it has been in the Bamenda Highlands.
Habitat: lower montane forest, often at edges; (1200–)1500–2000 m alt.
Threats: clearance of forest for wood and land for agriculture.
Management suggestions: the possibility of protecting the subpopulation at Mbingo should be investigated.

RUBIACEAE (M. Cheek)

Chassalia laikomensis Cheek
CR A1c
Range: Nigeria (Mambilla Plateau (1 collection)); Cameroon (Mwanenguba (1 collection), Bamenda Highlands (several sites)).
The assessment above was made in Cheek & Csiba (2000), listed as having been assessed by Cheek *et al.* (2000) in IUCN (2003) (www.redlist.org). That assessment is maintained here. A new location to the south, that of Mwanenguba is added here. Mwanenguba has seen similar forest loss to that in the Bamenda Highlands. This taxon has also been assessed for Cheek *et al.* (in press).
Habitat: montane evergreen forest; 1650–2000(–2400)m alt.
Threats: about 95% of the original forest cover of the Bamenda Highlands has been lost to e.g. agriculture (Cheek *et al.* 2000; Cheek & Csiba 2000) and there have been similar losses at Mambilla and Mwanenguba.
Management suggestions: more information is needed on the numbers of individuals at the known sites and levels of regeneration. Enforcement of existing protected area boundaries would help protect a significant portion of the surviving population.

Ixora foliosa Hiern
VU A2c+A3c
Range: Nigeria (Chappal Waddi and Chappal Hendu); Cameroon (Mt Cameroon (numerous collections), Mt Kupe, Mwanenguba, Bamenda Highlands (numerous collections from many sites)).
About half the area where this characteristic tree of wet montane forest occurred was in the Bamenda Highlands. This is now destitute of natural forest except for a very few exceptions. It is estimated that over 30% of the habitat of this tree has been lost over the last century and that over 30% of that remaining will be lost in the next century. This taxon has also been assessed for Cheek *et al.* (in press).
Habitat: montane forest.
Threats: forest clearance for agriculture and wood, especially in the Bamenda Highlands, once probably the main area for this species. In one study area of these highlands, 25% of forest was lost between 1987–1995 (Moat in Cheek *et al.* 2000).
Management suggestions: implementation and policing of protected area boundaries would ensure the survival of this species in most of its range.

Pavetta hookeriana Hiern var. *hookeriana*
VU A2c+A3c
Range: Cameroon (Mt Cameroon (numerous collections), Mwanenguba (c. 3 collections), Bamenda Highlands (numerous collections at several sites).
It is estimated that over 30% of the habitat of this species has been lost in the last century and that over 30% of that remaining will be lost in the next century. This taxon has also been assessed for Cheek *et al.* (in press).
Habitat: montane forest; 1900–2000(–2400)m alt.
Threats: secure from threat at Mt Cameroon, *P. hookeriana* is threatened by forest clearance for agriculture and wood throughout the extensive Bamenda Highlands, probably once the main area for the species. Study of one large area in the highlands between 1987–1995 showed that 25% of the surviving forest was lost (Moat in Cheek *et al.* 2000).
Management suggestions: improved policing of existing forest reserve boundaries could prevent extinction of this species in the Bamenda Highlands, where its survival is precarious, except at Kilum-Ijum. At Mwanenguba and Bamboutos Mts (presence inferred) it may not survive for much longer. The species is most secure at Mt Cameroon, where the narrowly endemic variety *pubinervia* also occurs.

Pentas ledermannii Krause
(syn. *P. pubiflora* S.Moore subsp. *bamendensis* Verdc.)
VU B2ab(iii)
Range: SE Nigeria (Obudu (1 collection)); Cameroon (Mt Oku area (3 collections), Bamboutos (1 collection), Mt Kupe-Mwanenguba (2 sites), Bali Ngemba F.R. (2 collections), Santa, Lake Aweng (1 collection each)).
At Mt Kupe known from 4 collections on the grassy summits (*Cable* 108, 1992; *Cheek* 7573, 1995, *Sebsebe* 5086, 1995), also known from Mwanenguba (*Leeuwenberg* 9970, 1972). Mt Kupe is the southernmost point for *Pentas ledermannii*: it does not appear to extend to Mt Cameroon. Mt Oku and the Ijim Ridge appear to be the northern extreme of its range. This taxon has also been assessed for Cheek *et al.* (in press). The only change in the assessment has been the addition of Bali Ngemba F.R. to the range.
Habitat: montane forest–grassland edge; 1000–2060 m alt.
Threats: forest clearance is a major threat in the Bamenda Highlands, with 25% loss in one area between 1987 and1995 (Moat in Cheek *et al.* 2000). This rate of loss probably continues and probably also extends to the 1800–2000 m altitudinal range in Obudu, Bamboutos and Mwanenguba. Frequent human-set fires in grasslands in these same areas probably also adversely effect the grassland-forest interface as a habitat for this taxon, although occasional natural fires may aid its regeneration.
Management suggestions: research to resolve these uncertainties would aid in management planning for the conservation of this and other species in this habitat in the Cameroon uplands. For the moment, only the summit area of Mt Kupe lacks the two threats outlined above, and so is alone in offering a secure base for *Pentas ledermannii.*

Psydrax dunlapii (Hutch. & Dalziel) Bridson
VU A2c; B2ab(iii)
Range: Bioko (1 collection); Cameroon (Mt. Cameroon (numerous collections). Bamenda Highlands: Bafut Ngemba and Tonagahi-Acha (1 collection each); Bali Ngemba F.R. (6 collections)).
This submontane tree, where it occurs, is so common (e.g. on Mt. Cameroon) that it characterises its band of submontane forest. Strangely it is not known from areas of apparently suitable habitat between its two subpopulations in Cameroon, this despite intensive collection in these habitats in, e.g. Mt. Kupe.
Habitat: understorey of evergreen forest; 1400–1500 m.
Threats: clearance of forest for agriculture and timber, principally in the Bamenda Highlands (secure at Mt. Cameroon) where in one sample area, 25% of forest cover was lost between 1987–1995 (Moat in Cheek *et al.* 2000).
Management suggestions: no action is required at Mt. Cameroon, but data on the status of the population at Bioko is advisable. In the Bamenda Highlands, forest has almost disappeared and high priority should be given to protecting surviving areas, particularly Bali Ngemba F.R.

SAPINDACEAE (M. Cheek)

Allophylus bullatus Radlk.
VU A1c
Range: SE Nigeria; Cameroon; Príncipe & São Tomé (Cameroon line mountains).
This understorey tree of upper submontane to montane forest, while secure on Mt Cameroon and on Mt Kupe, has lost large tracts of its habitat in recent decades in the Bamenda Highlands. Over 30% of its overall habitat is estimated to have been lost in the last 100 years. This taxon has also been assessed for Cheek *et al.* (in press).
Habitat: upper submontane and montane forest; 1600–2400 m alt.
Threats: clearance of forest for agriculture and wood, particularly in the Bamenda Highlands of Cameroon, once probably the main area for *A. bullatus*. Study of one area here (Moat in Cheek *et al.* 2000) showed that 25% of forest was lost between 1987–1995.
Management suggestions: improved policing of the existing protected areas could secure the future of this species.

Allophylus conraui Gilg ex Radlk.
EN B2ab(iii)
Range: Nigeria (Mambilla-Gembu, Mayo Naga-Njawe (1 collection each)); Cameroon (Kebo (unlocated precisely), Bali Ngemba F.R. (3 collections)).
This 6 m tree is unusal in its genus for being spiny. Published in 1908, it is only known from four sites, one of which, Kebo, has not been located, but may be a Germanic version of the modern Batibo with the usual Bantu place name prefix Ba- omitted. One of the three specimens referred to in FWTA under this name (*Olorunfemi* in FHI 30611) is discordant, respresenting an undescribed species.
Habitat: evergreen forest; c. 1400 m alt.

Threats: forest clearance for agriculture (e.g. Bali Ngemba F.R. where farming has increased inside the forest year by year) and in the Mambilla plateau area of Nigeria (H. Chapman pers. comm.).
Management suggestions: the status of forest at the two Nigerian sites needs to be ascertained. This would best be done by reference to Chapman and Olsen, who are conducting surveys in the area. In Cameroon, survey work should focus on Bali Ngemba where several individuals of *A. conraui* have been seen recently (pers. obs.).
A definitive enumeration of the subpopulation and assessment of regeneration and demography would provide the data needed to formulate a management plan for this species.

SIMAROUBACEAE (M. Cheek)

Quassia sanguinea Cheek & Jongkind ined.
(syn. *Hannoa ferruginea* Engl.)
VU A2c
Range: Nigeria (Obudu Plateau); Cameroon (SW and NW Provinces).
Restricted to the submontane forests of the Cameroon Highlands, from Mt Cameroon in the south to the Bamenda Highlands in the north, with an extension into Nigeria at Obudu, this small tree is distinctive in its red leaf axes. This taxon has also been assessed for Cheek *et al.* (in press).
Habitat: submontane forest; 800–1750 m alt.
Threats: forest clearance for wood, followed by agriculture, particularly in the northern part of its range, the Bamboutos Mts and the Bamenda Highlands. In the latter, a remote sensing study of one area over eight years (1987–1995) by Moat (in Cheek *et al.* 2000) showed 25% loss of forest. By extrapolation, it is here estimated that over 30% of its overall population has been lost due to habitat destruction over the last three generations, or sixty years (estimating one generation at twenty years).
Management suggestions: although forest loss in W and NW Provinces, Cameroon, has seriously reduced the population of *Q. sanguinea* in those areas, it seems relatively secure at Mt Cameroon and at Mt Kupe and the Bakossi Mts in SW Province. So long as these areas remain protected, no further action is currently needed to ensure the survival of the species. However, data on generation duration and other aspects of demography, together with data on densities, are desirable.

STERCULIACEAE (M. Cheek)

Dombeya ledermannii Engl.
CR A1c
Range: Nigeria (Mambilla Plateau (4 collections), Jos Plateau (2 collections)); Cameroon (Bamenda Highlands (6 collections)).
It is easily counted since, in April, trees are clearly visible from a distance on account of their white flowers. Flowering can occur in November (pers. obs.). The habitat of this species is highly threatened in all its known localities. Seyani (1982) reports that *D. ledermannii* is characteristic of forest edges and the early stages of forest succession, but that on exposed rocky slopes is a normal component of stunted, more open forest. In the Bamboutos Mts, it is propagated by cuttings and planted to form hedges (*Letouzey* 201, cited by Seyani 1982). This tree has previously been assessed in Cheek *et. al* (2000). The only change in the assessment has been the addition of Bali Ngemba F.R. to the range.
Habitat: woodland; (700–)1220–1980 m alt.
Threats: clearance for agricultural land, over-exploitation for bast fibre.
Management suggestions: more information is needed on the extent, distribution and threats to *D. ledermanniii* within the Kilum-Ijim area.

MONOCOTYLEDONAE

CYPERACEAE (I. Darbyshire, B.J. Pollard & M. Cheek)

Bulbostylis densa (Wall.) Hand.-Mazz. var. *cameroonensis* Hooper
VU D2
Range: Cameroon (SW Province: Mt Cameroon (1 site, 2 collections); Mt Kupe (1 collection). Littoral Province: Mt Mwanenguba (1 collection). NW Province: Bali Ngemba F.R. (1 collection)).

Until the 1990s, this variety was known only from the type collection (*Mann* 1360b) on Mt Cameroon, believed to be from the Mann's Spring area. It was then rediscovered there in 1992 (*Thomas* 9407), and has subsequently been recorded on three other montane grassland sites, at Mt Kupe, Mwanenguba and the Bali Ngemba F.R. It has not been recorded elsewhere in the Bakossi Mts. Although the extent of occurrence is now known to be much greater than when first assessed in Cable & Cheek (1998), the four known sites are highly isolated and limited in size, thus they are still vulnerable to local stochastic change, such as lava flow on Mt Cameroon or natural fires, thus the vulnerable status is still valid. Care must be taken not to confuse this plant with the sympatric typical variety. This taxon has also been assessed for Cheek *et al.* (in press).
Habitat: montane grassland; 1800–3000 m alt.
Threats: see above.
Management suggestions: more data on the size of each population are required to assess further its vulnerability to local stochastic events. Further botanical inventory work on montane grassland sites in e.g. the Bamenda Highlands, or Mt Nlonako in Littoral Province, may reveal further populations, in which case the conservation status would be downgraded.

Scleria afroreflexa Lye in press (see Fig. 8)
EN B2ab(iii)
Range: Cameroon (SW Province: Bakossi Mts (1 collection). NW Province: Boyo (1 collection); Bali Ngemba F.R. (2 collections).
This conservation assessment was originally made by B.J.Pollard in Lye & Pollard (in press., Nordic Journal of Botany), but here we add the additional site of the Bali Ngemba F.R. from where this taxon was collected twice in 2001. It was first collected in 1999 in two disjunct locations, 170 km apart, in western Cameroon: between Laikom and Fundong in Boyo Division, NW Province, and near Kodmin in the Bakossi Mts, SW Province. It was recorded as locally common at the latter site. Further subpopulations are likely to exist in locations that have not yet been included in botanical inventories, given the large areas of grassland at altutudes of around 1500 m in the Bamenda Highlands, Bamboutos and, to a lesser extent, in SW Province (Pollard pers. obs.). This taxon has also been assessed for Cheek *et al.* (in press).
Habitat: montane grassland and grassland patches in submontane forest; 1450–1550 m alt.
Threats: deliberate burning of the montane grasslands in NW province may result in losses of subpopulations or may lead to long-term habitat changes which do not favour this taxon. The site near Kodmin appears relatively unthreatened though may be lost to forest encroachment if human disturbance remains low.
Management suggestions: studies of the subpopulations in the Northwest Highlands could be made to better understand the ecology of this taxon, including its tolerance of fire and human disturbance. This species should also be searched for in suitable habitat elsewhere in western Cameroon; discovery of further subpopulations could lead to a downgrading of its conservation status.

ERIOCAULACEAE (M. Cheek)

Eriocaulon asteroides S.M.Phillips (see Plate 7(E))
VU D2
Range: Nigeria (Mambilla Plateau (Chappal Waddi, 1 collection)); Cameroon (Bamenda Highlands, Mount Oku and the Ijim Ridge area, Bali Ngemba F.R. (6 collections)).
This minute annual often co-occurs (in 3 of the 7 known sites) with the superficially similar *Eriocaulon parvulum*. The second species is not known to occur without the first. Indeed, both species were originally described (Phillips 1998) from an unwittingly mixed collection of the two species (*Zapfack* 1205, November 1996) from the Kumbo-Oku road (Iwooketele Mbai). At the time of its description, *Eriocaulon asteroides* was only known from that site and from Chappal Wadi, the highest mountain in Nigeria, situated on the border with Cameroon. Subsequently, in December 1998, a large colony of the two species was found on a headland of the Laikom spur of Ijim Ridge. This site was studied in some detail (see Cheek *et al.* 2000, vegetation chapter: basalt pavement). 1–2 km along the path from this site to the Ardo of Ijim's compound another site for this species was found. The basalt pavement area here was wetter than on the headland, perhaps accounting for the absence of *E. parvulum*. Finally at about 1700 m alt., between Laikom and Fundong, below the Fulani settlement, another colony of *E. asteroides* was encountered by Tadjouteu of our party, in November 1999. Again, the species was associated with an outcrop of wet basalt pavement, and again *E. parvulum* was absent, although about ten of us spent an hour searching for it. Finally, in listing the *Eriocaulon* holdings of YA for the monographer of the African species of this genus, Dr Phillips, another mixed collection of the two species was encountered. This was collected in September 1975 (*De Wilde* 8633) from km 21 on the Bamenda-Jakiri road at the southern boundary of our area. This taxon was assessed in Cheek *et al.* 2000). The only change in the assessment has been the addition of Bali Ngemba F.R. to the range.

Habitat: basalt pavement, i.e. thin, peaty, seasonally waterlogged soil in the cracks between blocks of basalt; with *Utricularia scandens*, *Loudetia simplex* and *Scleria interrupta*; c. 1700–2500 m alt.
Threats: unknown. However, too much trampling by cattle might cause damage to these small annual *Eriocaulon* plants by dislodging from the basalt substrate the thin layer of peaty soil in which they grow. Conversely, lack of grazing or of intermittent grassland fires might permit the build up of enough soil on the pavement to allow a *Sporobolus*-based community to encroach upon the basalt pavement and smother or compete with the *Eriocaulon*.
Management suggestions:
1. A survey of basalt pavement should be made in the Kilum-Ijim area. When areas are located, these species should be searched for and vouchered if found. A rough estimate of the area of occupation and total number of plants should be made. This will allow more complete mapping of the species and a more comprehensive understanding of their population size.
2. Consideration should be given to using experimental means to examine the effects of the possible threats mentioned above on these two rare annual species. Several square metres of one population could be fenced off and protected from fire and grazing. The effect of this could be monitored on an annual basis. Another area could be subjected to cattle or horse trampling to look at the effect of this on the soil that hosts these species. Results of this experimentation could then be used to guide management of the habitat of these species.

ORCHIDACEAE (B.J. Pollard)

Bulbophyllum nigericum Summerh.
VU A2c; B2ab(iii)
Range: Nigeria: (5 collections, at 5 sites); Cameroon: (Mt Kupe (1 collection), Bakossi Mts (1 collection), Bali Ngemba (1 collection), unlocated (2 collections)).
This species was described from *King* 124 collected in October 1958 from Plateau Province, Nigeria, from where King made 3 additional collections. A further record, from the Mambilla Plateau in November 1993 (*Sporrier* 18) remains unconfirmed, the specimen being labelled '*Bulbophyllum nigericum ?*'. The specimen cited in FWTA from the Ivory Coast is now referred to *B. bidenticulatum* J.J.Verm. subsp. *bidenticulatum* (Vermeulen 1987: 167). It was first collected in Cameroon on the southern side of Mt Kupe in (?)1970, *Letouzey* 408, but has not since been recorded there. Two additional specimens are recorded from western Cameroon by Vermeulen (1987: 92), but no site locations are listed. Recent intensive surveys in Cameroon have revealed only two additional sites, at Enyandong in the Bakossi Mts (*Salazar* 6322, Oct. 2001) and at Bali Ngemba F.R. (*Plot voucher* BAL25, Apr. 2002).
Habitat: an epilith or epiphyte in submontane and montane forest; c. 800–2050 m alt.
Threats: the Nigerian sites are threatened by continued extensive clearance of forest to high elevations; one or more of these subpopulations are likely lost. The plant at Enyandong was found growing in largely cleared forest, on a tree in the village. It is therefore likely to occur in the surrounding forest, some of which is being cleared for small-holder farming, thus threatening this subpopulation. In all, a loss of over 30% of the population is estimated over the past 3 generations, which we here estimate to be 10 years, much of this loss being irreversible.
Management suggestions: as this species is found within the village of Enyandong, on a tree in front of the house of the Chief of the village, this is an ideal location for promoting community-based conservation. Local residents here could be encouraged to search for this species in the surrounding forest, perhaps using a species conservation poster as an aid to identification, and to promote protection of any locations where it is found.

Polystachya anthoceros la Croix & P.J.Cribb
EN B2ab(ii,iii)
Range: Nigeria (1 Collection); Cameroon (3 Collections).
Polystachya anthoceros was described in 1996, known only from *Spurrier* N17, Nigeria, Mambilla Plateau, Ngel Nyaki Forest, 31 Oct. 1993, fl. in cultivation, Aug. 1994 (K holotype). In Oct. 2001, two collections were made within a week of each other, *Etuge* 4302r from Baba II forest, near Bali Ngemba F.R., and *Pollard* 696 from inside the Reserve. Another Bali Ngemba collection (*Simo* 158 (subject to confirmation)) was made in April 2004.
Habitat: epiphyte in montane forest or forest patches in grassland; 1500–2100 m alt.
Threats: conversion of forest to small-scale agriculture or plantations.
Management suggestions: this species should be introduced into cultivation, perhaps at the nearby Savanna Botanic Garden.

PTERIDOPHYTA

VITTARIACEAE (Y.B. Harvey & P.J. Edwards)

Vittaria guineensis Desv. var. *camerooniana* Schelpe
VU 2ab(iii)
Range: Cameroon (Mt Cameroon (4 collections), Bali Ngemba F.R. (1 collection)); Bioko (Moca, Lagoa Biao, Pico, Mte Baká (8 collections)).
This epiphyte is restricted to Mt Cameroon and Bali Ngemba F.R. in Cameroon, and 4 close locations in Bioko, and is not known to be locally abundant in any location.
Var. *camerooniana* differs from the more frequently encountered var. *guineensis* in its rhizome scales being longer (3–5 mm long), subulate with a long apical hair, entire, distinctly clathrate and lighter in colour.
Habitat: submontane forest, secondary forest and forest–grassland transition; 1200–1600 m alt.
Threats: forest clearance for agriculture (e.g. Bali Ngemba F.R. where farming has increased inside the forest year by year).
Management suggestions: a survey of the site at Bali Ngemba F.R. and Mt Cameroon should be made to assess the size and demography of the subpopulation and will provide the data needed to formulate a management plan for this species in Cameroon. Subpopulations in Bioko are fairly secure, lacking threats.

REFERENCES

Cable, S. & Cheek, M (1998). The Plants of Mount Cameroon, A Conservation Checklist. RBG, Kew, UK. lxxix + 198 pp.

Cheek, M., Onana, J.-M. & Pollard, B.J. (2000). The Plants of Mount Oku and the Ijim Ridge, Cameroon, A Conservation Checklist. RBG, Kew, UK. iv + 211 pp.

Cheek, M., Pollard, B.J., Darbyshire, I., Onana, J.-M. & Wild, C. (eds) (in press). The Plants of Kupe, Mwanenguba & the Bakossi Mts, Cameroon, A Conservation Checklist. RBG, Kew, UK.

de Wilde, J.J.F.E. (2002). *Begonia* section *Tetraphila* A.DC., a taxonomic revision. In de Wilde, J.J.F.E. (ed) (2002). Studies in Begoniaceae: 7. Leiden, Backhuys. 5–258 pp. (Wageningen University Papers; 2001–2).

IUCN (1994). IUCN Red List Categories. IUCN, Gland Switzerland. 21 pp.

IUCN (2001). IUCN Red List Categories and Criteria. Version 3.1. IUCN, Gland, Switzerland. 30 pp.

IUCN (2003). The IUCN Red List of Threatened Species. Available at http://www.redlist.org (accessed Sept. 2004).

Phillips, S.M. (1998). Two new species of *Eriocaulon* from West Africa. Kew Bull. 53: 943–948.

Pollard, B.J. (2002). Cameroon's First Ecological Reserve. Kew Scientist 22: 4. Available at http://kew.org/kewscientist/ks_22.pdf (accessed Nov. 2004).

Seyani, J.H. (1982). A Taxonomic Study of *Dombeya* Cav. (Sterculiaceae) in Africa. D.Phil. thesis, Univ. Oxford.

Vermeulen, J.J. (1987). A taxonomic revision of the continental African Bulbophyllinae. Orchid Monogr. 2: 300 pp.

BIBLIOGRAPHY

BirdLife International (2003). BirdLife's online World Bird Database: the site for bird conservation. Version 2.0. Cambridge. Available at http://www.birdlife.org (accessed Sept. 2004).

Borrow, N. & Demey, R. (2001). Birds of western Africa. Christopher Helm, London.

Brummitt, R.K. (1992). Vascular Plant Families and Genera. RBG, Kew, U.K. 804 pp.

Brummitt, R.K. & Powell, C.E. (eds) (1992). Authors of Plant Names. RBG, Kew, U.K. 732 pp.

Cable, S. & Cheek, M (1998). The Plants of Mount Cameroon, A Conservation Checklist. RBG, Kew, UK. lxxix + 198 pp.

Champluvier, D. (1990). Une nouvelle sous-espèce de *Magnistipula butayei* De Wild. (*Chrysobalanaceae*) de l'Ituri (Zaïre). Bull. Jard. Bot. Natl. Belg. 60: 393–403.

Cheek, M., Onana, J.-M. & Pollard, B.J. (2000). The Plants of Mount Oku and the Ijim Ridge, Cameroon, A Conservation Checklist. RBG, Kew, UK. iv + 211 pp.

Cheek, M., Pollard, B.J., Darbyshire, I., Onana, J.-M. & Wild, C. (eds) (in press). The Plants of Kupe, Mwanenguba & the Bakossi Mts, Cameroon, A Conservation Checklist. RBG, Kew, UK.

Chiron, G.R. & Guiard, J. (2001). Etude et conservation des orchidées de Bali Ngemba (Cameroun). Richardiana 1(4): 153–186.

Collar, N.J. & Stuart, S.N. (1988). Key Forests for Threatened Birds in Africa. International Council for Bird Preservation, Monograph No. 3. ICBP, Cambridge, UK. 102 pp.

Cook, F.M. (1995). Economic Botany Data Collection Standard. RBG, Kew, UK. 146 pp.

Courade, G. (1974). Commentaire des cartes. Atlas Régional. Ouest I. ORSTOM, Yaoundé.

de Wilde, J.J.F.E. (2002). *Begonia* section *Tetraphila* A.DC., a taxonomic revision. In de Wilde, J.J.F.E. (ed) (2002). Studies in Begoniaceae: 7. Leiden, Backhuys. 5–258 pp. (Wageningen University Papers; 2001–2).

Gartlan, S. (1989). La Conservation des Écostystèmes forestièrs du Cameroun. IUCN Switzerland & UK. 186 pp.

Hawkins, P. & Brunt, M. (1965). The Soils and Ecology of West Cameroon. 2 Vols. FAO, Rome. 516 pp, numerous plates and maps.

Hepper, F.N. & Neate, F. (1971). Plant collectors in West Africa; a biographical index of those who have collected herbarium material of flowering plants and ferns between Cape Verne and Lake Chad, and from the coast to the Sahara at 18° N. International Bureau for Plant Taxonomy and Nomenclature - Utrecht, Oosthoek. [(Regnum vegetabile, v. 74) xvi, 89 pp. 8 p. of photos.]

Holmgren, P.K., Holmgren, N.H. & Barnett, L.C. (1990). Index Herbariorum. 8th ed. New York Botanical Garden. 693 pp.

IUCN (1994). IUCN Red List Categories. IUCN, Gland Switzerland. 21 pp.

IUCN (2001). IUCN Red List Categories and Criteria. Version 3.1. IUCN, Gland, Switzerland. 30 pp.

IUCN (2003). The IUCN Red List of Threatened Species. Available at http://www.redlist.org (accessed Sept. 2004).

Ivens, G.W., Moody, K. & Egunjobi, J.K. (1978). West African Weeds. Ibadan, Oxford Univ. Press. x + 225 pp.

Keay, R.W.J. & Hepper, F.N. (eds) (1954–1972). Flora of West Tropical Africa, 2nd ed., 3 vols. Crown Agents, London.

Keay, R.W.J. (1958). Flora of West Tropical Africa: Volume 1. Second edition. Crown Agents, London, UK.

Lebrun, J. (1935). Les essences forestières des régions montagneuses du Congo oriental. Institut national pour l'étude agronomique du Congo Belge, Bruxelles, Belgium. 263pp.

Letouzey, R. (1968). Les Botanistes au Cameroun. Flore du Cameroun: 7. Museum National d'Histoire Naturelle, Paris. 110 pp.

Letouzey, R. (1985). Notice de la carte phytogéographique du Cameroun au 1: 500,000. IRA, Yaoundé, Cameroon.

Letouzey, R. & White, F. (1978). *Chrysobalanaceae*. Flore du Cameroun 20: 3–138. Museum National d'Histoire Naturelle, Paris.

Mabberley, D.J. (1997). The Plant Book. Second edition. Cambridge University Press, U.K.

Mackay, C.R. (1994). Survey of Important Bird Areas for Bannerman's Turaco *Tauraco bannermani* and Banded Wattle-eye *Platysteira laticincta* in North West Cameroon, 1994. Interim report. BirdLife Secretariat.

MacKinnon, J. & MacKinnon, K. (1986). Review of the Protected Areas System in the Afrotropical Realm. IUCN Switzerland & UK. xviii + 259 pp.

Nkembi, L.& Atem, T. (2003). A report of the biological and socio-economic activities conducted by the Lebialem Highlands Forest Project, South West Province, Cameroon. Environment and Rural Development Foundation (ERuDeF), Cameroon.

Okezie Akobundu, I. & Agyakwa, C.W. (1987). A Handbook of West African Weeds. International Institute of Tropical Agriculture, Ibadan. 521 pp.

Phillips, S.M. (1998). Two new species of *Eriocaulon* from West Africa. Kew Bull. 53: 943–948.

Pollard, B.J. (2002). Rediscovery. Kew Scientist 22: 4. Available at http://kew.org/kewscientist/ks_22.pdf (accessed Nov. 2004).
Prance, G.T. & Sothers, C.A. (2003). Chrysobalanaceae 2: *Acioa* to *Magnistipula*. Species Plantarum: Flora of the World, Part 10: 1–268.
Prance, G.T. & White, F. (1988). The genera of Chrysobalanaceae. Philos. Trans., Ser. B, v. 320, no. 1197: 161.
Secretariat of the Convention on Biological Diversity (2002). Convention on Biological Diversity: Text and Annexes. United Nations Environment Programme, Montreal, Canada. 34 pp.
Seyani, J.H. (1982). A Taxonomic Study of *Dombeya* Cav. (Sterculiaceae) in Africa. D.Phil. thesis, Univ. Oxford.
Stattersfield, A.J., Crosby, M.J., Long, A.J. & Wege, D.C. (1998). Endemic Bird Areas of the world: priorities for biodiversity conservation. BirdLife International, Cambridge.
Stuart, S.N. (ed) (1986). Conservation of Cameroon Montane Forests. International Council for Bird Preservation, Cambridge, UK.
Terry, P.J. (1983). Some Common Crop Weeds of West Africa and their control. USAID, Dakar. 132 pp.
Tye, H. (1986). Geology and landforms in the Highlands of Western Cameroon: 15–17. *In* Stuart, S.N. (ed) (1986). Conservation of Cameroon Montane Forests. International Council for Bird Preservation, Cambridge, UK.
Vermeulen, J.J. (1987). A taxonomic revision of the continental African Bulbophyllinae. Orchid Monogr. 2: 300 pp.
White, F. (1976). The taxonomy, ecology and chorology of African Chrysobalanaceae (excluding *Acioa*). Bull. Jard. Bot. Natl. Belg. 46: 265–350.
White, F. (1983). The vegetation of Africa: a descriptive memoir to accompany the Unesco/AETFAT/UNSO vegetation map of Africa. Paris, UNESCO. 356 pp. – maps.
Wild, C., Morgan, B. & Fotso, R. (in press). The Vertebrate Fauna. *In* Cheek, M., Pollard, B.J., Darbyshire, I., Onana, J.-M. & Wild, C. (eds). The Plants of Kupe, Mwanenguba & the Bakossi Mountains, Cameroon, A Conservation Checklist. RBG, Kew, UK.
World Bank (1993). Ecologically Sensitive Sites in Africa. Vol. 1: Occidental and Central Africa. The World Bank, Washington, USA. 128 pp.
Yana Njabo, K. & Languy, M. (2000). Surveys of selected montane and submontane areas of the Bamenda Highlands in March 2000. Cameroon Ornithological Club, Yaoundé. Cyclostyled.

READ THIS FIRST!: EXPLANATORY NOTES TO THE CHECKLIST

Iain Darbyshire & Yvette Harvey
Herbarium, Royal Botanic Gardens, Kew, Surrey, TW9 3AE, U.K.

Before using this checklist, the following explanatory notes to the conventions and format used should be read.

The checklist is compiled in an alphabetical arrangement: species within genera, genera within families and families within the groups Dicotyledonae, Monocotyledonae, Pinopsida, Lycopsida (fern allies) and Filicopsida (true ferns), following Kubitzki, in Mabberley (1997: 771–781).

Identifications and descriptions of the species were carried out on a family-by-family basis by both family specialists and general taxonomists; these authors are credited at the head of each account. As a general rule, if two authors are listed, the primary author is responsible for the determinations and the second author for the compiling of the account, including writing of descriptions, and distributional data. Where someone is responsible for only one or two genera within a larger family, this is noted in the author accreditation.

As the incomplete Flore du Cameroun (1963–) is a particularly relevant source of information on the plants of the checklist area, a reference to the volume and year of publication is listed at the head of each family account where available.

The families and genera accepted here follow Brummitt (1992), with recent updates on the R.B.G., Kew Vascular Plant Families and Genera database, with one exception: in *Cyperaceae* we follow K.Lye in sinking *Pycreus* Pal. and *Kyllinga* Rottb. into *Cyperus* L.

Within the checklist each taxon in the family account is treated in the following manner:

Taxon name

The species name adopted follows the most recent taxonomic work available. Author citations follow the standard forms Brummitt & Powell (1992).

Not all names listed are straightforward binomials with authorities. A generic name followed by 'sp.' generally indicates that the material was inadequate to name to species, e.g. *Dichapetalum* sp. (*Dichapetalaceae*). Use of 'sp. 1', 'sp. 2' and 'sp. A', 'sp. B' etc. generally indicate unmatched specimens which may be new to science or may prove to be variants of a currently accepted species; for example *Psychotria* sp. A and sp. B (*Rubiaceae*), or new taxa for which sufficient material is available, but which are awaiting formal description, e.g. *Deinbollia* sp. 1 (Sapindaceae). Unless otherwise stated, or inferred from the distribution, these provisional names are applicable only to the current checklist, thus 'sp. 1' indicates 'sp. 1 of the Bali Ngemba checklist'. The use of 'sp. nov.' is a firm statement that the taxon is new to science but awaiting formal description; sufficient material may or may not be available for this process. A generic name followed by '*cf.*': indicates that the specimens cited should be compared with the associated specific epithet, e.g. *Simo* 202 (*Diaphananthe* cf. *bueae*) should be compared with *Diaphananthe bueae* (Schltr.) Schltr. This is an indication of doubt (sometimes due to poor material), suggesting that the taxon is close to (but may differ from) the described taxon. The terms 'aff.', indicating that the taxon has affinity to the subsequent specific epithet, and 'vel. sp. aff.', indicating that the specimen refers to the taxon listed or a closely allied entity, are applied in a similar fashion. These uncertainties are generally explained in the taxon's 'Notes' section (see below).

Taxon reference

The majority of species referred to within the checklist are found in the 2nd edition of Flora of West Tropical Africa (FWTA: Keay 1954–58; Hepper 1963–72), the standard regional flora. Only species names which do not occur in FWTA are given a reference here; if no reference is cited the taxon name is currently accepted and occurs in FWTA. The references listed are not necessarily the place of first publication of the name; rather, we have tried where possible to use widely accessible publications which provide useful information on that taxon, such as a description and/or distribution and habitat data. The reference is recorded immediately below the taxon name. In the case of scientific journals, we list the journal name, volume and page numbers, and date of publication, with a record of the volume part number where it aids in access to the publication. In the case of books, we list the surname of the author(s), the book title, the page number for the taxon in question and the year of publication. Journal and book titles are often abbreviated in the interest of economy of space. Several notable publications are:

Fl. Cameroun	Flore du Cameroun (1963–). Muséum National d'Histoire Naturelle, Paris, France & MINREST, Yaoundé, Cameroon
Fl. Gabon	Flore du Gabon (1961–). Muséum National d'Histoire Naturelle, Paris, France.
F.T.E.A.	The Flora of Tropical East Africa (1952–). Crown Agents, London & A.A.Balkema, Lisse, Netherlands.

Synonyms

In some instances, names used in FWTA have been superseded and are thus reduced to synonymy; these are listed below the accepted name, with the prefix 'Syn.' Names listed in synonymy in FWTA are not recorded here. Other important synonyms are, however, recorded.

Taxon description

The short descriptions provided for each taxon are based primarily upon the material cited, in order to provide the most accurate representation for field botanists working within the checklist area. However, where necessary, they are supplemented by extracted details from the descriptions in FWTA, Fl. Cameroun and the cited taxonomic works. The descriptions are not exhaustive or necessarily diagnostic; rather, they aim to list the key characters to enable field identification of live or dried material, thus microscopic or complex characters are referred to only when they are essential for identification. Where two or more taxa closely resemble one another, a comparative description may be used, by for example stating 'Tree... resembling *Strombosia scheffleri*, but ...'; such comparisons are only made to other taxa occurring within the checklist area.

Several abbreviations are used in the descriptions, most notably d.b.h. (referring to 'diameter at breast height', being a standard measure of the diameter of a tree trunk), the use of 'c.' as an abbreviation for 'approximately', '±' meaning 'more or less'. In addition, in the Menispermaceae account, where male and female flowers and/or inflorescences are repeatedly referred to, the symbols '♂' and '♀' are applied for 'male' and 'female' respectively.

Habitat

The habitat, recorded at the end of the description, is derived mainly from the field notes of the cited specimens and therefore does not necessarily reflect the entire range of habitats for that taxon; rather, those in which it has been recorded within the checklist area. Habitat information is taken from published sources only where field data are not available, for example, where the only specimens recorded were not available to us, but are cited in FWTA Altitudinal measurements are derived from barometric altimeters carried by the collectors. In one case (Jean-Michel Onana in 2001) an altimeter was mis-calibrated, at 400 m below actual. These readings have been corrected in the book by Harvey. Altitudinal ranges, listed together with habitat, are generated directly from the database of specimens from the checklist area, and thus do not necessarily reflect the entire altitudinal range known for the taxon. In some cases no altitude is given for a taxon because this data is not given on the specimen.

Distribution

For the sake of brevity, country ranges are generally recorded for each taxon rather than listing each separate country, for example 'Sierra Leone to Uganda' is taken to include all or most of the intervening countries. Only where taxa are recorded from only two or three, rarely more, countries within a wide area of occurrence are the individual countries listed, for example 'Sierra Leone, Cameroon & Uganda'. For more widespread taxa, a more general distribution such as 'Tropical Africa' or 'Pantropical' is recorded. Where a species is alien to the checklist area, its place of origin is noted, together with its current distribution. Several country abbreviations are used:

Guinea (Bissau):	former Portuguese Guinea.
Guinea (Conakry):	the Republic of Guinea, or former French Guinea.
CAR:	Central African Republic, of former Oubangi-Shari.
Congo (Brazzaville):	the Republic of Congo or former French Congo.
Congo (Kinshasa):	the Democratic Republic of Congo, or former Zaïre.

Abbreviations for parts of the country are also used (N: north, S: south, E: east, W: west, C: central). Where appropriate, Equatorial Guinea is divided into Bioko, Annobon (both islands) and Rio Muni (mainland), and the Angolan enclave of Cabinda north of the Congo River is recorded separately from Angola itself (south of the river).

In addition to country range, a chorology, largely based upon the phytochoria of White (1983) but with modifications to reflect localised centres of endemism in W Africa, is recorded in square brackets. The main phytochoria used are:

Upper Guinea:	broadly the humid zone following the Guinean coast from Senegal to Ghana.
Lower Guinea:	separated from Upper Guinea by the 'Dahomey Gap', which is an area of drier savanna-type vegetation, that reaches the Atlantic coast. Lower Guinea represents the humid zone from Nigeria to Gabon, including Rio Muni, Cabinda, and the wetter parts of western Congo (Brazzaville).
Congolian:	the basin of the River Congo and its tributaries, from eastern Congo (Brazzaville) and southern CAR, through Congo (Kinshasa) and to Uganda, Zambia and Angola.
Afromontane:	a series of vegetation types restricted to montane regions, principally over 2000 m alt.
W(estern) Cameroon Uplands:	a subdivision of the Afromontane phytochorion, used for taxa restricted to the mountain chain running from the Gulf of Guinea islands (Annobon, São Tomé, Príncipe and Bioko) to western Cameroon and southeast Nigeria.
Cameroon Endemic:	for those taxa restricted to Cameroon, a subdivision of Lower Guinea. Taxa endemic to montane western Cameroon are however recorded under W Cameroon Uplands unless they are endemic to the checklist area, when they are listed as a Narrow Endemic (see below).
Narrow Endemic:	for those taxa restricted to the checklist area, a subdivision of W Cameroon Uplands.

These phytochoria are variously combined where appropriate, for example 'Guineo-Congolian (montane)' refers to an Afromontane taxa restricted to the mountains of the Upper and Lower Guinea and Congolian phytochoria. Taxa with ranges largely confined to the Guineo-Congolian phytochorion, but with small outlying populations in wet forest in, for example, west Tanzania or northern Zimbabwe, are here recorded as Guineo-Congolian rather than Tropical African, as the latter would provide a more misleading representation of the taxon's true phytogeography. A range of other chorologies are used for more widespread species, including 'Tropical Africa', 'Tropical & southern Africa', 'Tropical Africa & Madagascar', 'Palaeotropics' (taxa from tropical Africa and Asia), 'Amphi-Atlantic' (taxa from tropical Africa and S America), 'Pantropical' and 'Cosmopolitan'. If these terms are used in the distribution, no separate phytochorion is listed.

For some taxa, such as those native to one area of the tropics, but widely cultivated elsewhere, the chorology is difficult to define and is thus omitted. Both distribution and chorology are omitted for taxa where there is uncertainty over its identification.

Conservation assessment

The level of threat of future extinction on a global basis is assessed for each taxon that has been fully identified and has a published name, or for which publication is imminent, under the guidelines of the IUCN (2001). Under the heading 'IUCN:', each taxon is accredited one of the following Red List categories:

CR:	Critically Endangered	NT:	Near-threatened
EN:	Endangered	LC:	Least Concern
VU:	Vulnerable	DD	Data Deficient

Those taxa listed as VU, EN or CR are treated in full within the chapter on Red Data species, where the criteria for assessment are recorded. Those listed as LC or NT are not treated further in this publication, but it is recommended that further investigation of the threats to those taxa recorded as NT are made. Undescribed taxa, or those with an uncertain determination, are not assessed. In addition, we do not consider it appropriate to assess taxa from the poorly-known genera *Beilschmiedia* (*Lauraceae*) and *Aframomum* (*Zingiberaceae*) for which species delimitation is currently poorly understood and thus for which conservation assessments would be somewhat meaningless; these genera should be revisited once full taxonomic revisions have been completed.

Specimen citations

Specimens from the checklist area are recorded below the distribution and conservation assessment, with the following information:

Location: Bali Ngemba F.R. is a rough indicator as to the actual location, which can be provided by reference to the cited specimen.

Collector and number: within each location, collectors are ordered alphabetically and, together with the unique collection number, are underlined. Only the principal collector is listed here, thus for example, many collections listed under 'Zapfack' were originally recorded as 'Zapfack, Onana, Nana V., Nana F., Ela J., Nangmo & Kom'; this alteration has been made for the sake of brevity. Specimens listed under 'Plot voucher', with the prefix 'BAL' refer to collections made within a 25 m × 25 m plot in representative forest in the Bali Ngemba Forest reserve in April 2002 by *Zapfack et al.*

Phenology: this information is derived directly from the Cameroon specimen database at R.B.G., Kew and is thus dependent upon recording of such information at the time of collection or at the point of data entry onto the database; if this was not done, no phenological information is listed. Collections are recorded as flowering (fl.), fruiting (fr.) or sterile (st.), where applicable.

Date: the month and year of collection is recorded for each specimen; within each collector from each location, collections are ordered chronologically.

The specimens cited are of herbarium material derived from a variety of sources; the chapter on collectors in the checklist area provides further information. In addition, a very few confident sight records of uncollected taxa are also included. Owing to time constraints, only the new taxon additions from the 2004 Bali Ngemba inventory have been incorporated into this checklist. This also applies to the 2002 plot voucher specimens.

Notes

Items recorded in the notes field at the end of the taxon account include:

- Taxonomic notes. For example, in the case of new or uncertain taxa, notes are given to explain how they differ from closely related species.
- Notes on the source of specimen data, for example for those specimens recorded in Fl. Cameroun, or for specimens not seen by the author(s) of the family account.

Occasionally, notes are placed at either the head of a family account (for example, in Labiatae, Verbenaceae and Orchidaceae) which explain particular points of taxonomy, for example, where the species within a genus are poorly delimited, or where we have used broad generic or species concepts.

Ethnobotanical information

Local names and uses are listed for each taxon where appropriate; these are derived largely from local residents and field assistants and are reproduced here with their consent. Local names are in the Bali language unless otherwise stated; English names are listed only where they are widely used in Cameroon. For each local name or use listed, a source is attributed, usually by reference to the collector and number of the specimen from which the information was derived, including the source of the information where recorded, for example, the local name listed for *Hibiscus noldeae* (*Malvaceae*), 'Kakachong' was recorded by *Biye* (collection number 95). In circumstances where the ethnobotanical information was provided verbally with no attached specimen, or where it is known by the author(s) of the family account, the terms 'fide', 'pers. comm.' (personal communication) or 'pers. obs.' (personal observation) are used to attribute the information to a source.

The layout of the information on local uses follows the convention of Cook (1995), listing level 1 state categories of use in capital letters, followed by the level 2 state categories, then the specific use. In order to comply with guidelines set by the Secretariat of the Convention on Biological Diversity (2002), the detailed uses of medicinal plants are not listed here; only the level 1 state 'MEDICINES' is recorded.

REFERENCES

Brummitt, R.K. (1992). Vascular Plant Families and Genera. RBG, Kew, U.K. 804 pp.
Brummitt, R.K. & Powell, C.E. (eds) (1992). Authors of Plant Names. RBG, Kew, U.K. 732 pp.
Secretariat of the Convention on Biological Diversity (2002). Convention on Biological Diversity: Text and Annexes. United Nations Environment Programme, Montreal, Canada. 34 pp.
Cook, F.M. (1995). Economic Botany Data Collection Standard. RBG, Kew, UK. 146 pp.
IUCN (2001). IUCN Red List Categories and Criteria. Version 3.1. IUCN, Gland, Switzerland. 30 pp.
Keay, R.W.J. & Hepper, F.N. (eds) (1954–1972). Flora of West Tropical Africa, 2nd ed., 3 vols. Crown Agents, London.
Mabberley, D.J. (1997). The Plant Book. Second edition. Cambridge University Press, U.K.
White, F. (1983). The vegetation of Africa: a descriptive memoir to accompany the Unesco/AETFAT/UNSO vegetation map of Africa. Paris, UNESCO. 356 pp. – maps.

ANGIOSPERMAE

DICOTYLEDONAE

ACANTHACEAE

I. Darbyshire (K) & K.B. Vollesen (K)

Acanthus montanus (Nees) T.Anderson
Erect herb to 50cm; stems becoming woody at base, hirsute at nodes and apices; leaves subcoriaceous, oblanceolate, 25–30 × 6.5–13.5cm, ± pinnatisect, spinosely toothed, upper surface variegated, scabrid below; petiole 8mm; inflorescence a dense terminal raceme, to 14cm; bracts and calyx lobes ovate, spinose, glabrescent, 2.5cm; corolla lobes form a lip below androecium, 3.5 × 3cm, partially deflexed, white with purple streaking; stamens 4; anthers densely white-haired on one side, hairs to 3mm. Forest and secondary growth; 1570m.
Distr.: Benin to CAR, Angola & NW Zambia [Guineo-Congolian].
IUCN: LC
Bali Ngemba F.R.: Darbyshire 382 fr., 4/2004.
Note: field determination.

Asystasia gangetica (L.) T.Anderson
Straggling herb to 1m; stems sparsely pubescent; leaves ovate, c. 7 × 3cm, base attenuate, then abruptly rounded, apex acuminate, margin subentire, laminae sparsely pubescent or glabrous; petiole to 1.5cm; inflorescence an axillary or terminal spike to 14cm, 6–11-flowered; bracts lanceolate, c. 1mm, ciliate; calyx lobes subulate, c. 4mm, pubescent; corolla tube 1.3 × 0.8cm, puberulent, white with purple markings at base of throat; stamens 4; ovary and base of style pubescent; fruits 2 × 0.3cm, puberulent, green. Farmbush & forest edges; 1700m.
Distr.: widespread in palaeotropics, introduced to neotropics [Palaeotropics].
IUCN: LC
Bali Ngemba F.R.: Biye 98 11/2000.

Brillantaisia madagascariensis T.Anderson ex Lindau
Herb to 1.5m; stems puberulent towards apices; leaves ovate, to 17(–24) × 9.5cm, apex acuminate, base rounded then cuneate, merging into winged petiole, margin crenulate, raphides numerous, most-visible through upper surface; petiole to 4cm; inflorescence a terminal panicle, few-branched, to 14.5cm long; bracts ovate to obovate, ciliate, 1.2 × 0.8cm; calyx lobes linear, to 2cm; corolla white to blue or violet; fruit minutely puberulent. Forest & secondary growth; 1500m.
Distr.: Guinea to Congo (Kinshasa), E Africa & Madagascar [Tropical Africa & Madagascar].
IUCN: LC
Bali Ngemba F.R.: Ghogue 1037 11/2000.

Brillantaisia owariensis P.Beauv.
Syn. *Brillantaisia nitens* Lindau
Erect herb to 2.5m; stems robust, glabrous; leaves broadly ovate, 18–21 × 9–11cm, base acute to truncate, with a cuneate extension grading into the winged petiole, apex acuminate, margin serrate, laminae sparsely pilose, raphides numerous, most-visible through upper surface; petiole to 4cm, winged towards apex, tomentose, hairs to 2mm; panicles lax, to 20cm long, many-flowered, glandular-hairy throughout; bracts caducous; bracteoles linear to ovate, c. 7mm long; calyx lobes linear, uneven, longest 1.2cm; corolla bilabiate, c. 3cm, blue-violet with white throat; fruit linear, 2.2cm long, pubescent. Forest & forest-grassland transition; 1400–1500m.
Distr.: Nigeria to S Sudan and W Tanzania [Lower Guinea & Congolian].
IUCN: LC
Bali Ngemba F.R.: Cheek 10470 11/2000; Ghogue 1020 11/2000; Tadjouteu 403 11/2000.

Brillantaisia vogeliana (Nees) Benth.
Erect herb to 1.2m; similar to *B. owariensis*, but leaf-base truncate to subcordate, raphides abundant, visible on both leaf surfaces; petiole to 9cm, less clearly winged, cuneate extension of leaf-base may be absent; bracts usually persistent, ovate, to 2.2 × 1cm, becoming narrower towards the apex; bracteoles oblanceolate, 0.5cm long; calyx lobes linear, only 1cm long. Forest & forest margins; 1490–1900m.
Distr.: Ghana to Congo (Kinshasa), Sudan to Kenya [Tropical Africa].
IUCN: LC
Bali Ngemba F.R.: Ghogue 1007 11/2000; 1039 11/2000; Onana 1850 10/2001.

Dicliptera laxata C.B.Clarke
Suberect herb to 0.5m; stems glabrous; leaves ovate to elliptic, 6.5–11 × 2.5–5cm, base acute, apex acuminate, margin subcrenulate, upper surface sparsely pilose, raphides visible on under surface; petiole to 2.5cm, puberulent; flowers in axillary umbels, 2 per axis; peduncle to 1cm; pedicels to 0.4cm, puberulent; paired leafy, obovate, glabrous bracteoles, 0.8 × 0.4cm, enclose the flowers; corolla partially exserted, white. Farmbush, cultivation & forest margins; 1400m.
Distr.: Cameroon, Bioko, Congo (Kinshasa) & E Africa [Tropical Africa].
IUCN: LC
Bali Ngemba F.R.: Cheek 10471 11/2000.

Eremomastax speciosa (Hochst.) Cufod.
Fl. Gabon 13: 30 (1966).
Syn. *Eremomastax polysperma* (Benth.) Dandy
Syn. *Clerodendrum eupatorioides* Baker
Herb to 3m; stems robust, tomentose towards apex, hairs to 1.5mm; leaves ovate, 5–7 × 3–5cm, base truncate to subcordate, apex acuminate, margin crenate, surfaces pubescent along veins; petiole to 5.7cm; inflorescence a many-flowered panicle; bracts obovate, 1 × 0.5cm, finely pilose; calyx lobes linear, c. 1.5cm, tomentose; corolla 3–5cm long with neck 3cm, pale blue-lavender; 5 lobes all below the

androecium; stamens 4; anthers blue; fruit 1.7 × 0.3cm, pubescent. Forest & farmbush; 1560–1600m.
Distr.: W Africa to Uganda & NW Tanzania [Guineo-Congolian].
IUCN: LC
Bali Ngemba F.R.: Cheek 10441 11/2000; Etuge 4264 10/2001; 4808 11/2000.

Justicia paxiana C.B.Clarke

Bot. Jahrb. Syst. 20: 63 (1894).
Syn. ***Rungia paxiana*** (Lindau) C.B.Clarke
Herb or subshrub to 1.5m; stems glabrescent; leaves elliptic, 9.5–15 × 3.5–5cm, pairs unequal, base attenuate, apex acuminate, margin subentire, veins and margins sparsely pilose, raphides numerous, visible through both surfaces; petiole to 4.2cm, puberulent; inflorescence an axillary spike to 5.5cm; bracts obovate, 0.8 × 0.5cm, apex rounded to emarginate, ciliate, green with a 2mm hyaline white margin; calyx included within bracts; corolla bilabiate, c. 0.7cm long, pink; stamens 2. Forest & grassland; 1400–1560m.
Distr.: Guinea (Conakry) to Uganda & NW Tanzania [Guineo-Congolian].
IUCN: LC
Bali Ngemba F.R.: Cheek 10456 11/2000; Pollard 612 10/2001; Tadjouteu 402 11/2000.

Justicia striata (Klotzsch) Bullock subsp. *occidentalis* J.K.Morton

Kew Bull.: 502 (1932).
Suberect or creeping herb to 50cm tall, branching; stems puberulent; leaves more or less anisophyllous, ovate to elliptic, 4.5–7 × 1.4–3.3cm, acuminate, base acute to cuneate, margin subcrenulate, surfaces pilose, raphides more or less numerous, visible through upper surface; petiole 0.7–2.8cm; flowers in subsessile axillary clusters of 2–3, puberulent throughout; bracts oblanceolate, 0.5–0.8 × 0.2cm, pilose; calyx lobes lanceolate, 0.4cm; corolla bilabiate, 0.7–1cm long, lower lip trilobed, white with purple lines. Forest & forest edges; 1400–1950m.
Distr.: Ghana to CAR [Upper & Lower Guinea].
IUCN: LC
Bali Ngemba F.R.: Cheek 10457 11/2000; Ghogue 1121 fl., 11/2000.

Justicia unyorensis S.Moore

J. Bot. xlix. 308 (1911).
Straggling herb to 60cm; stems pubescent; leaves ovate, 2.6–5.5 × 0.6–1.9cm, base attenuate, apex acuminate or acute, margin subentire, surfaces pilose; petiole to 1.3cm, pilose; inflorescence a subsessile axillary cluster, 2–3-flowered, pubescent throughout; bracts linear, to 0.9cm; calyx lobes lanceolate, c. 2.5mm; corolla bilabiate, 0.7cm long, pink with purple and white markings on lower lip; fruit 0.6cm long, glabrous. Grassland; 1700m.
Distr.: Nigeria to Tanzania [Afromontane].
IUCN: LC
Bali Ngemba F.R.: Biye 110 11/2000.

ALANGIACEAE

I. Darbyshire (K)

Alangium chinense (Lour.) Harms

Tree 5–20m, glabrescent; leaves alternate, ovate, c. 10 × 6cm, acuminate, obliquely rounded, entire, digitately 5-nerved, scalariform; inflorescences cymose, axillary, c. 20-flowered, 4cm; peduncle 1.5cm; flowers orange, 1.1cm; fruit ellipsoid, 1cm, fleshy. Forest margins and farmbush; 1400–1700m.
Distr.: palaeotropical, excluding upper Guinea [Palaeotropics].
IUCN: LC
Bali Ngemba F.R.: Cheek 10462 11/2000; Onana 2043 fl., 4/2002; Pollard 1042 fl., 4/2002; Zapfack 2044 fl., 4/2002.

AMARANTHACEAE

C.C. Townsend (K) & I. Darbyshire (K)

Fl. Cameroun 17 (1974).

Achyranthes aspera L. var. *pubescens* (Moq.) C.C.Towns.

F.T.E.A. Amaranthaceae: 102 (1985).
Branching herb to 1m, pubescent throughout, particularly in young shoots and leaves; leaves opposite, elliptic, 10–20 × 4–10cm, acuminate, base attenuate, margin subentire; petiole 1–1.5cm; inflorescence a narrow terminal spike, 5–35cm long; flowers solitary, subsessile, dense in upper section, more widely spaced below; peduncle pubescent; flowers becoming deflexed in fruit; bracteoles c. 0.5cm, spinulose with membranous wings to half the bracteole length, subattenuate at apex; tepals lanceolate, c. 0.6cm, green; stamens and ovary purplish. Farmbush & forest edges; 1310–1440m.
Distr.: pantropical.
IUCN: LC
Bali Ngemba F.R.: Biye 77 11/2000; Cheek 10474 11/2000; Etuge 4704 11/2000; Zapfack 2032 fr., 4/2002.

Achyranthes aspera L. var. *sicula* L.

Fl. Cameroun 17: 32 (1974).
Herb resembling var. *pubescens*, but generally less robust and less pubescent; leaves ovate, c. 5.5 × 2.7cm, apex acuminate, base acute or shortly attenuate; spikes shorter, max. 10cm in our specimens, with smaller flowers; tepals 3–4mm; bracteole wings truncate at apex. Farmbush; 1490–1560m.
Distr.: pantropical.
IUCN: LC
Bali Ngemba F.R.: Ghogue 1008 11/2000; Pollard 609 fl., 10/2001.

Celosia pseudovirgata Schinz

Fl. Cameroun 17: 13 (1974).
Syn. ***Celosia bonnivairii*** sensu Keay
Suberect herb to 0.4–1m; stems glabrescent or finely pubescent; leaves alternate, ovate, 4.5–10(–14.5) × 2.5–7cm, acuminate, base shortly attenuate, margin subentire to obscurely undulate; petiole 1–3.5cm; inflorescence spikes terminal or occasionally axillary, sometimes branched, 2.5–5cm; peduncle to 4cm, glabrescent or finely pubescent;

bracteoles and tepals brown; tepals ovate, c. 2mm. Forest & forest edges; 1200–1500m.
Distr.: Nigeria to Congo (Kinshasa) [Lower Guinea & Congolian].
IUCN: LC
Bali Ngemba F.R.: Cheek 10444 11/2000; Ghogue 1002 11/2000; 1024 11/2000; Onana 1808 10/2001; Ujor 30351 fl., fr., 5/1951; Zapfack 2024 fl., 4/2002.

Cyathula prostrata (L.) Blume var. *pedicellata* (C.B.Clarke) Cavaco
Fl. Cameroun 17: 46 (1974).
Syn. ***Cyathula pedicellata*** C.B.Clarke
Scrambling or decumbent herb to 70cm high; stems furrowed, ± pilose; leaves opposite, elliptic-rhombic, 2.8–4.5 × 1.6–2.5cm, apex acute to shortly acuminate, base attenuate, margin subentire, sparsely pilose; petiole 0.3–0.8cm; racemes terminal, 4–13cm; peduncle puberulent; 2–3-flowered cymules on peduncles to 1.5mm; bracts and tepals ovate, to 1.5cm long, green; hooked bristles of infertile flowers numerous, radiating, 1.5mm. Farmbush & forest; 1500m.
Distr.: Sierra Leone to Congo (Kinshasa) & Tanzania [Guineo-Congolian].
IUCN: LC
Bali Ngemba F.R.: Ghogue 1060 11/2000.

Cyathula prostrata (L.) Blume var. *prostrata*
Scrambling herb to 1m, resembling var. *pedicellata*, but cymules subsessile; hooked bristles of infertile flowers less radiating, stem and leaves more densely pilose throughout, hairs to 1mm, leaves to 6 × 3cm. Farmbush; 1500m.
Distr.: pantropical.
IUCN: LC
Bali Ngemba F.R.: Ghogue 1012 11/2000; 1025 11/2000.

ANACARDIACEAE

F.J. Breteler (WAG), M. Cheek (K) & Y.B. Harvey (K)

Mangifera indica L.
Tree c. 15m; slash with translucent mango-scented exudate; leaves simple, long elliptic-oblong, c. 24 × 7cm, subacuminate, obtuse; petiole c. 5cm; inflorescence a terminal, erect, diffuse panicle or thyrse, c. 30 × 20cm; flowers white, 4mm; fruit fleshy, 1-seeded, oblique, c. 9 × 5 × 4cm. Farmbush, cultivated; c. 1400m.
Distr.: native of India [Pantropical].
IUCN: LC
Bali Ngemba F.R.: Pa Foncham Foseleng 2 4/2002.
Local name: Mango. **Uses:** FOOD – infructescences – an important cash crop; MEDICINES (*Foseleng* 2).

Pseudospondias microcarpa (A.Rich.) Engl. var. *microcarpa*
Tree 5–30m; broken stems scented of mango; leaves often only 3–4-jugate, pinnate, to 30cm long, leaflets 4–12 pairs, to 20 × 8cm, oblique, acuminate, obtuse to rounded; petiolule 0.5cm; petiole to 8cm; inflorescence c. 20 × 12cm, highly diffuse; flowers 4mm, greenish yellow; stamens 8; fruit ellipsoid, 2–5cm, red. Farmbush and secondary forest; 1600m.
Distr.: Senegal to Malawi [Tropical Africa].
IUCN: LC
Bali Ngemba F.R.: Onana 2032 fr., 4/2002.
Uses: FOOD – infructescences – fruit edible (*Onana* 2302).

Trichoscypha acuminata Engl.
Tree 15m; leaves 1.2m, c. 15-jugate; middle leaflets leathery, oblong-lanceolate, c. 20–25 × 5–10cm, acuminate, rounded, c. 15 nerve pairs; petiole 15cm; inflorescence cauliflorous, 15–30 × 6cm; flowers 2–3mm, pink; fruit red, 3cm. Forest; 1500m.
Distr.: S Nigeria to Congo (Kinshasa) [Lower Guinea & Congolian].
IUCN: LC
Bali Ngemba F.R.: Etuge 5360 fr., 4/2004.
Local name: Monkey plum. **Uses:** FOOD – infructescences – fruits edible, harvested from the wild (fide *Etuge*).
Note: field determination.

Trichoscypha lucens Oliv.
F.T.A. 1: 444 (1868).
Tree to 20m; leaves 60cm, 11-jugate, leaflets oblong-elliptic, to 20 × 6cm, acuminate, obtuse, to 10-nerved; petiole 9cm; inflorescence terminal, glabrescent, c. 50 × 25cm, branches numerous, patent, to 13cm; flowers 2–3mm, white; stamens bright orange. Forest; 1300–1450m.
Distr.: Cameroon, Gabon to Congo (Kinshasa) [Lower Guinea & Congolian].
IUCN: LC
Bali Ngemba F.R.: Cheek 10509 11/2000; Etuge 4695 11/2000; Kongor 58 fl., 10/2001; Onana 1921 10/2001; Ujor 30407 5/1951; Zapfack 2037 fr., 4/2002.
Uses: MATERIALS – wood – timber (*Cheek* 10509).

ANNONACEAE

G. Gosline (K) & Y.B. Harvey (K)

Artabotrys aurantiacus Engl. & Diels
Notizbl. Königl. Bot. Gart. Berlin 2: 300 (1899).
Woody climber; leaves coriaceous, elliptic-oblong, 7.5–10 × 2.5–6cm; inflorescence a recurved hook, c. 2cm long, few-flowered; pedicels 7–10mm long; petals 15–30 × 4–9mm, tomentose, white to yellow-orange; monocarps red at maturity, few, ellipsoid, both extremities rounded, 15–30 × 7–13mm. Forest; 1600m.
Distr.: Cameroon to Congo (Kinshasa) [Lower Guinea & Congolian].
IUCN: LC
Bali Ngemba F.R.: Onana 2027 fl., fr., 4/2002.

Isolona hexaloba (Pierre) Engl. & Diels
Kew Bull. 41: 296 (1986).
Syn. ***Isolona pleurocarpa*** Diels subsp. ***nigerica*** Keay
Tree to 18m; leaves oblong-oblanceolate, 8–15 × 2–4.5cm, gradually acuminate, broadly cuneate to obtuse at base, with 9–13 pairs of lateral nerves prominent above and beneath, extending almost to the margin without looping; corolla green with red centre; lobes oblong, 5–10 × 3–5mm; tube to 10mm; fruit subovoid to depressed subglobose, lobulate, with irregular transverse ridges, 8–10 × 7–8cm. Forest; 1600m.
Distr.: S Nigeria, Cameroon, CAR, Gabon, Congo (Kinshasa), Tanzania and Angola [Tropical Africa].

IUCN: NT
Bali Ngemba F.R.: Onana 2030 fl., 4/2002.
Note: widespread but rare (only 12 sites known). Likely to prove Vulnerable under criterion A when better data on site threats are available.

Monanthotaxis sp. of Bali Ngemba

Climber to 5m; slash aromatic; leaves oblong-oblanceolate, to 10.5 × 4cm, acute, truncate; petiole to 5mm; flowers not seen; fruits with pedicel c. 3.5cm, monocarps c. 6, cylindrical, to 3 × 0.5cm, terminated by a sharp point, constricted between seeds. Forest patches; 1600m.
Distr.: Bali Ngemba F.R.
Bali Ngemba F.R.: Etuge 4810 11/2000.

Xylopia africana (Benth.) Oliv.

Tree to 12m; leaves obovate-elliptic, 9–16 × 4–8cm, strongly reticulate, nerves on underside dark red; flowers yellow-orange, subglobose, c. 1.5cm diam., remaining closed except for a small apical opening; fruits with pedicel c. 3cm, monocarps c. 10, cylindrical, 3–12 × 1cm, pointed, constricted between seeds, stipe c. 2cm. Forest; 1700–2000m.
Distr.: SE Nigeria, SW Cameroon & São Tomé [W Cameroon Uplands].
IUCN: VU
Bali Ngemba F.R.: Cheek 10527 11/2000; Onana 1825 10/2001; 1835 10/2001; Pollard 1032 fl., fr., 4/2002; Tadjouteu 410 11/2000.

APOCYNACEAE

H.J. Beentje (K) & Y.B. Harvey (K)

Funtumia elastica (Preuss) Stapf

Tree to 15m, glabrous; stem drying black, smooth; leaves elliptic, to 19 × 7cm, acumen cuneate, 1.5cm, base acute-decurrent, lateral nerves c. 10 pairs, domatial pits to 1.5mm, margin undulate; fascicles axillary, 2cm; sepals 3–5mm; corolla white, c. 9mm, twisted to right, lobes shorter than tube; fruit with paired follicles, patent, c. 17 × 2cm, obtuse, angled; seeds hairy. Forest; 1200m.
Distr.: Guinea to Uganda [Guineo-Congolian].
IUCN: LC
Bali Ngemba F.R.: Onana 2034 4/2002.

Landolphia landolphioides (Hallier f.) A.Chev.

Liana 5–30m high; trunk up to 30cm diam., bark fissured; leaves 6–26 × 2.8–10.5cm, coriaceous when dry, paler beneath with pale venation, lateral nerves in 7–16 pairs, obtuse or acuminate, with acumen to 1cm long, rounded or acute or slightly angustate at base; petiole 4–17mm long; inflorescence an axillary or terminal cyme, 2.5–6.5 × 1.9–4cm; peduncle 3–15mm long; pedicels 2–8mm long; corolla white, cream or yellow, often tinged orange, red, purple or green; tube 8–14mm long; lobes 6.5–17.5 × 2.5–4mm, rounded at apex; fruit yellow, pyriform to globose or ovoid, 3.5–8 × 2.5–7 × 2.5–7cm; seeds irregularly ovoid, 11–21 × 6–13 × 6–8mm. Forest; 1600m.
Distr.: Nigeria to Uganda [Lower Guinea & Congolian].
IUCN: LC
Bali Ngemba F.R.: Onana 2026 fl., fr., 4/2002.

Rauvolfia vomitoria Afzel.

Shrub or tree, 2–8m; stems greyish white; leaves in whorls of 3(–4), membranous-papery, elliptic, c. 20 × 7cm, subacuminate, cuneate, lateral nerves 12 pairs; petiole 2cm; panicles puberulent, 10cm; corolla white, 8mm; fruit with 2 ovoid berries, 8mm. Forest, savanna.
Distr.: Senegal to Uganda [Guineo-Congolian].
IUCN: LC
Bali Ngemba F.R.: Pa Foncham Foseleng 3 4/2002.
Local name: Tongmogi (*Foseleng* 3). **Uses:** MEDICINES (*Foseleng* 3).
Note: this is only a tentative determination, since the material is sterile.

Tabernaemontana pachysiphon Stapf

Shrub or small tree, 2–15m high, glabrous; leaves broadly to narrowly elliptic, 10–50 × 5–26cm, acuminate, cuneate or rounded base, c. 7–16 pairs of lateral nerves, margin revolute; petiole 6–19mm long; inflorescence long-pedunculate; peduncle 3–14cm long; pedicels 8–22mm long; corolla sweet-scented, white; tube 18–35(–42)mm; lobes 14–50 × 6–18(–27)mm, recurved; fruit of 2 separate mericarps, obliquely subglobose, 7–15 × 6–13 × 6–14cm, fruit wall as thick as seed cavity is wide; seed dark brown, obliquely ellipsoid, 11–14 × 5–6.5 × 4.5–7mm, with 6–7 longitudinal grooves at each side. Forest; 1700m.
Distr.: Ghana to Malawi [Tropical Africa].
IUCN: LC
Bali Ngemba F.R.: Tadjouteu 414 11/2000.
Note: *Tadjouteu* 414 has flowers in tight bud and no fruits so the determination is provisional.

Tabernaemontana sp. of Bali Ngemba

Tree to 10m, glabrous; leaves to 23 × 9.5cm, elliptic to oblong, acute to subacuminate, cuneate, c. 15–16 pairs of lateral nerves, margin revolute; petiole to 20mm. Disturbed forest; 1770m.
Distr.: Bali Ngemba F.R.
Bali Ngemba F.R.: Tadjouteu 422 11/2000.
Note: description based on sterile material.

Voacanga sp. 1

Tree 6(–9)m, glabrous; stems greyish white, white pustular lenticellate, puberulent when young; leaves, oblong or oblanceolate, to 24 × 7cm, acumen 0.5cm, cuneate, lateral nerves to 14 pairs; petiole 2.5cm; flowers not seen; fruiting peduncle 8–12cm, 1-fruited, pendulous, fruit with carpels united, only slightly bilobed before dehiscence (when to 4 × 5cm), drying black, finely ridged, puberulent, dehiscing along a single terminal suture, when 4.5 × 7 × 4cm. Forest; 1310–1900m.
Distr.: Kupe-Bakossi & Bali Ngemba[W Cameroon Uplands].
Bali Ngemba F.R.: Onana 1837 10/2001; 1992 fr., 4/2002; Pollard 1045 fr., 4/2002; Zapfack 2002 fr., 4/2002.
Note: probably a new species; flowers required.

ARALIACEAE

D.G. Frodin (K) & M. Cheek (K).

Fl. Cameroun 10 (1970).

Polyscias fulva (Hiern) Harms

Pioneer tree to 35m, felty-puberulent; leaves alternate, 40cm, imparipinnate, 6-jugate, elliptic-oblong, 4.5–8 × 2.5–4cm, subacuminate, rounded, densely pale brown-felty below; umbels of panicles, each 30–50cm; partial-peduncles racemose, to 5cm; fruits 5mm, purple. Forest; 1550m.
Distr.: Guinea (Conakry) to Kenya [Tropical Africa].
IUCN: LC
Bali Ngemba F.R.: Darbyshire 396 4/2004.
Note: field determination.

Schefflera mannii (Hook.f.) Harms

Tree 12–15m, initially an epiphyte, but ends up strangling and substituting for the host; leaves c. 5–9, digitately compound; leaflets elliptic-oblong, 4.5–8.5 × 12–20cm, subacuminate, 12–18 pairs of lateral nerves; petiolules 4–6cm; stipule intrapetiolar; inflorescence a terminal fascicle of 15–20 racemes, 30–40cm; peduncles 1–2cm; flowers sessile, 3mm; fruits ellipsoid to obovoid, 5 × 7mm. Forest; 1950m.
Distr.: Nigeria, Bioko, Cameroon [W Cameroon Uplands].
IUCN: VU
Bali Ngemba F.R.: Biye 123 11/2000.

ARISTOLOCHIACEAE

M. Cheek (K)

Pararistolochia sp. of Bali Ngemba

Climber to 4 m, glabrous; leaves alternate, drying green, blade broadly ovate, to 13 × 10cm, papery, acumen 1cm, base shallowly cordate, palmately 3-nerved, the laterals pedately branched at the base, tertiary nerves prominent, coarsely reticulate; petiole 5cm, twisting. Beside streams; 1400m.
Distr.: Bali Ngemba F.R.
Bali Ngemba F.R.: Etuge 4797 11/2000.
Note: fertile material required to confirm placement and to determine fully.

ASCLEPIADACEAE

D.J. Goyder (K)

Epistemma cf. *decurrens* H.Huber

Bull. Mus. Natl. Hist. Nat., B, Adansonia, Sér. 4 11(4): 447–452 (1989, publ. 1990).
Epiphytic shrub; old stems 5mm diam. (dried), purple, glabrous, with copious white exudate, leafy stems brown, matt, white-puberulent; leaves oblong-elliptic, c. 6.5 × 2.5cm, acumen 3mm, lateral nerves c. 6 on each side of the midrib; petiole 6–8mm; inflorescence axillary, nodding, 5-flowered; peduncle 5mm; pedicel 8mm; calyx 5-lobed, 2–3mm; corolla yellow-green, 1 × 1cm, divided by $^1/_2$ to $^2/_3$, corona lobes filiform, densely-hairy. Upper forest edge; 2110m.
Distr.: Cameroon [Afromontane].
Bali Ngemba F.R.: Darbyshire 378 fl., 4/2004.

Note: of the three species of *Epistemma* described, *Darbyshire* 378 most closely matches the epiphytic *Epistemma decurrens* H.Huber. Until now, this species has been known only from the type, collected in montane forest at 2000m, c. 30km NNW of Banyo, further along the same mountain chain.
In the same paper Huber (Bull. Mus. Natl. Hist. Nat., B, Adansonia, Sér. 4, 11(4): 447–452 (1989)) described a second new species *E. rupestre* H. Huber, this time growing on granite rock at lower altitude, and having somewhat more fleshy flowers with shorter corolla lobes and the corona lobes pubescent only towards the tips.
The third species in the genus, *E. assianum* D.V.Field & J.B.Hall, is epiphytic, with a more open corolla, again with fleshy ridges in the tube, and with somewhat shorter corona lobes than in the other two species. This species was described from Ivory Coast and Ghana.
However, there are so few collections of any of the species that it is difficult to assess the variation shown, and whether or not the three should be regarded as regional variants of a more broadly defined species.

Marsdenia angolensis N.E.Br.

Kew Bull.: 258 (1895); F.T.A. 4(1): 423 (1904).
Syn. *Gongronema angolense* (N.E.Br.) Bullock
Herbaceous or woody scrambler to 4m with white latex; leaves 5–12 × 3–7cm, broadly ovate, acuminate, deeply cordate, pubescent; petiole 2–6cm; inflorescences extra-axillary, about as long as the petioles of the adjacent leaf pair, with 2–3 principal branches terminated by subumbelliform clusters of flowers, pubescent; pedicels 4–10mm; sepals c. 1.5–2mm; corolla cream or yellowish green, united into a broadly cylindrical tube for about half its length; tube 1.5–2.5mm; lobes 1.5–2 × 1.5mm; corona lobes c. 2mm, fleshy; follicles occurring singly or paired, narrowly cylindrical and tapering to a slightly displaced tip, 8–12 × 0.5cm, densely pubescent with short,spreading hairs; seeds flattened, c. 7 × 2mm, oblong with a narrow yellowish margin. Forest understorey; 1450m.
Distr.: tropical Africa.
IUCN: LC
Bali Ngemba F.R.: Pollard 928 fl., 4/2002.

Pentarrhinum abyssinicum Decne. subsp. *ijimense* Goyder

Cheek, M. *et al.* (2000). The Plants of Mt Oku and the Ijim Ridge, Cameroon, A Conservation Checklist: 92.
Herbaceous climber to c. 6m tall, exudate clear; stems c. 2mm diam., glabrous; internodes on flowering stems c. 15–20cm; leaves opposite, membranous, glabrous; blades ovate, c. 9 × 6cm, apex acuminate, base broadly cordate; petiole 4–5cm long; stipule-like subsidiary leaves ovate, c. 1 × 0.6cm; inflorescence axillary; peduncle as long as, or longer than, the subtending petioles, bearing a cluster of 8–15 flowers; pedicels c. 5mm; flowers 5-merous, c. 5mm diam., white-flecked with pink; fruits with two follicles arranged in a line, follicles cylindrical, smooth, c. 4 × 0.6cm, apex acute. Forest; 1560m.
Distr.: Mount Oku and the Ijim ridge; Bali Ngemba F.R. [Endemic].
IUCN: CR
Bali Ngemba F.R.: Etuge 4267 10/2001.

Sphaerocodon caffrum (Meisn.) Schltr.
Suberect perennial, 30–90cm, branched from base; leaves 5–10cm long; inflorescence of lateral umbelliform cymes; flowers very dark purple. Grassland; 1200m.
Distr.: widespread in tropical Africa extending to Natal [Tropical Africa].
IUCN: LC
Bali Ngemba F.R.: Onana 2050 fl., 4/2002.

***Tylophora* sp. of Bali Ngemba**
Climber to c. 5m high, glabrous, with clear exudate; stem twining, c. 3mm diam.; leaves broadly elliptic to obovate, acuminate to subacute, base cuneate to rounded; petiole to 3cm; flowers and fruits not seen. Forest patches. 1600m.
Distr.: Bali Ngemba F.R.
Bali Ngemba F.R.: Etuge 4818 11/2000.
Note: material (Etuge 4818) inadequate to determine further.

BALSAMINACEAE

I. Darbyshire (K) & M. Cheek (K)

Fl. Cameroun 22 (1981).

Impatiens burtoni Hook.f.
Terrestrial herb, c. 30cm; leaves alternate, lanceolate, c. 8 × 5cm; single-flowered; pedicel c. 3cm; flower white, c. 3 × 1cm, face concave; sepals densely long-hairy; spur about as long as petals. Farmbush; 1400–1490m.
Distr.: Cameroon to Burundi [Lower Guinea & Congolian].
IUCN: LC
Bali Ngemba F.R.: Ghogue 1004 11/2000; Pollard 591 fl., 10/2001.
Note: common in farmbush, notable for having large white-hairy flowers.

Impatiens filicornu Hook.f.
Terrestrial or epilithic, rarely epiphytic, herb to c. 30cm; leaves drying dark brown above, alternate, lanceolate, c. 9 × 4cm; peduncle c. 11cm; flowers fascicled at apex; bracts ovate, 3–4mm; pedicel 1.5cm; flowers pink to white with purple markings, face flat, c. 1.5cm diam., spur c. 15mm. Forest; 1430–1700m.
Distr.: Bioko, Cameroon to Cabinda [Lower Guinea].
IUCN: LC
Bali Ngemba F.R.: Pollard 407 fl., fr., 11/2000; Zapfack 2053 fl., 4/2002.

Impatiens hians Hook.f. var. ***hians***
Terrestrial herb, 30–100cm, glabrous; leaves alternate, ovate, c. 12 × 6cm; peduncle c. 4cm; flowers 2–5 on a short rachis; pedicels 3cm; flower green, face concave, 5 × 2.5cm; spur red. Forest.
Distr.: Bioko to W Congo (Kinshasa) [Lower Guinea].
IUCN: LC
Bali Ngemba F.R.: Daramola 40476 fl., 2/1959.
Note: distinct in the gaping mouth.

Impatiens mackeyana Hook.f. subsp. ***mackeyana***
Grey-Wilson, C. (1980), *Impatiens* of Africa: 221.
Terrestrial herb, 30–40cm; leaves alternate, ovate, c. 11 × 7cm; peduncle < 0.5cm, usually single-flowered; pedicels 2.5cm; flower purple-pink, face concave, 4.5 × 3cm, glabrous; lateral sepals 6mm, dentate on one edge; spur 1cm, recurved. Forest.
Distr.: Nigeria to Gabon [Lower Guinea].
IUCN: NT
Bali Ngemba F.R.: Ujor 30394 fl., 5/1951.

Impatiens sakerana Hook.f.
Terrestrial herb, 0.3–1m; leaves in whorls of 3–4 or opposite, ovate-elliptic, c. 7.5–12 × 3.5–4cm, apex acuminate, margin crenate, pale brown-hairy below; racemes short, 3–4-flowered; peduncle c. 4cm; flowers 1.5–3cm long; lip red, gradually narrowing into short spur, greenish red, swollen at tip. Forest patches; 2100–2200m.
Distr.: Bioko & Cameroon [W Cameroon Uplands].
IUCN: VU
Bali Ngemba F.R.: Ghogue 1085 11/2000; Pollard 657 fl., 10/2001.

BEGONIACEAE

M. Cheek (K)

Begonia fusialata Warb. var. ***fusialata***
Terrestrial or epiphytic, often scandent, to 3m; resembling *B. oxyanthera*, but leaves highly oblique at base; fruit 3-angled, white. Forest; 1700m.
Distr.: Guinea (Conakry) to Congo (Kinshasha) [Guineo-Congolian].
IUCN: LC
Bali Ngemba F.R.: Biye 99 11/2000.

Begonia oxyanthera Warb.
Syn. ***Begonia jussiaeicarpa*** Warb.
Stoloniferous epiphyte; leaves elliptic, c. 10 × 4cm, acuminate, obtuse to rounded, entire or slightly dentate, lower margin and nerves red, lower surface glabrescent; petiole 1cm; inflorescence 1.5cm; perianth 0.8cm, cream-red; fruit cylindric, 4 × 0.4cm, red. Forest; 1800m.
Distr.: Bioko & Cameroon [W Cameroon Uplands].
IUCN: VU
Bali Ngemba F.R.: Pollard 954 fl., 4/2002.
Uses: FOOD – leaves eaten raw; MEDICINES (fide *Kenneth Cha* in *Pollard* 954).

BIGNONIACEAE

Y.B. Harvey (K) & I. Darbyshire (K)

Fl. Cameroun 27 (1984).

Kigelia africana (Lam.) Benth.
Tree to 15m, glabrous; leaves opposite, imparipinnate, to 50cm, 5–6-jugate, leaflets oblong-elliptic, 11–20 × 3.5–8cm, apex acuminate, base acute to rounded; petiole 8–25cm; petiolules 3–6mm; panicles pendent, lax, to 40cm, lateral branches to 7cm; peduncle to 20cm; pedicels 1.2cm; calyx cupular, 2.8 × 2.5cm, lobes triangular, 0.6–1cm; corolla campanulate, 6(–8)cm, dark red; fruit sausage-shaped, 20–50(–100)cm. Forest & farmbush; 1310–1600m.
Distr.: throughout tropical Africa, widely cultivated elsewhere [Tropical Africa].
IUCN: LC

Bali Ngemba F.R.: Ghogue 1035 11/2000; Tiku 30417 fr., 6/1951; Zapfack 2004 fl., fr., 4/2002.
Note: *Tiku* 30417 is placed here, based on the assumption that *K. acutifolia* Engl. is a synonym of the polymorphic *K. africana.*

BOMBACACEAE

M. Cheek (K)

Fl. Cameroun 19 (1975).

Bombax buonopozense P.Beauv.

Tree to c. 30m, spines extending to leafy stems; leaves alternate, digitately compound, c. 25cm; leaflets 5–7, obovate, to c. 17 × 5cm, acumen long and slender, 2cm, secondary nerves white in dried leaves. Forest; 1400m.
Distr.: Sierra Leone to Gabon [Upper & Lower Guinea].
IUCN: LC
Bali Ngemba F.R.: Cheek 10434 11/2000.
Notes: description based on a sapling voucher, fertile material needed to confirm, but *Ceiba* unlikely in view of the acumen and nerves.

BORAGINACEAE

R. Becker (RNG)

Cordia millenii Baker

Shrub to tree, 4–32m; stem brownish; leaves suborbicular-obovate, 15–33 × 10–25cm; flowers in cyme; corolla white with yellowish tube; fruit 2.5 × 1.5–2cm. Forest; 1650m.
Distr.: Guinea to Gabon, Congo (Kinshasa), Angola and Sudan [Tropical Africa].
IUCN: LC
Bali Ngemba F.R.: Etuge 4820 11/2000.

Cynoglossum coeruleum A.DC. subsp. *johnstonii* (Baker) Verdc. var. *mannii* (Baker & C.H.Wright) Verdc.

F.T.E.A. Boraginaceae: 110 (1991).
Syn. *Cynoglossum lanceolatum Forssk.* subsp. ***geometricum*** (Baker & C.H.Wright) Brand
Herb 0.3–1.2m tall; cauline leaves elliptic to lanceolate, 2–22 × 0.8–5.5cm, sessile or short-petiolate; radical leaves long-petiolate; nutlets with glochidia only on meridian and top surface. Forest edge, margins of cultivation; 1700m.
Distr.: Cameroon, Congo (Kinshasa), Rwanda, Burundi, East Africa, Southern East Africa to Zimbabwe [Tropical & subtropical Africa].
IUCN: LC
Bali Ngemba F.R.: Zapfack 2064 fl., fr., 4/2002.

BUDDLEJACEAE

Y.B. Harvey (K)

Nuxia congesta R.Br. ex Fresen.

Syn. *Lachnopylis mannii* (Gilg) Hutch. & M.B.Moss
Tree 2–8m; bole white, fibrous; stems orange-brown; leaves in whorls of 3, dimorphic: type 1 ovate-elliptic, c. 4 × 2.5cm, apex rounded, base cuneate, serrate, petiole 0.2cm; type 2 c. 11 × 5.5cm, entire, petiole 2cm; inflorescence terminal, 7 × 10cm, dense, many-flowered; flowers white, c. 7mm. Forest & forest edge; 1450m.
Distr.: Guinea (Conakry) to South Africa [Afromontane].
IUCN: LC
Bali Ngemba F.R.: Tadjouteu 426 11/2000.
Uses: FUELS – wood – 'utilisé pour les feux de bois' (*Tadjouteu* 426).

BURSERACEAE

I. Darbyshire (K) & J.-M. Onana (YA)

Canarium schweinfurthii Engl.

Tree to 40m; copious exudate, drying whitish; young branches and young leaves rusty-pubescent; leaves clustered towards branch apices, c. 30–60cm long, imparipinnate, 8–12-jugate; leaflets oblong-ovate, 5–20 × 3–5.5cm, base cordate, lateral nerves pubescent below; petiole flat on upper side towards base; panicles to 30cm long; pedicels c. 3mm; flowers cream-white, c. 1cm; fruits ellipsoid, c. 3.5 × 2cm, epicarp glabrous, purple-black, endocarp trigonous. Forest, farmbush; 1400m.
Distr.: Senegal to Sudan & Tanzania [Guineo-Congolian].
IUCN: LC
Bali Ngemba F.R.: Etuge 5370 fl., 4/2004.
Note: field determination.

Dacryodes edulis (G.Don) H.J.Lam

Tree to 20m; exudate brown; branchlets, petioles, inflorescence rachis and outside of flowers stellately brown-hairy; leaves imparipinnate, 4–7-jugate; leaflets oblong-ovate, 12.5–20.5 × 3.5–5.5cm, acumen to 1.8cm, base acute, oblique in lateral leaflets, undersurface sparsely stellate-hairy with long simple hairs along midrib; lateral nerves 11(–14) pairs; petiole 9cm; petiolules 0.5cm; panicles many-branched; partial peduncles to 14cm, grooved; flowers 0.4cm, cream; fruit ovoid-ellipsoid, 5.5 × 2.2cm. Forest, and planted near villages.
Distr.: S Nigeria to Angola [Lower Guinea & Congolian].
IUCN: LC
Bali Ngemba F.R.: sight record (*Darbyshire*, pers. comm.) 4/2004.
Note: common around Mantum (*Darbyshire*, pers.comm.).

Santiria trimera (Oliv.) Aubrév.

Tree to 20m with laterally-flattened stilt roots, glabrous; stem strongly smelling of turpentine when cut; leaves imparipinnate, 2–3-jugate, leaflets oblong-elliptic, 10.5–16.5 × 4–6.5cm, acumen narrow, to 2.2cm long, base acute to obtuse, lateral nerves 7–10 pairs; petiole 9.5cm; petiolules 1cm; panicles axillary, to 9cm; fruit green-black, asymmetrically oblate, 1.2 × 2 × 1.5cm, style-remains lateral; fruiting pedicel 0.8cm. Forest; 1430m.
Distr.: Sierra Leone to Congo (Kinshasa) [Guineo-Congolian].
IUCN: LC
Bali Ngemba F.R.: Etuge 4722 11/2000.

CACTACEAE

M. Cheek (K)

Rhipsalis baccifera (J.Miller) Stearn

Syn. ***Rhipsalis cassutha*** Gaertn.
Epiphyte, pendent, 45cm; stems terete, grey-green, succulent, glabrous, branching at intervals of c. 10cm; flowers red, 1cm; fruit globose, 8mm, white-tinged-red. Forest; 1200m.
Distr.: tropical America & Africa [Amphi-Atlantic].
IUCN: LC
Bali Ngemba F.R.: Onana 2053 fl., 4/2002.

CAMPANULACEAE

M. Thulin (UPS) & Y.B. Harvey (K)

Lobelia columnaris Hook.f.

Herb 2m tall, unbranched; stem 2cm diam. at base, glabrous; leaves oblanceolate-oblong, c. 30 × 8cm at base of stem, gradually diminishing in size towards the apical inflorescence, apex acute, margin inconspicuously serrate, softly-hairy; inflorescence a densely-flowered, unbranched spike occupying the apical half of the plant; corolla pale blue, lilac to white, 2–3cm long. Forest (edges) and grassland; 1200m.
Distr.: Bioko, Cameroon and Nigeria [Western Cameroon Uplands (montane)].
IUCN: NT
Bali Ngemba F.R.: Onana 2052 fl., 4/2002.

Lobelia neumannii T.C.E.Fr.

Notizbl. Bot. Gart. Berlin-Dahlem 8: 409 (1923).
Annual erect to decumbent herb, 3–30cm; stems often purple-tinged towards base; leaves 4–35 × 4–25mm, ovate or elliptic to suborbicular, narrowing towards apex, dentate; petiole 1.5–9mm; flowers in lax leafy racemes; pedicels 5–30mm; upper bracts linear, shorter than the pedicels; bracteoles 0.3–0.6mm near the base of the pedicel; hypanthium broadly obconical; calyx lobes narrowly triangular; corolla 3.5–5mm, blue; capsule obovoid; seeds elliptic, 0.4–0.5mm long, brown. Upland grassland, usually on bare or rocky ground; 2100m.
Distr.: Cameroon, Ethiopia, Sudan, Kenya [Afromontane].
IUCN: LC
Bali Ngemba F.R.: Pollard 674 fl., 10/2001.

Wahlenbergia ramosissima (Hemsl.) Thulin subsp. *ramosissima*

Symb. Bot. Upsal. 21: 189 (1975).
Syn. ***Lightfootia ramosissima*** (Hemsl.) E.Wimm. ex Hepper
Erect annual herb, to c. 10cm high; stems laxly-pubescent; leaves sessile, basal leaves broadly ovate, 5 × 4mm, stem leaves narrowly ovate, up to 10 × 4mm, margin undulate; inflorescence lax; bracts at base of pedicels, to 3mm long; pedicels 12–14mm; hypanthium obovoid, to 1mm long, expanding to 2mm in fruit; calyx lobes to 0.5mm; corolla blue, to 1mm long, split almost to base in linear lobes; seeds elliptic, to 0.6mm long. Grassland; 2100m.
Distr.: Nigeria & Cameroon [W Cameroon Uplands].
IUCN: VU
Bali Ngemba F.R.: Pollard 675 fl., 10/2001.

Wahlenbergia silenoides Hochst. ex A.Rich.

F.T.E.A. Campanulaceae: 12 (1976).
Syn. ***Wahlenbergia mannii*** Vatke
Perennial herb, up to 60cm; stems hirsute at base; leaves lanceolate to narrowly ovate, 5–25 × 1–7mm, margin sometimes undulate; inflorescence few-flowered, lax; pedicels up to 60mm; hypanthium narrowly obconical; calyx lobes 1.6–4mm long; corolla 4–6.5mm long, +/- white, deeply-split into narrowly elliptic lobes; capsule narrowly obconical, up to 17mm long; seeds 0.6–0.7mm long. Grassland & forest-grassland transition; 2100m.
Distr.: Nigeria, Bioko & W Cameroon, east to Sudan, Ethiopia and E Africa [Tropical Africa].
IUCN: LC
Bali Ngemba F.R.: Pollard 464 fl., fr., 11/2000.

CAPPARACEAE

M. Cheek (K)

Fl. Cameroun 29 (1986).

Cleome iberidella Welw. ex Oliv.

Fl. Cameroun 29: 61 (1986).
Syn. ***Cleome montana*** A. Chev. ex Keay
Herb; branches prostrate, reddish, radiating from taproot, c. 30cm; leaves to 5cm, trifoliolate; leaflets to 2.5cm; flowers numerous at stem apex; petals lilac with purple streaking, whitish at base; fruits cylindrical, puberulent, c. 4cm. Thicket; 1550m.
Distr.: Guinea Bissau to Zambia [Afromontane].
IUCN: LC
Bali Ngemba F.R.: Darbyshire 401 4/2004.
Note: tentative field determination.

Ritchiea albersii Gilg & Benedict

Shrub or tree to 11(–20)m; leaves simple to 5-foliolate; leaflets elliptic or oblong-elliptic, to 20 × 8cm, base cuneate; petiolule 2–5mm long; inflorescence terminal; pedicels 1.5–3.5cm; sepals lanceolate or ovate, 12–20 × 5–10mm; petals 4, to 3.5cm, clawed; gynophore 2–3cm long; fruits cylindrical, to 5 × 2.5cm, with 6 longitudinal grooves. Forest; 1310m.
Distr.: SE Nigeria to E Africa [Afromontane].
IUCN: LC
Bali Ngemba F.R.: Onana 1994 fr., 4/2002; Zapfack 2030 fr., 4/2002.

CARYOPHYLLACEAE

I. Darbyshire (K) & M. Cheek (K)

Drymaria cordata (L.) Willd.

Straggling, slightly viscid herb; stems to 90cm; leaves broadly ovate, 1–2.5cm long; inflorescence a terminal cyme; flowers few, white. Grassland; 1500m.
Distr.: pantropical.
IUCN: LC
Bali Ngemba F.R.: Ghogue 1068 11/2000.

Stellaria mannii Hook.f.

Straggling, sometimes sticky herb to 60cm; leaves long-petiolate; inflorescence a terminal cyme; flowers few, white. Forest edge; 1400m.
Distr.: pantropical.
IUCN: LC
Bali Ngemba F.R.: Cheek 10443 11/2000.

Uebelinia abyssinica Hochst.

Bull. Jard. Bot. Natl. Belg. 55: 435 (1985).
Syn. ***Uebelinia hispida*** **Pax**
Prostrate herb; stems straggling, hispid, to 45cm long; leaves elliptic to orbicular, 7–15mm long; flowers axillary, white. Grassland; 2100m.
Distr.: Cameroon to Ethiopia [Afromontane].
IUCN: LC
Bali Ngemba F.R.: Pollard 678 10/2001.

CELASTRACEAE

Y.B. Harvey (K)

Fl. Cameroun 19 (1975) & 32 (1990).

Maytenus gracilipes (Welw. ex Oliv.) Exell subsp. *gracilipes*

Symb. Bot. Upsal. 25(2): 51 (1985).
Syn. ***Maytenus ovatus*** (Wall. ex Wight & Arn.) Loes. var. ***argutus*** (Loes.) Blakelock
Tree or shrub to 5m; spiny; leaves 8–16 × 3–17cm, elliptic, base strongly attenuate, apex angled, shortly acuminate; lateral nerves 10–12 pairs; inflorescence a loose axillary cyme, 5–14cm long; flowers tiny, c. 2 × 2mm; fruit a tri-locular capsule, red-violet when fresh, 1.5–2 × 1cm; 2–4 seeds per locule. Forest; 1400–1550m.
Distr.: Cameroon to Angola [Lower Guinea].
IUCN: LC
Bali Ngemba F.R.: Etuge 4692 fl., fr., 8/2000; Etuge 5361 fl., 4/2004.
Note: *Etuge* 5361 is a field determination.

Salacia erecta (G.Don) Walp. var. *erecta*

Climber to 15m, no resinous threads; branchlets dark brown, smooth; leaf blade elliptic, 4–15 × 1.5–5cm, 7–10 secondary nerves, margin toothed; petioles to 5mm, channel-margins undulate; flowers in sessile axillary fascicles, sometimes on thin shoots appearing like peduncles; pedicels 3–7mm; buds ovoid; flowers c. 5mm diam., yellow; fruits globular, 1–3cm diam., orange-red. Forest; 1500–1900m.
Distr.: Guinea (Conakry) to Zambia [Tropical Africa].
IUCN: LC
Bali Ngemba F.R.: Onana 1849 10/2001; Zapfack 2018 fr., 4/2002.

Salacia sp. nr. *erecta* (G.Don) Walp.

Similar to *S. erecta* var. *erecta* except having thinner pedicels and smoother, smaller fruits. Forest; 1400–1500m.
Distr.: Bali Ngemba F.R.
Bali Ngemba F.R.: Etuge 4269r 10/2001; 4691 11/2000; Etuge 4798 fr., 11/2000.
Uses: MATERIALS – fibres – ropes – rope (*Etuge* 4691).

CHRYSOBALANACEAE

C.A. Sothers (K), G.T. Prance (K), B.J. Pollard (K) & M. Cheek (K)

Fl. Cameroun 20 (1978).

Magnistipula butayei De Wild. subsp. *balingembaensis* Sothers, Prance & B.J.Pollard

Tree to 15m; branches lenticellate; leaves elliptic; blades 9.9–13.4 × 3.4–5.4cm, glabrous on both sides (including midrib), apex acute to acuminate, base rounded to subcuneate, glands on lower surface, circular, near to base but also randomly on lamina; petiole 5(–10)mm, glabrous; stipules caducous (not seen); inflorescences branched and congested; flowers 7–8mm long; petals ciliate (minutely) at apex; stamens 7, on one side, fused regularly; receptacle gibbous; ovary and fruit glabrous. Forest; 1310–1950m.
Distr.: Bali Ngemba F.R. [Endemic].
IUCN: CR
Bali Ngemba F.R.: Ghogue 1081 11/2000; Nana F. 105 fl., 4/2004; Pollard 1028 fl., 4/2002; Zapfack 2038 fl., 4/2002.

Magnistipula sp. 1 of Bali Ngemba

Tree 8m; stems sparsely brown-scurfy, lenticels white, linear; leaves leathery, drying mid-brown above, pale brown below, elliptic to lanceolate, to 20 × 8cm, acumen 1.5cm, obtuse to acute, midrib sparsely-hairy, lateral nerves red, 16–17 pairs, finely reticulate, 3–4 pairs bullate glands above at base of midrib; petiole 1cm; stipules persistent, narrowly oblong or spathulate, 10 × 2mm, densely-hairy; flowers and fruit lacking. Forest; 1400m.
Distr.: Bali Ngemba F.R.
Bali Ngemba F.R.: Cheek 10433 11/2000.
Note: known only from Bali Ngemba but fertile material needed to place.

COMBRETACEAE

C.C.H. Jongkind (WAG) & Y.B. Harvey (K)

Fl. Cameroun 25 (1983).

Combretum molle R.Br. ex G.Don

Small tree 5–7(–17)m; bark dark grey to black, rough, reticulately fissured; branchlets with bark peeling in grey fibrous strips; leaves opposite; lamina narrowly elliptic to broadly ovate or obovate, up to 21 × 12.5cm, acute, cuneate, pubescent above, grey tomentose below; petiole 2–3mm; inflorescences of axillary spikes, 7–11cm long; peduncles 1–2cm; flowers yellow or yellowish green, fragrant; fruits 4-winged, subcircular to elliptic, 1.3–2.5 × 1.5–2.5cm. Woodland and grassland areas; 1400m.
Distr.: tropical and subtropical Africa, and Yemen [Tropical & subtropical Africa and Arabian Peninsula].
IUCN: LC
Bali Ngemba F.R.: Onana 1913 10/2001; Pollard 441 fr., 11/2000.

COMPOSITAE

H.J. Beentje (K), D.J.N. Hind (K) & Y.B. Harvey (K)

Acmella caulirhiza Delile

Fragm. Florist. Geobot. 36(1), Suppl.1: 228 (1991).
Syn. ***Spilanthes africana*** DC.
Syn. ***Spilanthes filicaulis*** (Schum. & Thonn.) C.D.Adams
Creeping annual herb, 5–30cm tall; leaves ovate, up to 5 × 2.5cm, obtuse, subentire to dentate; petiole 0.25–1cm; peduncles to 10cm; capitula conical; ray florets present, sometimes tiny, yellow to orange. Forest & stream-banks; 1350–1850m.
Distr.: tropical & subtropical Africa and Madagascar [Tropical & subtropical Africa].
IUCN: LC
Bali Ngemba F.R.: Biye 111 11/2000; Pollard 432 fl., 11/2000; 434 fl., fr., 11/2000.
Uses: ANIMAL FOOD – placed in chicken feed; MEDICINES (*Wirmum Clare* in *Pollard* 432).

Adenostemma caffrum DC.

Erect or semi-procumbent herb to 1.5m, rooting at nodes on wet ground; stem ± fleshy; leaves opposite, ovate, 5–16 × 0.5–8cm; capitula 6–10mm across; phyllaries 1–2-seriate, florets many, small, white; pappus of (2–)3–4(–5) pegs. Wet ground; 1420m.
Distr.: widespread in tropical Africa [Tropical Africa].
IUCN: LC
Bali Ngemba F.R.: Pollard 410 fl., fr., 11/2000.

Adenostemma mauritianum DC.

Erect annual herb c. 1m tall; leaves ovate, 14 × 8cm, apex acute, base cuneate, margin serrate; petiole c. 6cm; capitula 5–10 per inflorescence, c. 5mm across, discoid; with small white florets; pappus of club-like pegs, 1mm long. Forest; 1500–1900m.
Distr.: SE Nigeria to Zimbabwe & E Africa (Afromontane).
IUCN: LC
Bali Ngemba F.R.: Onana 1842 10/2001; Tadjouteu 400 11/2000.

Ageratum conyzoides L.

Annual herb 0.3–1m, tall; leaves ovate, 5 × 2.5cm, apex acute, base acute to truncate, margin serrate; petiole c. 1cm, white-pilose; capitula c. 6mm diam., numerous, in dense terminal aggregations; florets blue; pappus of 5 scales, 2mm long. Farmbush; 1350m.
Distr.: Senegal to Tanzania [Pantropical].
IUCN: LC
Bali Ngemba F.R.: Pollard 431 fl., 11/2000.
MEDICINES/SOCIAL USES – eat the young shoots to cure a hangover (*Wirmum Clare* in *Pollard* 431).

Aspilia africana (Pers.) C.D.Adams subsp. *africana*

Syn. ***Aspilia africana*** (Pers.) C.D.Adams var. ***minor*** C.D.Adams
Scrambling shrub or herb to 2m tall; leaves lanceolate, c. 7.5 × 2.8cm, apex acute, base rounded, margin serrate, upper and lower surfaces scabrid; capitula terminal, single or few, 2.5cm diam., radiate; ray florets orange, c. 6 × 3mm, apex bilobed. Farmbush; 1500m.
Distr.: Senegal to Cameroon [Upper & Lower Guinea (montane)].
IUCN: LC
Bali Ngemba F.R.: Ghogue 1065 11/2000.

Bidens barteri (Oliv. & Hiern) T.G.J.Rayner

Kew Bull. 48: 483 (1993).
Syn. ***Coreopsis barteri*** Oliv. & Hiern
Annual herb, 0.3–1m tall; stems purple; leaves lanceolate, c. 5 × 1.5cm, apex acute, base rounded, margin serrate, glabrous; petiole 1–2mm long; capitula few, 3cm diam., radiate; involucre biseriate, c. 10 phyllaries each, inner broader, black, outer green; ray florets yellow, clawed, 14 × 4mm; disc florets black. Farmbush; 1300–1950m.
Distr.: Ghana to Cameroon [Lower Guinea].
IUCN: LC
Bali Ngemba F.R.: Biye 103 11/2000; Onana 1874 10/2001; Pollard 503 11/2000.

Bidens camporum (Hutch.) Mesfin

Fragm. Florist. Geobot. 36(1) Suppl.1: 156 (1991).
Syn. ***Coreopsis camporum*** Hutch.
Annual herb, 0.3–1.2m tall; cauline leaves bipinnate, broadly ovate in outline, 4–5 × 3.5cm, glabrous; capitula 4.5cm diam., radiate; involucre biseriate, with more or less equal ranks of narrowly elliptic phyllaries, 9 × 2mm; ray florets oblong, 25 × 5mm. Grassland; 1300–2100m.
Distr.: Sierra Leone to Cameroon [Afromontane (Guinea)].
IUCN: LC
Bali Ngemba F.R.: Onana 1890 10/2001; Pollard 683 10/2001.

Bidens pilosa L.

Erect herb, c. 1m tall; leaves trifoliolate; leaflets ovate or lanceolate, to 6 × 2cm, apex acute, base obtuse, margin serrate, glabrous; petiole 1–1.5cm long; capitulum c. 1.5cm diam., radiate; involucre with inner phyllaries narrowly elliptic, with scarious margins, outer much shorter, linear; ray florets white, oblong, c. 5 × 2mm; disc florets yellow. Farmbush, grassland & forest edges; 1500m.
Distr.: pantropical.
IUCN: LC
Bali Ngemba F.R.: Ghogue 1066 11/2000.

Conyza attenuata DC.

Fragm. Florist. Geobot. 36(1), Suppl.1: 89 (1991).
Syn. ***Conyza persicifolia*** (Benth.) Oliv. & Hiern
Herb, 0.4(–3)m tall; leaves sessile, linear, 10 × 0.7cm, apex acute, base clasping the stem for half its diam., margin distantly serrate-dentate; capitula 6–10mm wide, discoid, clustered in dense aggregations; involucre of c. 25 lanceolate phyllaries, c. 4 × 1mm, margins scarious, florets white. Farmbush; 2200m.
Distr.: tropical Africa [Afromontane].
IUCN: LC
Bali Ngemba F.R.: Ghogue 1088 11/2000.

Conyza pyrrhopappa Sch.Bip. ex A.Rich.

F.T.A. 3: 318 (1877).
Syn. *Microglossa angolensis* Oliv. & Hiern
Shrub or woody herb to 3m, pubescent; leaves narrowly elliptic to oblanceolate, acute; capitula campanulate, 0.75–2cm wide; phyllaries linear-lanceolate, acute; corollas tiny, white or yellow; achenes hairy; pappus reddish or tawny. Fallow, grassland; 2100m.
Distr.: Nigeria to Sudan, south to Angola & Zambia [Lower Guinea & Congolian].
IUCN: LC
Bali Ngemba F.R.: Pollard 469 fl., fr., 11/2000; 966 fl., 4/2002.

Conyza stricta Willd.

Erect herb, woody below, to c. 60cm, branched and leafy above; leaves linear-spathulate, pinnatifid or dentate, apex and lobes usually mucronulate, hispid-pubescent, sessile, clasping at base, 1–2.5cm long; capitula widely campanulate, many-flowered, c. 0.5 × 0.5cm, on short,slender pedicels; ray florets yellow, with a linear ray equalling tube in length and breadth, shorter than style, sometimes obsolete; achenes somewhat pubescent; pappus tawny or subrufous. Grassland; 1900m.
Distr.: tropical Africa to India [Palaeotropics].
IUCN: LC
Bali Ngemba F.R.: Pollard 507 11/2000.

Conyza subscaposa O.Hoffm.

Perennial herb with a weak flowering shoot from 5–30cm high; leaves forming a basal rosette; phyllaries often purplish; florets dull yellow. Grassland; 2100m.
Distr.: W Cameroon to Congo (Kinshasa) & E Africa [Afromontane].
IUCN: LC
Bali Ngemba F.R.: Pollard 470 11/2000.

Crassocephalum bauchiense (Hutch.) Milne-Redh.

Herb, 0.3–1m tall; stems with white crisped hairs; cauline leaves pinnately lobed, c. 7 × 3.5cm, lobes 5, divided almost to base, margins denticulate; capitula 1cm diam., discoid; involucre of c. 20 phyllaries, as long as capitulum, and c. 20 short, patent, linear, basal calycular bracts, 2–3mm long; florets blue or purple. Forest edge; 1900m.
Distr.: N Nigeria & Cameroon [Lower Guinea (montane)].
IUCN: VU
Bali Ngemba F.R.: Onana 1843 11/2001.

Crassocephalum crepidioides (Benth.) S.Moore

Annual herb, to 1m, puberulous or glabrous, branched; leaves deeply lyrate-pinnatifid, sometimes not lobed, 2–25cm long, mostly petiolate; capitula oblong, c. 1cm long, on slender pedicels of 0.5–5cm, in a dense or lax corymbose cyme; phyllaries linear-subulate; corollas yellow or with purplish tips, appearing brownish red. Weed of farms and waste places. 1400–1900m.
Distr.: Guinea (Conakry) to Mascarenes, naturalised in tropical Asia & Pacific islands [Palaeotropics].
IUCN: LC
Bali Ngemba F.R.: Etuge 4718 11/2000; Ghogue 1052 11/2000; Onana 1844 11/2001; Pollard 440 fl., fr., 11/2000.

Crassocephalum montuosum (S.Moore) Milne-Redh.

Herb, 1.5m tall; cauline leaves elliptic in outline, 8.5 × 6cm, pinnately 5-lobed, apical lobe largest, lower lobes decreasing in size, lobes dentate, glabrous; capitula 10–20, in aggregations, c. 0.5mm diam., discoid; involucre as in *C. bauchiense*; florets yellow or brownish yellow. Forest & forest-grassland transition; 1560–1900m.
Distr.: W Cameroon to Congo (Kinshasa) to E Africa & Madagascar [Afromontane].
IUCN: LC
Bali Ngemba F.R.: Onana 1838 10/2001; Pollard 601 fl., 10/2001; Ujor 30332 fl., 5/1951.

Crassocephalum rubens (Juss. ex Jacq.) S.Moore

Syn. *Crassocephalum sarcobasis* (DC.) S.Moore
Annual herb, to c. 1.5m, puberulous; leaves irregularly dentate or lobed, 2–9cm, upper leaves sessile, lower leaves petiolate; capitula terminal, solitary, 1–2cm long, oblong-hemispherical; peduncles 15–30cm; phyllaries linear, subulate, glabrous; corollas tiny, purple. Fallow & rocky slopes in grassland; 2000m.
Distr.: Cameroon, Sudan to South Africa, Madagascar, Yemen [Afromontane].
IUCN: LC
Bali Ngemba F.R.: Pollard 451 fl., 11/2000.

Crassocephalum vitellinum (Benth.) S.Moore

Herb, 0.3–2m high, ascending from a procumbent base, with patent branches, subglabrous to puberulent; leaves ovate or lanceolate, 2.5–5cm long, irregularly dentate; petiole 0.25–4cm, auriculate at base; capitula 0.75–2cm long, subhemispherical, solitary; peduncles 6–30cm; phyllaries linear-subulate; florets yellow or orange, equalling or exceeding the pappus. Forest edge; 1500–2000m.
Distr.: SE Nigeria & W Cameroon to E Africa [Afromontane].
IUCN: LC
Bali Ngemba F.R.: Onana 1978 11/2001; Pollard 459 11/2000; Tadjouteu 397 11/2000.

Echinops giganteus A.Rich.

Erect herb, 1–7m; branches cottony above; leaves spiny, the upper obovate, 30 × 15cm, deeply pinnatifid, strigose-hispid above, whitish tomentose beneath, sessile, uppermost leaves elliptic, deeply toothed; capitula up to 3.5cm long, numerous, in a globose head up to 9cm diam.; flowers white or cream, fragrant; with pappus of laciniate scales; achenes glabrescent near apex. Grassland; 1400m.
Distr.: Cameroon to Sudan and Tanzania [Tropical Africa].
IUCN: LC
Bali Ngemba F.R.: Onana 1904 10/2001.
Uses: FOOD ADDITIVES – condiment (*Onana* 1904).

Echinops gracilis O.Hoffm.

Spiny herb, 0.6–0.9(–2)m tall; leaves dense, spiny, linear, stiff, 18 × 0.5–1mm, heads mauve, 5cm diam.; achenes with narrow scales. Grassland; 1350–2200m.
Distr.: N Nigeria, Cameroon, Sudan, Congo (Brazzaville), Uganda [Lower Guinea & Congolian (montane)].
IUCN: LC

Bali Ngemba F.R.: Ghogue 1096 11/2000; Onana 2040 fl., 4/2002; Pollard 429 11/2000; 935 fl., 4/2002.
Uses: FOOD – infructescences – fruit is eaten by people, and tastes like coconut; MATERIALS – fibres – tools – sieves – to filter palm wine; and to deters rats when tied in bunches near corn store. (*Wirmum Clare* in *Pollard* 429).

Elephantopus mollis Kunth

Perennial herb, 40–90cm; leaves oblanceolate, 6–16 × 2–5cm, the base clasping; capitula are united into glomerules, these in compound cymes; florets white; pappus of 5–6 setae. Grassland, forest edge; 1500–1560m.
Distr.: tropical America, introduced in Africa & Asia [Pantropical].
IUCN: LC
Bali Ngemba F.R.: Ghogue 1063 11/2000; Pollard 604 fl., 10/2001.

Emilia coccinea (Sims) G.Don

Annual herb, 15–120cm high; leaves ± subsucculent, 1–20 × 0.4–6cm; capitula in terminal corymbs; florets bright orange; pappus setae white. Forest & farmbush; 1350m.
Distr.: tropical Africa [Afromontane].
IUCN: LC
Bali Ngemba F.R.: Pollard 433 fl., fr., 11/2000.

Galinsoga parviflora Cav.

Annual herb, 10–75(–100)cm high, erect, spreading or decumbent; leaves opposite, ovate, 1–11 × 0.5–7cm; 3-veined from base; capitula in axillary cymes; ray florets white, small; disc florets yellow; pappus absent or of 15–20 small scales. Grassland; 1500m.
Distr.: S America, introduced in W Cameroon [Montane].
IUCN: LC
Bali Ngemba F.R.: Ghogue 1061 11/2000.

Gynura miniata Welw.

Erect, rosette-forming perennial herb to 1m; stem brown-purplish, subglabrous; leaves polymorphic, basal leaves petiolate, ovate or spathulate, 4–22 × 2.5–11cm, middle and upper leaves narrower, obovate, lobed to pinnatisect; inflorescence scapose with 1–7 campanulate capitula per shoot; involucre 13 × 13–15mm; florets 45–100, orange; achenes glabrous or pubescent. Grassland.
Distr.: Ghana, east to Ethiopia, south to Angola, Zambia and Malawi [Tropical Africa].
IUCN: LC
Bali Ngemba F.R.: Surgenor *s.n.* (photographic record, pers. comm.) 4/2004.

Gynura pseudochina (L.) DC.

Prodr. 6: 299 (1838).
Perennial herb 0.4–1m, unpleasantly aromatic; leaves in rosette or leafy to mid-point, pubescent or glabrous, green or purplish, basal leaves ovate or spathulate, 4–22 × 2.5–11cm, margins entire, middle and upper leaves elliptic, obovate or narrowly obovate, lobed to pinnatisect, 6–25 × 2–7cm, margins lobed; capitula terminal; involucre campanulate, 8–13 × 13–15mm; corolla 9–12.5mm; florets orange or yellow; achenes 3mm long; pappus 7–11mm long. Grassland; 1450m.
Distr.: Sierra Leone to Ethiopia, southwards to Angola and Zambia, and tropical Asia [Palaeotropics].
IUCN: LC
Bali Ngemba F.R.: Pollard 521 11/2000.

Helichrysum forskahlii (J.F.Gmel.) Hilliard & B.L.Burtt

Syn. ***Helichrysum cymosum*** (L.) Less. subsp. ***fruticosum*** (Forssk.) Hedberg
Syn. ***Helichrysum cymosum*** sensu Adams
Syn. ***Helichrysum helothamnus*** Moeser var. ***helothamnus***
Perennial herb or shrub, 0.2–1m high, densely leafy; leaves narrowly lanceolate, 0.4–3 × 0.1–1cm; capitula in clusters in corymbs; involucre and florets pale yellow; pappus of small white bristles. Forest-grassland transition; 2100m.
Distr.: Nigeria to Zambia, Sudan to Tanzania & Yemen [Afromontane & Arabian Peninsula].
IUCN: LC
Bali Ngemba F.R.: Pollard 473 11/2000.

Helichrysum nudifolium (L.) Less. var. *nudifolium*

Syn. Gen. Comp.: 299 (1832).
Herb with annual stems, to 1m; leaves forming a rosette, tomentose below, glabrescent above; capitula in clusters, with yellow or brown phyllaries and small yellow florets. Grassland; 1450–1600m.
Distr.: W Africa to Ethiopia and down to South Africa [Tropical Africa].
IUCN: LC
Bali Ngemba F.R.: Onana 2049 fl., 4/2002; Pollard 945 fl., 4/2002.

Laggera pterodonta (DC.) Sch.Bip. ex Oliv.

Stout erect herb, 0.6–1.5m, branched above, viscid; branches puberulous, winged; leaves alternate, elliptic-oblong or lanceolate, to 15 × 6cm, acute, dentate, puberulous; capitula broadly campanulate, many-flowered, to 1cm long; pedicels c. 1.5cm; phyllaries linear-laneolate; ray florets filiform, purple or greenish; achenes pilose; pappus uniseriate. Grassland; 1900m.
Distr.: tropical Africa and Asia [Pantropical].
IUCN: LC
Bali Ngemba F.R.: Tadjouteu 428 11/2000.
Note: *Laggera pterodonta* may be a synonym of *Laggera crispata*.

Mikania chenopodifolia Willd.

Fragm. Florist. Geobot. 36(1) Suppl.1: 460 (1991) (in FWTA 2: 286 (1963) as a synonym of *M. cordata* (Burm.f.) B.L.Robinson var. *cordata*).
Syn. ***Mikania capensis*** DC.
Shrub 5–9m; leaves opposite, ovate, 2–10 × 1–7cm, subsagittate to hastate, glandular-punctate, with 3 main veins; capitula in a dense, leafy corymb; phyllaries 4; florets 4, white; pappus of bristles. Forest & thicket; 1200–1500m.
Distr.: tropical Africa & Asia [Palaeotropics].
IUCN: LC
Bali Ngemba F.R.: Biye 80 11/2000; Etuge 4706 11/2000; Ghogue 1030 11/2000; Onana 1809 10/2001.

Pseudognaphalium luteo-album (L.) Hilliard & B.L.Burtt

Fl. Masc. Compositae: 74 (1993).
Syn. ***Gnaphalium luteo-album*** L.
Herb 15–50cm, sometimes woody at base; stem and branches grey-green-cottony; leaves spathulate, obtuse or linear, 2–7 × 0.25–2cm, upper smaller, entire or subcrenulate; capitula campanulate, many-flowered, c. 3mm long, subsessile, in crowded clusters at the ends of the stem and branches; phyllaries straw-coloured; florets dull yellow; achenes with white bristles. Sandy waste places and streamsides; 1790m.
Distr.: tropical and subtropical Africa [Afromontane].
IUCN: LC
Bali Ngemba F.R.: Cheek 10497 11/2000.

Pseudognaphalium oligandrum (DC.) Hilliard & B.L.Burtt

Bot. J. Linn. Soc. 82: 204 (1981).
Annual herb, 0.8–1m; leaves oblong, lanceolate or sublinear, 6–8 × 0.2–0.8cm, cordate-amplexicaul at base; capitula campanulate, c. 4mm diam.; phyllaries in 4 series, external ovate, inner sublinear; male florets white, numerous, 1–7mm long; female florets white; achenes ellipsoid. Forest edges and grassland; 1450m.
Distr.: Cameroon, Rwanda & Congo (Kinshasa) [Afromontane].
IUCN: LC
Bali Ngemba F.R.: Pollard 943 fl., 4/2002.

Pseudognaphalium undulatum (L.) Hilliard & B.L.Burtt

Bot. J. Linn. Soc. 82: 205 (1981).
Syn. ***Gnaphalium undulatum*** L.
Erect perennial, 20–60cm; stem leafy to the top, pubescent or glabrate; leaves ovate or lanceolate, acute, apiculate, upper sessile, ovate-lanceolate, base cordate, lower narrowed below the middle or suddenly contracted into a winged auricle, 2.5–9 × 1–4cm; capitula hemispherical, 0.5–1cm wide, in small dense rounded pedunculate or subsessile cymes, 2–5cm diam.; phyllaries linear-lanceolate; florets dull yellow; achenes minutely appressed-pubescent. Grassland; 1450m.
Distr.: Cameroon to Ethiopia, south to Natal & Madagascar [Tropical Africa & Madagascar].
IUCN: LC
Bali Ngemba F.R.: Pollard 940 fl., 4/2002.

Senecio lelyi Hutch.

Erect perennial herb, to c. 60cm, from a woody rootstock, subglabrous; lower leaves oblanceolate, upper leaves with narrowly rounded base, margins regularly minutely denticulate; capitula few, 1–2cm long; phyllaries 10–12mm long, glabrous; ray florets yellow; disc florets yellow. Grassland; 1450m.
Distr.: Guinea (Conakry), Ghana, Nigeria & Cameroon [Guinea montane].
IUCN: NT
Bali Ngemba F.R.: Pollard 936 fl., 4/2002.
Note: fairly rare, only 12 sites known.

Senecio ruwenzoriensis S.Moore

Perennial herb, to 0.6m; stems fleshy, glaucous; leaves ovate to obovate, to 8 × 1–3cm, glabrous, with pinnate veins; capitula several, corymbose, to 1cm long; phyllaries 8–10mm long; ray florets yellow; disc florets yellow, small; achenes glabrous. Grassland; 1450–2100m.
Distr.: Cameroon, Congo to Tanzania [Afromontane].
IUCN: LC
Bali Ngemba F.R.: Pollard 939 fl., 4/2002; 964 fl., 4/2002.

Solanecio mannii (Hook.f.) C.Jeffrey

Kew Bull. 41: 922 (1986).
Syn. ***Crassocephalum mannii*** (Hook.f.) Milne-Redh.
Shrub or soft-wooded tree, up to c. 7m high, subglabrous; leaves ovate, 10–20 × 2–5cm, acuminate, puberulous on midrib beneath; petiole 0.7–2.5cm; capitula oblong, c. 1cm long, on short,slender pedicels in dense panicled cymes, c. 1cm long; phyllaries 5, linear, cymbiform; florets about 6, tubular, yellow; achenes minutely-hairy. Forest; 1900m.
Distr.: Nigeria to Congo (Kinshasa), Sudan to E Africa & Zimbabwe [Afromontane].
IUCN: LC
Bali Ngemba F.R.: Pollard 1038 fr., 4/2002.

Sonchus oleraceus L.

Erect herb, to 1.5m; sparsely branched in the upper half, branches to 45cm, ascending; lower leaves to 28 × 9cm, oblanceolate in outline, often pinnately-lobed, upper leaves smaller, lanceolate, ± lobed; capitula stalked; involucres to c. 13 × 10mm, broadly cylindric, swollen in fruit; phyllaries from 3mm (outside) to 13mm (inside), narrowly lanceolate; florets numerous, all ligulate; corollas yellow, to 13mm; achenes to 3 × 1mm, light to reddish brown; pappus to 7mm.
Distr.: cosmopolitan weed (native of Eurasia & N Africa).
IUCN: LC
Bali Ngemba F.R.: Pollard 457 fr., 11/2000.

Vernonia doniana DC.

Syn. ***Vernonia conferta*** Benth.
Tree, 6–13m high; leaves elliptic to obovate, 10–90 × 8–26cm, glabrous above, tomentose beneath; capitula many, in a panicle to 1m long; florets small, white; pappus of small outer scales and longer inner bristles. Forest, near stream.
Distr.: Guinea (Conakry) to Bioko, Cameroon, Angola, Uganda & Sudan [Guineo-Congolian].
IUCN: LC
Bali Ngemba F.R.: Daramola FHI 40479 fl., 2/1959.

Vernonia glabra (Steetz) Vatke var. *hillii* (Hutch. & Dalziel) C.D.Adams

Perennial herb, with stems 0.3–3m, from a woody rootstock; leaves elliptic, 2–16 × 0.5–7cm, serrate; capitula many in dense corymbose cymes; florets blue; pappus of two rows of bristles, the outer smaller. Forest edge, thickets; 1450m.
Distr.: Cameroon & Kenya to South Africa [Afromontane].
IUCN: LC
Bali Ngemba F.R.: Pollard 439 fl., 11/2000.

Vernonia guineensis Benth. var. *cameroonica* C.D.Adams

Erect herb, with densely grey or brownish pubescent stems, 0.3–1.5m high; leaves lanceolate to narrowly obovate, to 19 × 5cm, woolly beneath, coarsely toothed; capitula about 2cm wide; tips of phyllaries papery and acute, mauve or pink, fading to brown, appendages acute; florets purple, fading to pale mauve. Grassland; 1450m.

Distr.: Nigeria and Cameroon [W Cameroon Uplands].
IUCN: VU
Bali Ngemba F.R.: Pollard 944 fl., 4/2002.

Vernonia hymenolepis A.Rich.

Kew Bull. 43: 237 (1988).
Syn. ***Vernonia leucocalyx*** O.Hoffm. var. ***acuta*** C.D.Adams
Syn. ***Vernonia leucocalyx*** O.Hoffm. var. ***leucocalyx***
Syn. ***Vernonia insignis*** (Hook.f.) Oliv. & Hiern
Woody herb or shrub, to 4m; leaves lanceolate, 5–20 × 1–7cm, serrate, densely-hairy beneath; capitula large; phyllaries with pink appendages; small mauve or purple flowers. Forest-grassland transition; 1200–1420m.
Distr.: Cameroon, Sudan, Ethiopia, Uganda & Kenya [Afromontane].
IUCN: LC
Bali Ngemba F.R.: Onana 1807 10/2001; Pollard 411 fl., fr., 11/2000.
Local name: Bitterleaf = 'sweet bitterleaf', as opposed to 'bitter bitterleaf', this one is less bitter. **Uses:** FOOD – leaves –take the leaves, squish and cut them into small pieces, wash them well, even thricefold, then cook and eat (*Nkankano Peter* in *Pollard* 411).

Vernonia nestor S.Moore

Perennial herb, to 1.8m; leaves lanceolate, 2–12 × 1–4cm, densely-hairy beneath, with small purple or mauve florets. Grassland; 1450–1600m.
Distr.: Guinea to Malawi [Montane].
IUCN: LC
Bali Ngemba F.R.: Etuge 4816 11/2000; Pollard 522 11/2000.

Vernonia purpurea Sch.Bip.

Perennial herb, to 2m; leaves elliptic or ovate, 3–10 × 1–3cm, sandpapery; capitula many, with small purple or mauve florets. Grassland; 1400–2100m.
Distr.: Senegal to Mozambique [Afromontane].
IUCN: LC
Bali Ngemba F.R.: Onana 1906 10/2001; Pollard 477 fl., fr., 11/2000.

Vernonia sp. aff. *rugosifolia* De Wild.

Erect herb, 0.6–1m; capitula many, c. 1cm long, shortly pedicelled; florets purple. Grassland; 1300–2250m.
Distr.: Cameroon [W Cameroon Uplands].
Bali Ngemba F.R.: Ghogue 1100 11/2000; Onana 1900 10/2001.
Note: although *V. rugosifolia* is applied to this taxon in FWTA, Lisowski (e.g. Kalenda and Lisowski, Fragm. Florist. Geobot. 40(2): 625 (1995)) excludes FWTA material from *V. rugosifolia* which he delimits as centred in Congo, whence the type originated. A new name may be required for the FWTA material once a more detailed investigation has been completed [M. Cheek].

Vernonia smithiana Less.

Perennial herb with annual shoots to 1m, from a woody rootstock; leaves lanceolate, 2–9 × 0.3–24cm, densely-hairy beneath; capitula c. 0.75cm wide, hemispherical; florets pinkish or reddish purple. Grassland; 1450m.
Distr.: W Africa to E Africa and south to Angola [Tropical Africa].
IUCN: LC
Bali Ngemba F.R.: Pollard 938 fl., 4/2002.

Vernonia stellulifera (Benth.) C.Jeffrey

Fl. Zamb. 6(1): 142 (1992).
Syn. ***Triplotaxis stellulifera*** (Benth.) Hutch.
Annual herb, 0.1–1.2m high; leaves ovate or elliptic, 2–10 × 1–3cm, remotely crenate; capitula in corymbose cymes, florets mauve to purple; pappus of a rim of scales with a few inner bristles. Forest and farmbush; 770m.
Distr.: Guinea (Conakry) to Zambia & Uganda (Guineo-Congolian [montane)].
IUCN: LC
Bali Ngemba F.R.: Pollard 632 10/2001.

Vernonia subaphylla Baker

Herb with annual stems, to 60cm; capitula few, with small mauve florets. Grassland; 1450m.
Distr.: Cameroon to Congo (Kinshasa) [Guineo-Congolian].
IUCN: LC
Bali Ngemba F.R.: Pollard 947 fl., 4/2002.

CONNARACEAE

Y.B. Harvey (K) & I. Darbyshire (K)

Agelaea pentagyna (Lam.) Baill.

Wageningen Agric. Univ. Pap. 89(6): 144 (1989).
Syn. ***Agelaea hirsuta*** De Wild.
Syn. ***Agelaea dewevrei*** De Wild. & T.Durand
Syn. ***Agelaea floccosa*** Schellenb.
Syn. ***Agelaea grisea*** Schellenb.
Syn. ***Agelaea obliqua*** (P.Beauv.) Baill.
Syn. ***Agelaea preussii*** Gilg
Syn. ***Agelaea pseudobliqua*** Schellenb.
Liana, 6(–25)m; branches furrowed, puberulent; leaves trifoliolate, 2.2–30 × 1.5–17cm; leaflets ovate, usually asymmetric, terminal leaflet 18–24 × 1.4–12.5cm, acumen 1.5cm, base rounded, finely tomentose on lower surface and veins; petiole 1–26cm; petiolule 1–9mm; panicles to 35cm, glabrous to tomentose; sepals 2.5–5mm, with fringing multi-cellular hairs; petals 3–5.5mm; seeds 10–15 × 5–10mm. Forest; 1700m.
Distr.: Senegal to Mozambique & Madagascar [Tropical Africa & Madagascar].
IUCN: LC
Bali Ngemba F.R.: Plot voucher BAL8 4/2002; Plot voucher BAL42 4/2002.

CONVOLVULACEAE

P. Wilkin (K) & M. Cheek (K)

Ipomoea involucrata P.Beauv.

Climber resembling other *Ipomoea* species, distinct in the entire, sheathing, involucral bract, c. 1.5 × 7cm wide; flowers 5–15, capitate; corolla pink, 3–4cm. Farmbush; 1850m.
Distr.: Senegal to Kenya [Tropical Africa].
IUCN: LC
Bali Ngemba F.R.: Biye 91 11/2000.
Note: a very common weed, identified by the sheathing involucral bract.

Ipomoea tenuirostris Choisy
Slender climber; stems pilose; leaves lanceolate, c. 4–8 × 1.6–4.5cm, appressed-hairy; inflorescence subracemose, 3–12cm, rhachis as long as peduncle, 3–10-flowered; calyx lobes c. 0.5cm; corolla 1.5cm, white, purple throat; fruit globose, 0.5cm. Forest edge; 1710m.
Distr.: Sierra Leone to Cameroon [Upper & Lower Guinea].
IUCN: LC
Bali Ngemba F.R.: Biye 105 11/2000.

CRASSULACEAE

Y.B. Harvey (K)

Kalanchoe crenata (Andrews) Haw.
Syn. ***Kalanchoe laciniata*** sensu Hepper
Erect, unbranched, succulent herb to 1m; leaves elliptic, c. 7 × 4.5cm, rounded, obtuse, crenate; petiole 1cm; inflorescence corymbose, 6cm across, c. 20-flowered; pedicels glandular-pubescent; flowers pink, yellow or red, 1.5cm long; petals 5. Farmbush; 1700m.
Distr.: Guinea (Conakry) to Cameroon [Upper & Lower Guinea].
IUCN: LC
Bali Ngemba F.R.: Tadjouteu 406 11/2000.

CUCURBITACEAE

I. Darbyshire (K)

Fl. Cameroun 6 (1967).

Coccinia barteri (Hook.f.) Keay
Herbaceous climber; stems ridged, glabrous, angular; tendrils bifid; leaves variable, palmatisect, (3–)5-lobed to sublobed, 12–20 × 10–15cm, base ± deeply cordate, lobe apices acute or acuminate, margins ± serrate; dioecious, both sexes on racemes; peduncle 2–3cm, pubescent; flowers subsessile; males with 5–10 flowers; females 3–5; flowers c. 1.8 × 1.5cm; petals yellow; fruits ellipsoid or suborbicular c. 4 × 2.5cm, green, drying blackish, smooth; seeds numerous, c. 4 × 2mm, white, smooth, with narrow rim. Farmbush & forest; 1310m.
Distr.: widespread in (sub)tropical Africa [Tropical Africa].
IUCN: LC
Bali Ngemba F.R.: Zapfack 2017 fl., 4/2002.

***Coccinia* cf. *barteri* of Bali Ngemba**
Climber; stems glabrous; tendrils bifid; leaves 13 × 11cm, deeply 3-lobed, lateral lobes with secondary lobing, margin with mucros at the vein termini, upper surface scabridulous, lower surface punctate; differing from *C. barteri* in the long petioles, to 7.5cm. Forest edge; 1500m.
Distr.: Bali Ngemba F.R.
Bali Ngemba F.R.: Ghogue 1046 11/2000.
Note: specimen cited is sterile, fertile material required for further investigation.

Momordica foetida Schum. & Thonn.
Syn. ***Momordica cordata*** Cogn.
Herbaceous climber; stems deeply-ridged, finely-pubescent or glabrous; leaves ovate, 9–14 × 8–12cm, base deeply cordate, apex acuminate, margin shallowly-dentate, pubescent along veins on abaxial lamina; petioles c. 7cm; monoecious, male flowers in umbels; peduncle c. 10cm, finely-pubescent, sometimes kinked; pedicels 1–5cm, subtended by a leafy bract, c. 4mm; sepals rounded at apex, pubescent, drying black; petals obovate, 1.5 × 0.5cm; stamens 2 with sinuous anthers; female flowers solitary, pedicel c. 7cm; fruits densely-hispid when immature, pale orange with sparser bristling when mature, ovoid to ellipsoid, c. 5.5 × 3cm, seeds red. Secondary forest & farmbush; 1490–1700m.
Distr.: tropical & S Africa [Tropical & subtropical Africa].
IUCN: LC
Bali Ngemba F.R.: Ghogue 1003 11/2000; Tadjouteu 413 11/2000.

Raphidiocystis phyllocalyx C.Jeffrey & Keraudren
Fl. Cameroun 6: 102 (1967).
Herbaceous climber to 5m; stems, petioles and veins of lower leaf surface densely reddish tomentose; leaves ovate in outline, 10–14 × 7.5–10cm, apex rounded, base deeply cordate, margin highly sinuate, troughs with an apiculum at vein terminus; flowers simple or fasciculate, subsessile, c. 1cm long; sepals pinnatipartite, 1cm; fruit oblong-ellipsoid, densely red-hairy. Forest; 1700m.
Distr.: Cameroon to Sudan & Congo (Kinshasa) [Lower Guinea & Congolian].
IUCN: LC
Bali Ngemba F.R.: Cheek 10515 11/2000.

Zehneria minutiflora (Cogn.) C.Jeffrey
Fl. Cameroun 6: 37 (1967).
Syn. ***Melothria minutiflora*** Cogn. var. ***hirtella*** Cogn.
Syn. ***Melothria minutiflora*** Cogn. var. ***minutiflora***
Syn. ***Melothria minutiflora*** Cogn. var. ***parviflora*** Cogn.
Herbaceous climber; stems c. 1mm, ridged, minutely-pubescent; leaves c. 4.5 × 4cm, shallowly 3-lobed, base cordate, apex acute or acuminate, margin shallowly-dentate, surfaces with white hairs to 1mm; petiole 2–5cm, pubescent; dioecious; male inflorescence not observed; female flowers solitary; pedicels 2.5–3cm long, thin; petals white (2–3mm); ovary ellipsoid, c. 3mm long; fruit ellipsoid, c. 2 × 0.8cm, orange-red, smooth; seeds visible through thin epicarp. Disturbed vegetation including farmbush; 1440m.
Distr.: tropical & South Africa [Tropical & subtropical Africa].
IUCN: LC
Bali Ngemba F.R.: Biye 78 11/2000.

DICHAPETALACEAE

M. Etuge

Fl. Cameroun 37 (2001).

***Dichapetalum* sp.**
Woody climber, to 10m high; stem woody with short, brownish hairs; leaves to 15 × 8cm, ovate; petioles to 8mm; fruits c. 1cm diam., hairy, brownish. Secondary forest patch; 1350m.
Bali Ngemba F.R.: Etuge 5443 fr., 4/2004.

DROSERACEAE

M. Cheek (K)

Drosera pilosa Exell & Laundon

Bol. Soc. Brot., Sér. 2, 30: 213 (1956).
Carnivorous acaulous (?annual) herb; leaf rosette 3cm diam., leaves to 20, appressed, spathulate, blade subcircular, 5 × 5mm, rounded-truncate, base cuneate-decurrent; petiole 1cm; scape 8cm, pubescent with angled hairs, not glandular; flowers 5mm. Grassland in damp peaty soil; 1450m.
Distr.: Cameroon, Rwanda, Kenya, Tanzania [Afromontane].
IUCN: VU
Bali Ngemba F.R.: Pollard 531 11/2000; 1044 4/2002.
Note: extremely rare though widespread. Pollard 1044 is almost certainly this taxon although fertile material is needed for confirmation.

EBENACEAE

Y.B. Harvey (K)

Fl. Cameroun 11 (1970).

Diospyros zenkeri (Gürke) F.White

Tree, to 20m; bole black and smooth; slash black outside, pink with a yellow inner edge inside; leaves drying black, elliptic to ovate, to 10 × 5cm, acumen to 2.5cm, with 4–5 pairs lateral veins looping far from edge; male flowers in clusters of 1–9 in leaf axils or on older twigs, c. 1cm long, cylindric; calyx with 3 short lobes; female flowers in clusters of 1–3, shorter and wider than males; fruits globose, yellow, turning black, 2.5cm, glabrous, calyx not accrescent. Forest; 1500m.
Distr.: Nigeria to Congo (Kinshasa) [Lower Guinea & Congolian].
IUCN: LC
Bali Ngemba F.R.: Ghogue 1059 11/2000.

ERICACEAE

Y.B. Harvey (K) & M.Cheek (K)

Fl. Cameroun 11 (1970).

Agarista salicifolia G.Don

Syn. ***Agauria salicifolia*** (Comm. ex Lam.) Hook.f. ex Oliv.
Shrub or small tree, to 13m; leaves oblong-lanceolate, acute at ends; inflorescence a short, axillary raceme; flowers pink. Forest edge; 1950–2200m.
Distr.: Cameroon to Madagascar [Afromontane & Madagascar].
IUCN: LC
Bali Ngemba F.R.: Biye 119 11/2000; Ghogue 1095 11/2000; Pollard 670 10/2001.

Erica mannii (Hook.f.) Beentje

Utafiti 3: 13 (1990).
Syn. ***Philippia mannii*** (Hook.f.) Alm & Fries
Heath-like shrub, to 4m; leaves ascending to erect, in whorls of 3; few purplish flowers at end of each branchlet. Forest edge; 2100m.
Distr.: Cameroon to Kenya [Afromontane].
IUCN: LC
Bali Ngemba F.R.: Pollard 661 fl., 10/2001.

Erica tenuipilosa (Engl. ex Alm & T.C.E.Fr.) Cheek subsp. *tenuipilosa*

Kew Bull. 52: 753 (1997).
Syn. ***Blaeria spicata*** Hochst. ex A.Rich. subsp. ***mannii*** (Engl.) Wickens
Syn. ***Blaeria mannii*** (Engl.) Engl.
Syn. ***Blaeria spicata*** Hochst. ex A.Rich. var. ***mannii*** (Engl.) Letouzey
Syn. ***Blaeria spicata*** Hochst. ex A.Rich. var. ***fakoensis*** Letouzey
Syn. ***Blaeria spicata*** Hochst. ex A.Rich. var. ***nimbana*** (A.Chev.) Letouzey
Heath-like undershrub, 0.3–0.8m; leaves in whorls of 3; inflorescence a spicate panicle; flowers purplish. Forest-grassland transition; 1950–2100m.
Distr.: Ivory Coast to Bioko and W Cameroon [Guineo-Congolian (montane)].
IUCN: LC
Bali Ngemba F.R.: Biye 120 11/2000; Pollard 684 fl., 10/2001.

EUPHORBIACEAE

M. Cheek (K), Petra Hoffmann (K) & G.L. Challen (K)

Alchornea laxiflora (Benth.) Pax & K.Hoffm.

Shrub, to 3.5m; leaves to 12.5 × 6cm, elliptic-lanceolate to oblong-oblanceolate, acuminate or cuneate, shallowly crenate-serrate, sparingly pubescent on the midrib and main nerves at first, soon glabrescent except in the domatia, reddish when young; petioles to 2cm; male inflorescences up to 12cm long, axillary; bracts 1.5–5 × 1–2mm, ovate; female inflorescences usually not more than 10cm long, terminal, spicate, few-flowered; bracts 2–3mm long, ovate-lanceolate; fruits 5–7 × 7–8mm, dark green, brown or black. Secondary forest patch; 1420m.
Distr.: Nigeria eastwards to Ethiopia and south to South Africa.
IUCN: LC
Bali Ngemba F.R.: Etuge 5378 4/2004.

Antidesma vogelianum Müll.Arg.

Tree or shrub, 3–10m, pubescent; leaves elliptic, rarely slightly obovate, c. 15 × 5.5cm, acumen c. 2cm, base acute; petiole c. 5mm; stipule lanceolate, c. 6 × 2mm, entire; infructescence non-interrupted, pendulous, racemose, c. 15cm; fruits ellipsoid, 7mm, red, fleshy. Forest; 1400–1600m.
Distr.: Nigeria to Tanzania [Lower Guinea & Congolian].
IUCN: LC
Bali Ngemba F.R.: Cheek 10430 11/2000; Onana 2033 fl., 4/2002; Ujor 30410 fr., 6/1951.

Bridelia sp.

Tree to 10m; much-branched; leaves with prominent yellowish midrib, secondary nerves extend to and from the

leaf margin; fruits abundant along leafy branches, green, c. 8mm, ellipsoid. Degraded forest and farmbush; 1880m.
Bali Ngemba F.R.: Darbyshire 358 fr., 4/2004.
Note: field determination.

Clutia kamerunica Pax

Shrub 2–4m, stems and petioles white-downy-hairy; stems densely and evenly leaved, internodes 1cm; blades oblong-elliptic, to 10 × 2cm, acute, lateral nerves 10 pairs; petiole 1cm; inflorescences axillary, sessile, 1–2-flowered; pedicels 1.5cm; flowers 2mm; fruit globose, 5mm, dehiscent. Forest and farmbush; 1900–2000m.
Distr.: Nigeria and Cameroon [W Cameroon Uplands].
IUCN: EN
Bali Ngemba F.R.: Cheek 10523 11/2000; Onana 1851 10/2001; Pollard 512 11/2000.

Croton sp.

Weak shrub, sprawling on dense herbaceous vegetation; pubescent thoughout; leaves with yellow-brown scales; fruits 3-locular, grey-green, with yellow-brown scales; styles persistent, 3, each lobed; stipules brown-scaled. Streamside; 1550m.
Bali Ngemba F.R.: Darbyshire 403 fr., 4/2004.
Note: field determination.

Discoclaoxylon hexandrum (Müll.Arg.) Pax & K.Hoffm.

F.T.E.A. Euphorbiaceae (1): 280 (1987).
Syn. ***Claoxylon hexandrum*** Müll.Arg.
Pithy tree, 3–5(–10)m, minutely puberulent; leaves elliptic, to 28 × 12cm, short acuminate, acute, margin serrate; petiole c. 18cm; inflorescences numerous, one per axil, pendent, spike-like, c. 12cm; flowers minute. Forest; 1310–1830m.
Distr.: Liberia to Uganda [Guinea-Congolian].
IUCN: LC
Bali Ngemba F.R.: Ghogue 1047 11/2000; Ujor 30350 fr., 5/1951; Zapfack 2000 fl., fr., 4/2002.

Drypetes sp. 3

Tree 15m, 20cm dbh; leafy stems minutely brown puberulous, soon glabrescent, white; leaves drying black, oblong-elliptic, 16 × 6.5cm, acumen blunt, 0.5cm, base assymmetric, obtuse, lateral nerves 7 pairs, drying pale brown below; petiole 3mm; stipules caducous. Forest; 1310–1700m.
Distr.: Bali Ngemba [W Cameroon Uplands].
Bali Ngemba F.R.: Cheek 10735 10/2001; Onana 1983 fl., 4/2002; Plot voucher BAL62 fl. 4/2002; Tadjouteu 404 11/2000; Zapfack 2020 fl., 4/2002.
Note: appears close to *D. calvescens* Pax. & Hoffm. of Adamoua highlands, extending to Congo (Kinshasa), but fertile material need to confirm.

Drypetes sp. aff. *leonensis* Pax of Bali Ngemba

Tree 6–15m, 10cm dbh, glabrous; leafy stems ridged; leaves elliptic, to 8 × 3.5cm, leathery, drying pale grey-green, acumen to 1cm, base acute, assymmetric, lateral nerves 5 pairs; stipules caducous; fascicles axillary, to 14-flowered; pedicels 8mm; sepals 4, pubescent on both sides; stamens five, outside a funnel-like, glabrous disc; ovary 4mm, densely white-hairy; styles 2mm; stigmas flat, peltate; fruit ellipsoid, 2.5 × 2cm, grey pubescent; pedicel 1cm. Forest; 1400–1900m.
Distr.: Bali Ngemba [W Cameroon Uplands].
Bali Ngemba F.R.: Cheek 10736 10/2001; Pollard 1034 fl., 4/2002; 1041 fl., fr., 4/2002; Zapfack 2067 fr., 4/2002.
Note: probably a species new to science, further study needed.

Erythrococca sp. cf. *hispida* (Pax) Prain = Daramola 41161

Shrub 1.5–2m; stem, petioles and lower blade thinly covered in curved hairs; spines absent; larger leaves lanceolate-ellipic, to 15 × 7cm (atypical?), acuminate, obtuse to rounded, lateral nerves 5 pairs; petiole 1–2cm; female inflorescence 4cm, 6-flowered, involucre cup-like; ovary glabrous; styles spreading; fruit 7mm diam. Forest; 1450–1600m.
Distr.: Bamenda Highlands [W Cameroon Uplands].
Bali Ngemba F.R.: Onana 2044 fr., 4/2002; Pollard 927 fl., 4/2002; Pollard 1029 fr., 4/2002.
Note: not matched in FWTA, possibly new.

Euphorbia schimperiana Hochst. ex A.Rich. var. *schimperiana*

Herb, 0.6–1.2m; leaves alternate, 4–10cm long, narrowly oblanceolate, with a whorl of 4–10 below the inflorescence; inflorescence an umbel; glands of involucre crescent-shaped; capsule 3–4mm. Forest edges, grassland; 1850–2200m.
Distr.: Cameroon, Congo (Kinshasa), Somalia, Ethiopia, Uganda, Kenya, Tanzania, Malawi and into Zimbabwe [Afromontane].
IUCN: LC
Bali Ngemba F.R.: Biye 88 11/2000; Ghogue 1084 11/2000.

Macaranga occidentalis (Müll.Arg.) Müll.Arg.

Tree, 3–25m; trunk often spiny, with red or clear exudate, glabrous; leaves suborbicular, 30–50cm diam., shallowly 3–5-lobed, lobes acuminate, base cordate, lower surface often bluish white below and with minute red glands; petiole c. 30cm; stipules 4 × 2cm; inflorescence axillary, pendent, paniculate; bracts 5–10cm, deeply dentate-sinuate, densely pubescent; female inflorescences to 18cm; fruit 1–2-lobed, 8mm. Forest edge; 1400–1500m.
Distr.: SE Nigeria & Cameroon [W Cameroon Uplands].
IUCN: NT
Bali Ngemba F.R.: Cheek 10738 10/2001; Ghogue 1033 11/2000.

Margaritaria discoidea (Baill.) G.L.Webster var. *discoidea*

J. Arnold. Arbor. 48: 311 (1967).
Syn. ***Phyllanthus discoideus*** (Baill.) Müll.Arg.
Deciduous tree, 10–25m, glabrous; leaves ovate–elliptic, 7–9 × 3.5cm, subacuminate, acute to obtuse, bluish white below, drying green, secondary nerves 10 pairs, arcuate, quaternary nerves reticulate, conspicuous; petiole 8mm; fruit 3-lobed, glossy, 7mm. Farmbush & forest edge; 1560m.
Distr.: widespread in tropical Africa [Tropical Africa].
IUCN: LC
Bali Ngemba F.R.: Pollard 600 fr., 10/2001.

Neoboutonia mannii Benth. var. *mannii*

Tree, to 23m, stellate-hairy; leaves orbicular-cordate, 10–20cm long, lower surface with long, soft, spreading hairs as well as scurf; petiole 7cm; stipules ovate, 3mm; inflorescence terminal, much branched, 20 × 20cm; male calyx puberulent; flowers minute, yellowish green; styles narrow with 2 linear lobes. Forest edge & farmbush; 1350m.
Distr.: S Nigeria to Cameroon [Lower Guinea].
IUCN: NT
Bali Ngemba F.R.: Etuge 5435 fl., fr., 4/2004.
Note: field determination.

Phyllanthus mannianus Müll.Arg.

Erect subshrub, to 1.5m; leaves obovate or obovate-elliptic, 1–2.5cm; flowers dioecious, males 1–3 in each leaf axil, females solitary; sepals white with broad green central line. Forest-grassland transition; 1560–2100m.
Distr.: Guinea, Ivory Coast and W Cameroon [Guineo-Congolian (montane)].
IUCN: LC
Bali Ngemba F.R.: Biye 129 11/2000; Pollard 593 fl., 10/2001; 656 fl., 10/2001.

Plukenetia conophora Müll.Arg.

Flora 47: 530 (1864).
Syn. ***Tetracarpidium conophorum*** (Müll.Arg.) Hutch. & Dalziel
Liana to 15m, glabrous; leaves ovate-elliptic, c. 10 × 5.5cm, acumen 0.5cm, base obtuse, crenate-serrate, 3-nerved at base; petiole 8cm; inflorescences axillary, paniculate, c. 6cm; flowers dull-white, 2mm; fruit capsule 4-lobed, 7cm diam., winged and ridged. Secondary forest and plantations; 1400m.
Distr.: Sierra Leone to Congo (Kinshasa) [Guineo-Congolian].
IUCN: LC
Bali Ngemba F.R.: Etuge 5424 fl., fr., 4/2004.
Local name: Nga (Mantum) (*Etuge* 5424).
Note: field determination.

Pseudagrostistachys africana (Müll.Arg.) Pax & K.Hoffm. subsp. *africana*

Tree, 5–20m tall; bark whitish with green spots, slash brown to red; leaves leathery, elliptic, 30 × 11–18cm, subacuminate, base rounded then slightly decurrent, with a pair of flat glands, ± serrate, nerves 21–24 pairs, venation scalariform; petiole 2.5–5cm; stipule single, 3cm, caducous, scar completely encircling stem. Forest; 1700m.
Distr.: Ghana, Bioko, São Tomé, Cameroon [Upper & Lower Guinea].
IUCN: VU
Bali Ngemba F.R.: Pollard 510 11/2000.

Shirakiopsis elliptica (Hochst.) H.-J.Esser

Kew Bull. 56(4): 1018 (2001).
Syn. ***Sapium ellipticum*** (Krauss) Pax
Deciduous shrub or tree, 3–25m; white exudate, glabrous; leaves drying black, leathery, elliptic, c. 9 × 4cm, obtuse, acute, finely serrate; petiole 3mm; inflorescence terminal spike, 6cm; flowers numerous, 1mm, green; fruit globose or bilobed, fleshy, 8mm; styles 2, coiled. Forest or forest edge; 1450–1800m.
Distr.: tropical & S Africa [Tropical Africa].
IUCN: LC
Bali Ngemba F.R.: Onana 1821 10/2001; Pollard 926 4/2002.

Tragia benthamii Baker

Climber, to 3m; stems long-hairy (stinging); leaves ovate, c. 6.5 × 3.5cm, acuminate, cordate, serrate, densely-pubescent below; petiole 2cm; female calyx lobes 6, lacking foliaceous apex, deeply pectinate, the lobes setose. Forest and scrub; 1950m.
Distr.: tropical and subtropical Africa [Afromontane].
IUCN: LC
Bali Ngemba F.R.: Biye 125 11/2000.

Vernicia montana Lour.

F.T.E.A. Euphorbiaceae 1: 178 (1987).
Syn. ***Aleurites montana*** Wilson
Tree, 12m, no exudate, glabrous; leaves broadly ovate, dimorphic, entire, c. 12 × 10cm, acuminate, cordate, basal glands connate on upper surface of petiole apex, or 3-lobed, c. 18 × 18cm, lobes c. 6cm deep, with a gland at the base of the sinuses; inflorescences terminal, paniculate, c. 20-flowered, 15cm long, in fascicles from flushing, often leafless stems; flowers c. 3cm wide; calyx 2-lobed; petals 5, white; anthers 5+5. Farms, shade tree; 1400m.
Distr.: SE Asia, planted in tropical Africa [Palaeotropics].
IUCN: LC
Bali Ngemba F.R.: Pollard 1047 fr., 4/2002; 1048 fl., 4/2002.

FLACOURTIACEAE

M. Cheek (K)

Dasylepis racemosa Oliv.

Tree, (4–)6–15m; bark smooth, green-grey with brown plaques; leaves elliptic or oblong, c. 19 × 9cm, acuminate, obtuse, undulate-dentate; petiole 1cm; inflorescence axillary, spicate, 5–6cm; flowers 1–2cm diam.; sepals 4, pink-red, 0.7cm; petals c. 8, white, stigmas 3; fruit 3-valved, globose, 1.5–2cm diam., thick-walled. Forest & *Aframomum* thicket; 1450m.
Distr.: SE Nigeria, Cameroon, Congo (Kinshasa) & Uganda [Guineo-Congolian (montane)].
IUCN: LC
Bali Ngemba F.R.: Cheek 10511 11/2000; Ujor 30344 fr., 5/1951.

Oncoba lophocarpa Oliv.

F.T.A. 1: 117 (1868).
Syn. ***Caloncoba lophocarpa*** (Oliv.) Gilg
Tree, (6–)12–25m; leaves elliptic or elliptic-ovate, c. 14 × 7cm, long-acuminate; inflorescences cauliflorous, c. 2cm long, or on long, whip-like, mostly leafless branches radiating 3–5m from the trunk on the ground; flowers 5–7cm diam.; sepals 3, 1.5cm; petals white, 8, c. 3cm; fruit ovoid, c. 5 × 4cm, strongly 8–12-winged. Forest; 1850m.
Distr.: Cameroon [W Cameroon Uplands].
IUCN: VU
Bali Ngemba F.R.: Biye 90 11/2000.

Oncoba sp. nov.

Tree 5–7m with spines 2–4cm long; leaves papery, elliptic, c. 6cm long, serrate; flowers c. 4cm diam.; petals white; staminal-mass yellow; fruits globular, with equidistant longitudinal lines, c. 5cm diam. Forest; 1790–1900m.
Distr.: Cameroon [Cameroon Endemic].
Bali Ngemba F.R.: Cheek 10489 11/2000; Pollard 952 fl., 4/2002; 1037 4/2002.
Local name: Tamti. **Uses:** none known (*Darbyshire*, pers. comm.).

GENTIANACEAE

Y.B. Harvey (K) & I. Darbyshire (K)

Sebaea brachyphylla Griseb.

Erect herb, 20(–50)cm; stems glabrous, few-branched, dichotomous; leaves sessile, orbicular, 8mm; cymes terminal, dense, 5–20-flowered; pedicels c. 1mm; sepals oblong, 3mm, with a prominent central nerve; corolla 6mm, inflated at base, yellow; stamens exserted. Grassland; 2000m.
Distr.: tropical Africa [Afromontane].
IUCN: LC
Bali Ngemba F.R.: Pollard 460 fl., 11/2000.

Swertia mannii Hook.f.

Erect herb, to 15cm; lateral branches to 8cm; leaves linear-lanceolate to narrowly elliptic, to 1–3cm long; cymes more lax, 6–8-flowered; pedicels to 1.8cm; sepals lanceolate, 3mm; petals 5, lanceolate, 5.5mm, white with a purple stripe externally. Grassland; 1300–2100m.
Distr.: Guinea (Conakry) to Cameroon [Upper & Lower Guinea].
IUCN: LC
Bali Ngemba F.R.: Onana 1883 10/2001; Pollard 465 fl., fr., 11/2000; 692 fl., 10/2001.

Swertia quartiniana Hochst. ex A.Rich.

Perennial, 0.15–0.6m high; simple, or sparsely branched; leaves narrowly lanceolate to elliptic, 1.5–5cm long, acute; inflorescence a dense cyme; corollas mauve; lobes oblong-obovate, 10–13 × 5mm. Grassland; 2100m.
Distr.: W Africa, Ethiopia, E Africa and southern tropical Africa [Tropical & subtropical Africa].
IUCN: LC
Bali Ngemba F.R.: Pollard 965 fl., 4/2002.

GERANIACEAE

I. Darbyshire (K) & M. Cheek (K)

Geranium arabicum Forssk. subsp. *arabicum*

Notes Roy. Bot. Gard. Edinburgh 42: 171 (1985).
Syn. ***Geranium simense*** Hochst. ex. A.Rich.
Straggling herb, 6–20cm; leaves orbicular in outline, 2–3cm diam., deeply palmately lobed, lobes 5, with 2–3 lateral lobes, sparingly pubescent; petiole 3cm; inflorescence 1–2-flowered; peduncle c. 8cm; flowers 1cm, pale pink (-white) with dark veins. Grassland-forest boundary, roadsides; 2040m.
Distr.: Nigeria to Kenya [Afromontane].
IUCN: LC
Bali Ngemba F.R.: Pollard 449 11/2000.

GESNERIACEAE

P. Bhandol (K)

Fl. Cameroun 27 (1984).

Streptocarpus elongatus Engl.

Erect fleshy herb to 1m; leaves ovate-elliptic, acuminate, rounded or subcordate at base, upper surface shortly-pubescent, lower surface subglabrous; cymes lax, pedunculate; corolla tubular, white, 1cm long; fruit bright-green, 5cm long, twisted, glabrous. Forest; 1420–1560m.
Distr.: Sierra Leone & Cameroon [Upper & Lower Guinea].
IUCN: LC
Bali Ngemba F.R.: Ghogue 1038 11/2000; Pollard 414 11/2000; 595 fl., 10/2001.
Uses: none known (*Nkankano Peter* in *Pollard* 414).

GUTTIFERAE

I. Darbyshire (K) & M. Cheek (K)

Allanblackia gabonensis (Pellegr.) Bamps

Bull. Jard. Bot. Natl. Belg. 39: 347 (1969).
Syn. ***Allanblackia*** **sp. of FWTA**
Tree to 10–30(–45)m; leaves obovate, c. 12 × 5cm, acuminate, obtuse to rounded, lower surface matt, lateral nerves c. 15 pairs, resin canals inconspicuous, midrib pinkish red; petiole 1.5cm, axillary cup; male inflorescence terminal, 3–15-flowered; flowers pale yellow or pink, 4.5cm diam.; petal apex rounded; staminal phalanges 5, with anthers on both upper and lower surface; central disc 5-lobed, slightly undulate; fruit ovoid, c. 15cm. Forest; 1310–1500m.
Distr.: Cameroon & Gabon [Lower Guinea].
IUCN: VU
Bali Ngemba F.R.: Ghogue 1058a 11/2000; Onana 1988 fl., 4/2002; 1995 fl., 4/2002; Ujor 30327 fl., 5/1951.
Local name: Mbibabdon. **Uses:** MEDICINES (*Etuge*, pers. comm.).
Note: flowering abundantly within the reserve, April 2004 (pers. obs.). Flowers pink to red in the Bali Ngemba subpopulation.

Garcinia smeathmannii (Planch. & Triana) Oliv.

Syn. ***Garcinia polyantha*** Oliv.
Tree, 4–15m; leaves thickly leathery, drying pale brown below, narrowly oblong-elliptic, c. 20 × 8cm, obtuse, subacuminate, base obtuse, secondary nerves c. 20 pairs; petiole c. 1.5cm; inflorescence sessile, axillary, on leafy stems, umbellate-fascicled, 15–20-flowered; pedicels c. 15mm; flowers white, 1cm diam.; anthers with free filaments inserted on ligules, as long as petals; fruits globose, 2cm; stigmas 2; pedicels 4cm. Forest; 1310–1900m.
Distr.: Guinea (Bissau) to Zambia [Guineo-Congolian].
IUCN: LC
Bali Ngemba F.R.: Cheek 10447 11/2000; 10487 11/2000; 10734 fl., 10/2001; Ghogue 1058b 11/2000; Onana 1927 10/2001; 1991 fr., 4/2002; Ujor 30370 fr., 5/1951; Zapfack 2036 fr., 4/2002.

Local name: Tehsusong (*Onana* 1991). **Uses:** MATERIALS – branches used as toothbrush (*Onana* 1991); MEDICINES (*Cheek* 10447).
Note: Ghogue 1058b and Onana 1991 are tentatively placed in this taxon. Fruits/flowers are required to fully identify.

Harungana madagascariensis Lam. ex Poir.
Shrub or small tree, 3–6m, glabrescent; leaves ovate, c. 12 × 5.5cm, acuminate, base rounded, nerves c. 10 pairs; petiole 2cm, producing bright orange exudate when broken; inflorescence a dense terminal panicle, 7–15cm; flowers white, 2mm, petals hairy; berries orange, 3mm. Farmbush; 1430m.
Distr.: tropical Africa & Madagascar [Tropical Africa].
IUCN: LC
Bali Ngemba F.R.: Etuge 4700 11/2000.

Hypericum peplidifolium A.Rich.
Herb; stems prostrate, to 60–90cm long; leaves elliptic or obovate, 6–15mm long; flowers yellow; fruit fleshy and indehiscent. Grassland; 2100m.
Distr.: tropical Africa [Afromontane].
IUCN: LC
Bali Ngemba F.R.: Pollard 969 fr., 4/2002.

Hypericum revolutum Vahl subsp. *revolutum*
Webbia 22: 239 (1967); Bull. Jard. Bot. Natl. Belg. 41: 438 (1971); Kew Bull. 33: 581 (1979).
Syn. ***Hypericum lanceolatum*** Lam.
Shrub or tree to 12m; leaves narrowly elliptic, 1–2(–3.5)cm long; petals yellow, 2–3cm long. Grassland & forest-grassland transition; 2100m.
Distr.: tropical Africa [Afromontane].
IUCN: LC
Bali Ngemba F.R.: Pollard 659 fl., 10/2001.

Hypericum roeperianum Schimp. ex A.Rich.
F.T.E.A. Hypericaceae: 3 (1953).
Shrub, to 3m; leaves elliptic, ovate or lanceolate, 3–8cm long; petals yellow 2.5–3.5cm long. Forest edge; 2100m.
Distr.: Cameroon, Ethiopia to Tanzania [Afromontane].
IUCN: LC
Bali Ngemba F.R.: Pollard 667 10/2001.

Psorospermum aurantiacum Engl.
Shrub or small tree, to 2(–5)m; young stems rusty-tomentose; leaves bullate, elliptic, 3.5–7 × 1.8–3.5cm, apex shortly acuminate, base acute, brown-green and sparsely tomentose above, venation impressed, densely rusty-tomentose below, obscuring the venation; petiole 4–5mm, tomentose; inflorescence terminal on lateral branches, pubescent throughout, 10–many-flowered; sepals acute, 2.5–3mm; petals cream, 4.5mm, pubescent within; fruit ovoid, wine-red. Grassland and forest edge; 2100m.
Distr.: Nigeria & Cameroon [W Cameroon Uplands].
IUCN: VU
Bali Ngemba F.R.: Darbyshire 363 fl., fr., 4/2004; Pollard 967 fl., 4/2002.
Note: locally common within its range and appearing tolerant of limited human disturbance (pers. obs.). *Darbyshire* 363 is a field determination.

Psorospermum densipunctatum Engl.
Shrub or small tree, to 2(–3)m; young stems sparsely pubescent; leaves densely bullate, elliptic, 3–5 × 1.8–2.8cm, apex shortly acuminate, base acute, upper surface dark green, glossy, venation deeply impressed, lower surface pubescent only on midrib and ± on lateral nerves; petiole 3–6mm, ± pubescent; inflorescence terminal on lateral branches, cymes subumbellate, c. 10–20-flowered, puberulent throughout; sepals acute, c. 2.5mm; petals white to cream, pubescent within; fruit ovoid, wine-red. Grassland and forest edge; 2100m.
Distr.: Sierra Leone, Nigeria, Cameroon [Upper & Lower Guinea (montane)].
IUCN: NT
Bali Ngemba F.R.: Darbyshire 380 fr., 4/2004.
Note: widespread only in the Cameroon highlands, where it is perhaps threatened by intensification of agriculture in the Northwest Highlands. Field determination.

Psorospermum febrifugum Spach
Fl. Zamb. 1(2): 387 (1961).
Syn. ***Psorospermum febrifugum*** Spach var. ***ferrugineum*** (Hook.f.) Keay & Milne-Redh.
Shrub or small tree, to 3m; bark cork-like; young stems rusty pubescent; leaves coriaceous, subsessile, broadly ovate-elliptic, 5–10 × 3–6.5cm, apex shortly acuminate, base rounded to subcordate, upper surface glaucous, lower surface stellately rusty-hairy; inflorescences terminal on lateral branches, subumbellate, many-flowered; pedicels 2–3mm, rusty-pubescent; sepals acute, 3mm; petals 5–6mm, cream, pubescent within; fruit ovoid, wine-red. Wooded grassland; 1600m.
Distr.: Sierre Leone to Mozambique [Tropical Africa].
IUCN: LC
Bali Ngemba F.R.: Darbyshire 407 fl., fr., 4/2004.
Note: field determination.

Psorospermum cf. *tenuifolium* Hook.f.
Shrub or climber, to 4m; lateral shoots puberulent; leaves in 2–3 pairs on lateral branches, opposite or rarely subopposite, papery, concolorous, elliptic-obovate, c.6.5 × 3.4cm, apex shortly acuminate, base cuneate, margin subcrenulate, lower surface densely punctate; petiole c. 5mm, puberulent; cymes terminal on lateral branches, umbellate, 8–30-flowered; peduncle 1.7cm; pedicels 0.4cm, puberulous; sepals acute, c. 2mm long, puberulent; petals elliptic c. 4mm long, white, dark veined, densely white-pubescent within; fruit ovoid, wine-red.. Forest; 1450m.
Distr.: Bali Ngemba F.R.
Bali Ngemba F.R.: Darbyshire 353 fr., 4/2004; Pollard 929 fl., 4/2002.
Note: differing from *P. tenuifolium* in the puberulent young stems, petioles, pedicels and calyx, and the generally fewer-flowered, less dense inflorescences. *Pollard* 929 matches *Daramola* in FHI 40648 from Bamenda. Several plants of this taxon were recorded along the lower boundary of the reserve in April 2004 (pers. obs.), *Darbyshire* 353 is a field determination.

HERNANDIACEAE

Y.B. Harvey (K) & M. Cheek (K)

Illigera pentaphylla Welw.

Climber, 5m; leaves shortly-pubescent, alternate; petioles 12cm, twisting; leaflets 5, elliptic, to 9 × 6cm, subacuminate, sub- 5-nerved, nervation scalariform; petiolules 2.5cm; inflorescences axillary, 30cm, 1–5-branched; flowers 7mm; corolla white and purple; fruit 3 × 9cm, 1-seeded, 2-winged. Fallow forest & *Hypselodelphys* thicket; 1450m.
Distr.: Ivory Coast to Uganda [Guineo-Congolian].
IUCN: LC
Bali Ngemba F.R.: Cheek 10508 11/2000.
Note: easily confused with *Dioscorea*, when sterile.

ICACINACEAE

M. Cheek (K)

Fl. Cameroun 15 (1973).

Apodytes dimidiata E.Mey. ex Arn.

F.T.A. 1: 355 (1868).
Tree 15m, glabrous; leaves drying black above, alternate, simple, ovate-elliptic, to 13 × 6cm, acumen 0.5cm, base obtuse-rounded, entire, lateral nerves 6–7 pairs; petiole 1.5cm; stipules nil; inflorescence terminal, minutely brown-puberulent, c. 10 × 6cm; old flowers 3mm; fruits immature, strongly asymmetric, style single. Forest; 1920m.
Distr.: Nigeria to South Africa [Afromontane].
IUCN: LC
Bali Ngemba F.R.: Etuge 5334 fr., 4/2004.
Local names: Canda stick (*Etuge* 5334)

Leptaulus daphnoides Benth.

Tree or shrub, to 15m; leaves elliptic, 6.5–16 × 2.5–6.5cm, base attenuate, base obtusely acuminate, 6–7 pairs lateral nerves ascendant, anastomosing, glabrous; inflorescence a short cyme, joined to the leaf; flowers 5-merous, white, 11 × 1.5mm, corolla tubular, linear; lobes 1mm long; anther filaments joined to the petals; style glabrous; fruit an ellipsoid drupe, 12 × 8 × 5mm, papillose. Forest.
Distr.: Sierra Leone to Sudan & Tanzania [Guineo-Congolian].
IUCN: LC
Bali Ngemba F.R.: Tiku 30418 fl., 6/1951.

Rhaphiostylis beninensis (Hook.f. ex Planch.) Planch. ex Benth.

Woody climber, to 10–12m; leaves elliptic to lanceolate-elliptic, 6–15 × 2–7cm, 5–7 pairs lateral nerves; flowers rather numerous in axillary facicles, pentamerous; petals free, 7 × 1mm, drying black; stamens 5, white; pedicels 7–8mm; fruits red, reniform, 1.3 × 1–2 × 1cm, reticulate. Forest; 1790m.
Distr.: Senegal to Tanzania [Guineo-Congolian].
IUCN: LC
Bali Ngemba F.R.: Cheek 10492 11/2000.

LABIATAE

B.J. Pollard (K)

Note: here follows the treatment of Labiatae *sensu lato*, adding *Clerodendrum* L. and *Vitex* L., which were previously included in Verbenaceae. See Harley, R.M., Atkins, S., Budantsev, A., Cantino, P.D., Conn, B., Grayer, R.J., Harley, M.M., De Kok, R., Krestovskaja, T., Morales, A., Paton, A.J., Ryding, O. and Upson., T., (2004). Labiatae. *In*: Kadereit, J.W. (ed.). The Families and Genera of Vascular Plants, vi: (Lamiales). Springer, Berlin.

Achyrospermum africanum Hook.f. ex Baker

An erect herbaceous undershrub, to about 3m; leaves ovate, c. 15 × 8cm; terminal inflorescences 8–12 × 1.5cm, the laterals smaller; bracts broadly ovate, c. 0.8cm; calyx teeth twice as long as broad; corolla exceeding calyx, purplish pink with white throat; stamens ascending. Forest; 1200–1560m.
Distr.: Guinea (Conakry) to Cameroon [Guineo-Congolian (montane)].
IUCN: LC
Bali Ngemba F.R.: Ghogue 1021 fl., 11/2000; Onana 1806 fl., 10/2001; 1922 fl., 10/2001; Pollard 437 fl., 11/2000; 505 fl., 11/2000; Pollard 605 fl., 10/2001.

Achyrospermum oblongifolium Baker

An erect, herbaceous, little-branched, undershrub, 30–70cm; stems little-branched, tomentose; leaves 10–18 × up to 8cm, narrowly obovate-elliptic, acuminate; inflorescence terminal, 3–5(–10) × 2cm; bracts broadly ovate, ciliate; calyx teeth as broad as long, margin ciliate; corolla greenish white. Forest; 1500m.
Distr.: Guinea (Conakry) to Bioko, Cameroon, São Tomé, Angola (Cabinda) [Upper & lower Guinea)].
IUCN: LC
Bali Ngemba F.R.: Pollard 504 fl., fr., 11/2000.

Aeollanthus angustifolius Ryding

Syn. *Aeollanthus repens* sensu Morton
Annual herb, 10–50cm; stems decumbent or erect; leaves linear to narrowly-lanceolate, (6–)10 × 1.5–5mm, usually more than 6 × as long as broad, ± pubescent; margins revolute; inflorescence of dense spikes, to 2(–3.5)cm, secund; corolla blue, pink or violet; lower lip 3-lobed, with narrow teeth; calyx two-lipped, dehiscent (circumscissile), 3–4mm in fruit, basal part undulate. On rocks or shallow soil, in wet depressions; 1300–1650m.
Distr.: Nigeria, Cameroon, CAR [Lower Guinea & Congolian].
IUCN: LC
Bali Ngemba F.R.: Onana 1879 fl., fr., 10/2001; Pollard 519 fl., fr., 11/2000; 699 fl., fr., 10/2001.

Aeollanthus trifidus Ryding

Symb. Bot. Upsal. 26: 136 (1986).
Subshrub, to 30–50cm; stem indumentum to 1.5mm; leaves fleshy, shortly petiolate, broadly elliptic, to 45–65 × 22–36mm; apex subacute; base attenuate; margin crenate; veins prominent below; inflorescence paniculate, lax, conspicuously bracteate; bracts elliptic, to 4–6.5 × 2–3.5mm; fruiting calyx circumscissile; corolla white; lower lip tinged purple distally; upper lip white, speckled purple; tube much-

broadened towards the throat, 5.5–8mm; upper lip 4-lobed; lower lip trifid. Rocks or shallow soil; 1500m.
Distr.: SE Nigeria (Mambilla Plateau), W Cameroon (NW Province) [Uplands of Western Cameroon].
IUCN: VU
Bali Ngemba F.R.: Pollard 698 fl., fr., 10/2001.

Clerodendrum capitatum (Willd.) Schum. & Thonn. var. *capitatum*

Erect or scrambling shrub, or climber with long petiole-derived spines; internodes hollow; leaves ovate or elliptic, 7–20 × 4–10cm, with 3–6 pairs of lateral nerves; inflorescence terminal or occasionally lateral on young branches, conspicuously bracteate; bracts ovate-acuminate or ovate-lanceolate; corolla white; tube 4–8cm; calyx fimbriate on margins. Forest; 1500–1900m.
Distr.: tropical and subtropical Africa [Afromontane].
IUCN: LC
Bali Ngemba F.R.: Etuge 4268 10/2001; Onana 1848 11/2001.

Clerodendrum formicarum Gürke

Woody climber, or climbing shrub; internodes hollow; leaves elliptic, ternate, 4–10 × 2–5cm, glabrous, acuminate, shortly cuneate; inflorescence terminal; axis short; peduncles long and conspicuously horizontal; flowers small; calyx 2–3mm; corolla tube very short, 4–5mm, yellowish white. Forest; 1830m.
Distr.: tropical Africa.
IUCN: LC
Bali Ngemba F.R.: Biye 100 11/2000.

Clerodendrum silvanum Henriq. var. *buchholzii* (Gürke) Verdc.

Mem. Mus. Natl. Hist. Nat. B. Bot. 25: 1555 (1975);
Syn. ***Clerodendrum buchholzii*** Gürke
Woody climber, to 10m; stems with petiolar spines; leaves elliptic or ovate, glabrous, 8–20 × 3–10cm, often confined to the canopy; inflorescence an elongate, leafless panicle, frequently cauliflorous; rachis 5–30cm long; calyx enlarged, 8–10mm; corolla white, fragrant; tube (1.5–)1.7–2.5cm; fruits red. Forest; 1400m.
Distr.: tropical and subtropical Africa.
IUCN: LC
Bali Ngemba F.R.: Etuge 4297r 10/2001.

Clerodendrum violaceum Gürke

Straggling or climbing shrub; young shoots quadrangular; leaves distinctly petiolate, thinly membranaceous, ovate, elliptic or oblong, 5–12 × 3–10cm; inflorescence paniculate; calyx lobes obtuse; flowers about 2.5cm, a conspicuous violet, violet and white, or greenish. Forest; 1600m.
Distr.: Guinea (Conakry) to Cameroon, Congo (Kinshasa), Zimbabwe [Tropical Africa].
IUCN: LC
Bali Ngemba F.R.: Onana 2051 fr., 4/2002.

Haumaniastrum caeruleum (Oliv.) J.K.Morton

Perennial herb, 0.2–1m; stems one to several, sparsely pubescent; leaves almost glabrous to densely pubescent, leaf-base sometimes clasping stem; petioles 0–2mm long; upper leaves subtending heads, 5–25mm long, caudate or apiculate, white or bluish, apex green, often inrolled; corolla white, pink, blue or purple; stamens declinate. Grassland; 1500m.
Distr.: tropical Africa [Tropical Africa].
IUCN: LC
Bali Ngemba F.R.: Pollard 524 11/2000; 702 fl., 10/2001.

Haumaniastrum sericeum (Briq.) A.J.Paton

Kew Bull. 52(2): 332 (1997).
Perennial suffrutex, 0.2–1.2m; stems several, faintly aromatic or not, densely pubescent or sericeous; leaves sessile, densely pubescent to sericeous; blades linear, narrowly elliptic or obovate, sometimes folded along midrib, 8–65 × 1.5–7mm, entire or remotely serrate; veins parallel or narrowly diverging from midvein, prominent below; inflorescence heads, 3–10 × 3–8mm, corymbose; flowers pale pink, blue or purple, rarely white; fruiting calyx 3–5mm. Damp areas in grassland and woodland; 2100m.
Distr.: Cameroon, Congo (Kinshasa), Mozambique, Zambia, Zimbabwe, Angola, Namibia [Afromontane].
IUCN: LC
Bali Ngemba F.R.: Pollard 475 fl., fr., 11/2000.

Isodon ramosissimus (Hook.f.) Codd

Bothalia 15: 8 (1984); Fl. Rwanda 3: 311 (1985).
Syn. ***Homalocheilos ramosissimus*** (Hook.f.) J.K.Morton
An erect or straggling herb, to 4m; stems hollow, strongly quadrangular, pilose; leaves ovate, up to 7 × 4cm; inflorescence an axillary panicle of many-flowered dichotomous cymes; calyx tube declinate, ventricose, teeth subequal; corolla 5mm long, white, speckled purple in throat; upper lip very small, recurved; stamens declinate. Forest margins; 1400–2100m.
Distr.: Sierra Leone to Bioko, Cameroon, Sudan, Uganda, Zimbabwe [Afromontane].
IUCN: LC
Bali Ngemba F.R.: Biye 136 fl., 11/2000; Pollard 467 fl., fr., 11/2000.

Leucas deflexa Hook.f.

A straggling or semi-erect aromatic herb, to 2m; leaves lanceolate, cuneate at base, serrate, with an entire, acute tip, petiolate; inflorescence a densely globose axillary whorl with numerous linear-subulate bracteoles; corolla white; stamens ascending; anthers often conspicuously-hairy, orange.
Forest, forest margins, savanna; 1440–1800m.
Distr.: Ghana, Bioko, Cameroon, Angola [Guineo-Congolian (montane)].
IUCN: LC
Bali Ngemba F.R.: Onana 1831 fl., 10/2001; Pollard 419 fl., fr., 11/2000.
Uses: MEDICINES (*Nkankano Peter* in *Pollard* 419).

Leucas oligocephala Hook.f.

Herb, to 1.2m; stems slender, branched, pilose; leaves linear-lanceolate to elliptic, 2.5 or more × 1cm; inflorescence of several verticils of dense, globose, axillary whorls; calyx with longest tooth on the lower side, densely ciliate; flowers densely-hairy and purple tinged in bud, white when open; stamens ascending; anthers orange or red. Grassland and forest-grassland transition; 2100m.
Distr.: W Africa to E Africa and South Africa [Afromontane].
IUCN: LC

Bali Ngemba F.R.: Pollard 474 fl., fr., 11/2000.
Note: Sebald's (Stuttgarter Beitr. Naturk. A. 341: 1–200 (1980) subspecies are not recognised here because there are major overlaps between his distinguishing characters.

Ocimum gratissimum L. subsp. *gratissimum* var. *gratissimum*

A branched, erect, pubescent shrub, to ± 3m, emanating an aroma similar to that of cloves (*Syzygium aromaticum*); leaves ovate to obovate, 6–12 × 3cm, cuneate, acutely acuminate; inflorescence of several dense spikes, >1cm; calyx dull, densely lanate, horizontal or slightly downward-pointing in fruit; corolla small, greenish white; stamens declinate. Forest, woodland, savanna; 1410m.
Distr.: widespread in the tropics from India to W Africa, south to Namibia & South Africa; naturalised in tropical S America [Pantropical].
IUCN: LC
Bali Ngemba F.R.: Pollard 514 fl., 11/2000.
Uses: FOOD ADDITIVES – flavourings (Pollard 514).

Platostoma rotundifolium (Briq.) A.J.Paton

Syn. ***Geniosporum rotundifolium*** Briq.
A stout woody perennial, to ± 2m; stems grooved, densely ferrugineous-pubescent; leaves subrotund to broadly lanceolate, 2–5 × 1–3cm, crenulate; inflorescences several, dense, cylindrical, 2.5–10cm long; bracts broadly ovate, conspicuously white or mauve-tinged; calyx tubular, c. 4mm long at maturity, 4-toothed; corolla twice as long as calyx. Forest edge, grassland; 1350–2100m.
Distr.: Sierra Leone to Cameroon, Congo (Kinshasa), E Africa, Angola [Afromontane].
IUCN: LC
Bali Ngemba F.R.: Nkeng 192 fl., 10/2001;Onana 1901 fl., 10/2001; Pollard 435 fl., fr., 11/2000; 671 fl., 10/2001.

Plectranthus decumbens Hook.f.

Syn. ***Solenostemon decumbens*** (Hook.f.) Baker
Epilithic or terrestrial (at Mt Cameroon) herb, 0.2–0.8m; stems usually annual, arising from a perennial tuber; lamina broadly ovate-triangular to triangular, (5–)20–40(–55) × (4–)15–35(–45)mm, acute, broadly cuneate, deeply crenate at maturity; leaves petiolate at maturity; petiole 5–32mm; inflorescences lax, to 44 × 35mm, with 10–20 verticillasters; fruiting calyx 3–4mm. Grassland, on rocks in grassland; 1500–1900m.
Distr.: SE Nigeria, Cameroon (Mt Cameroon, Mt Kupe, Bakossi Mts, Bamenda Highlands, Mt Oku) [W Cameroon Uplands].
IUCN: NT
Bali Ngemba F.R.: Pollard 511 fr., 11/2000; 713 10/2001.

Plectranthus glandulosus Hook.f.

A coarse, scrambling to erect, often robust, glandular and strongly aromatic herb, to ± 3.5m; leaves to 15cm long, glandular-punctate, margin with very uneven, rather small, double or treble crenations; inflorescence of copious loose panicles to ± 65cm long; mature calyx 9mm long; corolla violet; stamens declinate. Forest; 1420–1900m.
Distr.: Mali to Bioko, Cameroon [Upper & Lower Guinea].
IUCN: LC
Bali Ngemba F.R.: Onana 1845 fl., 11/2001; Pollard 409 fl., fr., 11/2000.
Local name: Watzamam kob. **Uses:** NON VERTEBRATE POISONS – when a chicken hatches, put the leaves around its sleeping place to deter insects, especially lice/fleas (*Nkankano Peter* & *Achu John* in *Pollard* 409).

Plectranthus kamerunensis Gürke

Straggling, densely-woolly, herb, to 1m; leaves ovate, to 11 × 9cm, acutely acuminate, cordate, coarsely crenate; inflorescences little-branched; mature calyx 8mm long, with a long white-pubescence; lower teeth lanceolate, acuminate; corolla violet; stamens declinate. Forest, forest margins; 2000–2100m.
Distr.: SE Nigeria, W Cameroon, E Africa [Afromontane].
IUCN: LC
Bali Ngemba F.R.: Pollard 455 fl., 11/2000; 668 fl., 10/2001.

Plectranthus occidentalis B.J.Pollard, in press

Syn. ***Solenostemon mannii*** (Hook.f.) Baker
A herbaceous, or somewhat woody, perennial herb or shrub, to c. 1m; stems climbing or erect; leaves ovate, 4–15cm, acutely acuminate, crenate, long-petiolate, often purplish tinged; inflorescence a copiously-flowered, dense raceme, up to 25 × 3–4cm or more in fruit; calyx 4–5mm; corolla rich bluish purple. Forest, woodland. 1420–1500m.
Distr.: Sierra Leone to Bioko & W Cameroon [Upper & Lower Guinea].
IUCN: LC
Bali Ngemba F.R.: Ghogue 1036 fl., 11/2000; Pollard 428 fl., fr., 11/2000.
Local name: Banbayin. **Uses:** MEDICINES (*Nkankano Peter* in *Pollard* 428).

Plectranthus tenuicaulis (Hook.f.) J.K.Morton

Syn. ***Plectranthus peulhorum*** (A.Chev.) J.K.Morton
Slender, branched, annual herb, to ± 1m; stems pubescent; leaves shortly petiolate, lanceolate, acute, 0.5–6cm long, a third as broad, crenate, pubescent; inflorescence a panicle with lateral racemose branches; mature calyx 3–5mm; corolla 8–10mm, pale blue; stamens declinate. Forest, forest margins; 1440–1560m.
Distr.: tropical Africa.
IUCN: LC
Bali Ngemba F.R.: Pollard 420 fr., 11/2000; Pollard 596 fl., 10/2001; 626 fl., 10/2001.
Uses: none known (*Nkankano Peter* in *Pollard* 420).

Satureja pseudosimensis Brenan

A slender, weak, straggling, perennial herb, to 10–60cm; stems softly pubescent and pilose, branched from the base; leaves subsessile, rounded at base, up to 2cm long, crenulate; inflorescence of distant axillary whorls; calyx broadly tubular, 6–8mm in fruit; corolla about 1.2cm; stamens ascending. Grassland; 2100m.
Distr.: W Cameroon, Bioko, Congo (Kinshasa), Sudan, E Africa [Afromontane].
IUCN: LC
Bali Ngemba F.R.: Tadjouteu 430 fl., 11/2000.

Satureja robusta (Hook.f.) Brenan

An erect, robust, strongly aromatic, perennial, 0.8–1.4m; stems branched; leaves ovate-rotund, crenate; inflorescence broad, of many dense terminal sessile spikes; mature calyx 4–

5mm; corolla white with mauve marks on the lip; stamens ascending. Grassland, woodland, forest edges; 2000m.
Distr.: W Cameroon [Cameroon Endemic].
IUCN: LC
Bali Ngemba F.R.: Pollard 479 fl., 11/2000.
Uses: MEDICINES (*Nkankano Peter* in *Pollard* 479).
Note: although this taxon is of limited distribution, it is quite common within its range, and so does not warrant assignation of an IUCN category of threat.

Vitex doniana Sweet
Tree, 10–20m; branches glabrous; leaves coriaceous, 5-foliolate; leaflets obovate to elliptic, middle ones 5–16 × 4–10cm; petiolule 1–2.5cm; inflorescence axillary or axillary and with terminal cymes, congested; peduncle 2–8cm long; fruits obovoid to subglobose, c. 1cm. Savanna; 1400m.
Distr.: tropical Africa and the Comores Is. [Tropical Africa].
IUCN: LC
Bali Ngemba F.R.: Pollard 592 fr., 10/2001.
Local name: Tambe. **Uses:** MATERIALS – wood – timber, furniture & roofing for houses (*Pollard* 592).

LAURACEAE

M. Cheek (K)

Fl. Cameroun 18 (1974).

Beilschmiedia sp. 1 of Bali Ngemba
Tree, 12–25m, aromatic when wounded; stems and leaves glabrous; leaves alternate, elliptic-oblong or slightly oblanceolate, to 20 × 8cm, acumen to 2cm, base acute, midrib drying slightly yellowish brown below, finely ridged below, yellow above when live, lateral nerves 5–8-pairs; petiole purple live, to 1.5cm; inflorescences dense, axillary, patent-puberulent, held above branches, c. 10 × 10cm, orange-green; flowers 3–4mm wide, C3+3, A3+3, G1; fruit ellipsoid, 1-seeded, fleshy, blue-black, c. 3 × 1.5cm. Forest; 1400–1900m.
Distr.: Cameroon [Cameroon Endemic].
Bali Ngemba F.R.: Cheek 10486 11/2000; 10522 11/2000; Etuge 4720 11/2000; Ujor 30409 6/1951.
Local name: Feshik (*Cheek* 10486). **Uses:** FOOD – infructescences – fruits harvested in dry season (Dec.–Jan.), and can be stored for one year: used for soup, for baking of accra. According to *Tadjouteu*, not known to the Bamilike tribe. According to *Clement Toh*, used by the Kom tribe, mostly harvested from trees in compounds; FOOD ADDITIVES – used in corn fufu and a special corn beer to make it slippery and as a flavour enchancer, not imparting a particular taste, but noticable if absent. Also used for baking Accora (Guinea corn beer); 3 seeds would sell for for 10 CFA some years ago, old women cherish it a lot; MATERIALS – wood – Fulanis also use it a lot for timber, furniture and especially roofing (*Cheek* 10486); timber (*Etuge* 4720).
Note: further analysis of the Bali Ngemba material is needed to see if it matches any of the 41 existing taxa documented in Cameroon (FDC 18: 1974). It may well be new since no other species in the Flore du Cameroun are recorded from such a high altitude (1900m). *Beilschmiedia acuta* and *B. congestiflora* are both known from Tschappe Pass (1400m), otherwise *B. lancilimba* is known from Babanki, at 1300m. Unfortunately these taxa are mostly known from single collections and are not represented at K.

LEEACEAE

Y.B. Harvey (K)

Fl. Cameroun 13 (1972).

Leea guineensis G.Don
Erect or suberect, soft-wooded, shrub, to 7m; leaves bipinnate; leaflets opposite, imparipinnate, oblong-elliptic, to 18cm long; flowers bright yellow, orange or red; fruits brilliant red, turning black. Forest and forest gaps; 1400m.
Distr.: tropical Africa.
IUCN: LC
Bali Ngemba F.R.: Cheek 10439 11/2000.

LEGUMINOSAE-CAESALPINIOIDEAE

B.A. Mackinder (K)

Fl. Cameroun 9 (1970).

Caesalpinia decapetala (Roth) Alston
F.T.E.A. Leguminosae: Caesalpinioideae: 36 (1967).
Spiny shrub or scrambler, to 6m; leaves bipinnate; leaflets 9–10 pairs, (obovate-)oblong, to 1.8 × 0.7cm; inflorescence an erect raceme; flowers pale yellow, 1.5cm diam., pod to 8.5cm. Secondary forest; 1400m.
Distr.: Asia, widely cultivated in Africa.
IUCN: LC
Bali Ngemba F.R.: Etuge 5416 fl., fr., 4/2004.

Chamaecrista kirkii (Oliv.) Standl. var. *kirkii*
Lock, M. (1989). Legumes of Africa: 31.
Erect shrub, to 1.5m; leaflets 25–30 pairs, oblong-linear, 9–17 × 1.5–4mm; petiole with an apical sessile gland; flowers yellow, solitary or 2–3; fruits 6–7.5cm long. Grassland; 1300m.
Distr.: tropical Africa.
IUCN: LC
Bali Ngemba F.R.: Kongor 55 fl., 10/2001.

Chamaecrista mimosoides (L.) Greene
Lock, M. (1989). Legumes of Africa: 32.
Prostrate, or more commonly erect, herb or subshrub, to 1.5m; leaves paripinnate; leaflets 20–70 pairs, linear to linear oblong, 3–8 × 1–1.5mm; flowers yellow, 4–13mm; pod linear to linear-oblong, up to 8cm. Grassland; 1440m.
Distr.: widespread in the palaeotropics.
IUCN: LC
Bali Ngemba F.R.: Biye 81 11/2000.

Senna septemtrionalis (Viv.) H.S.Irwin & Barneby
Lock M., Legumes of Africa: 39.
Syn. ***Cassia laevigata*** Willd.
Shrub or small tree, to 3m; leaves paripinnate; leaflets 3–4 pairs, lanceolate to ovate, 4–11 × 2–4cm; flowers yellow, 1–

1.5cm; pod subterete, up to 10cm. Farmbush & villages; 1560m.
Distr.: pantropical.
IUCN: LC
Bali Ngemba F.R.: Pollard 602 fr., 10/2001.

Zenkerella citrina Taub.

Tree or shrub, to 20m; leaves 1-foliolate; leaflet narrowly-elliptic or elliptic, 8–12 × 4.5–6cm; flowers white and pink, 5–7mm; pod broadly oblong, compressed, somewhat asymmetric, up to 7cm, leathery. Forest; 1700m.
Distr.: SE Nigeria to Gabon [Lower Guinea].
IUCN: NT
Bali Ngemba F.R.: Ghogue 1101 11/2000.

LEGUMINOSAE-MIMOSOIDEAE

B.A. Mackinder (K)

Albizia gummifera (J.F.Gmel.) C.A.Sm. var. *gummifera*

Tree, to 30m; crown flat; leaves bipinnate; leaflets numerous, auriculate at base, up to 2.0 × 1.1cm; inflorescence capitate, calyx and corolla inconspicuous; stamens numerous, showy, up to 2.5cm long, greenish becoming red towards apex, fused into a tube, the free ends extending a further 5–7mm; pod compressed, coriaceous, glabrescent, becoming glossy, up to 18 × 3.2cm wide. Forest; 1310m.
Distr.: SE Nigeria, W Cameroon (Bamenda Highlands), Congo (Kinshasa), Sudan, Ethiopia south to Zimbabwe, Mozambique, Madagascar [Tropical and subtropical Africa & Madagascar].
IUCN: LC
Bali Ngemba F.R.: Daramola FHI 40478 2/1959; Zapfack 2043 fl., 4/2002.

Entada abyssinica Steud. ex A.Rich.

Tree, to 10m; leaves bipinnate, 4–16 pinnae pairs, each pinna with 25–50 leaflet pairs; leaflets narrowly-oblong, less than 1cm long, pubescent, midvein not central; flowers subsessile, pale yellow, up to 2mm long, sweetly-scented; pod compressed, the valves splitting transversely into 1-seeded papery segments leaving the persistent suture. Dry woodland and wooded grassland; 1300–1700m.
Distr.: widespread in tropical Africa [Tropical Africa].
IUCN: LC
Bali Ngemba F.R.: Etuge 4814 11/2000; Onana 1875 10/2001; Pollard 977 fl., 4/2002.

Newtonia camerunensis Villiers

Bull. Jard. Bot. Natl. Belg., 60(1–2): 123 (1990).
Tree, 50cm diam. at base, fluted; leaves c. 25cm long; petiole 1.2–1.5cm; pinnae 8–10 pairs, gland between each pair; leaflets 25 per pinnae, 15 × 3.5mm, oblong, apex rounded, base rounded to slightly retuse, margin ciliate, upper surface glossy, midrib prominent; fruits 19–30 × 1.8–2.3cm, elliptic-oblong, straight or gently curved, apex rounded-apiculate; seeds winged, 4–8 × 1.5–2.2cm, elliptic to ovate. Forest; 1700m.
Distr.: Cameroon (Bamenda Highlands and Bamboutos Mts).
IUCN: CR
Bali Ngemba F.R.: Plot voucher BAL52 4/2002.
Note: flowers unknown in this species.

LEGUMINOSAE-PAPILIONOIDEAE

B.A. Mackinder (K) & R.P. Clark (K)
(determinations also provided by R.M. Polhill (K), B.J. Pollard (K), B.D. Schrire (K) & B. Verdcourt (K))

Adenocarpus mannii (Hook.f.) Hook.f.

Shrub, to 5m; leaves 3-foliolate; leaflets very variable in shape, 5–8 × 1.5–3.5cm; flowers yellow, 9–14mm; pod oblong, up to 2.5cm, viscose-glandular indumentum. Grassland & forest-grassland transition; 2200m.
Distr.: Bioko, Cameroon, Congo (Kinshasa) & E Africa [Afromontane].
IUCN: LC
Bali Ngemba F.R.: Ghogue 1082 11/2000.

Aeschynomene baumii Harms

Erect, woody, herb or shrub, to 1m; stem ribbed; leaves 1-pinnate, paripinnate, c. 3.5cm long, 8–20(–32) leaflets; leaflets oblong to ovate, 4–20mm long, opposite, asymmetric at base, discolorous; stipules foliaceous; inflorescence an axillary or terminal raceme or panicle, 2–8cm long; pedicels 10–15mm long; paired bracts at base of calyx; flowers yellow/pinkish, to 2cm long; pods flat, 14–22 × 8–12mm, 1 (occasionally 2)-seeded with marginal suture, greenish red, lunate-elliptic. Grassland, often frequently/annually burnt, forest understorey; 1200–1450m.
Distr.: Nigeria east to Burundi & Rwanda, south to Equatorial Guinea, Tanzania and Zambia [Tropical Africa].
IUCN: LC
Bali Ngemba F.R.: Onana 2039 fl., fr., 4/2002; Pollard 933 fl., fr., 4/2002.

Antopetitia abyssinica A.Rich.

Slender herb, to 75cm; leaves imparipinnate; leaflets 3–4 pairs, narrowly oblanceolate, 1cm; umbels 2–4-flowered; fruits torulose. Grassland, disturbed ground; 2100m.
Distr.: Cameroon to E Africa [Afromontane].
IUCN: LC
Bali Ngemba F.R.: Pollard 450 fl., fr., 11/2000.

Crotalaria glauca Willd.

Erect or spreading herb, to 1.2m; leaves unifoliolate; leaflets linear to very narrowly oblanceolate, 2–4.2 × 0.2–0.7cm, glabrous or sparsely-hairy on lower surface; inflorescence a raceme; flowers yellow, 5–8mm long; standard veined brown; pod stipitate, cylindrical, inflated, up to 3cm long, glabrous. Grassland; 1300m.
Distr.: widespread in tropical Africa [Tropical Africa].
IUCN: LC
Bali Ngemba F.R.: Onana 1886 10/2001.
Note: a polymorphic widespread species. Many varieties have been described by not upheld by Polhill (*Crotalaria* of Africa, 1982).

Crotalaria incana L. subsp. ***purpurascens*** (Lam.) Milne-Redh.
F.T.E.A. Papilionoideae: 870 (1971).
Herb, 0.6–1.2m; stems coarsely-hairy; leaves obovate, 3–4cm long; flowers numerous, yellow with purple veins; fruit coarsely-hairy. Forest edge; 1950m.
Distr.: tropical Africa [Tropical Africa].
IUCN: LC
Bali Ngemba F.R.: Biye 121 11/2000.

Crotalaria ledermannii Baker f.
Lock, M. (1989). Legumes of Africa: 187.
Erect, well-branched, annual or short-lived perennial, 20–70cm; stems appressed puberulous; leaves 3-foliolate; leaflets oblanceolate, 7–20 × 1–5mm, apex rounded or truncate, apiculate; inflorescence 1.5–3cm long; flowers yellow, reddish veined, brown puberulous outside, 5.5–6.5mm long; pod ovoid, inflated. Grassland; 1900m.
Distr.: W Cameroon & N Nigeria [Lower Guinea].
IUCN: VU
Bali Ngemba F.R.: Biye 108 11/2000.
Note: *C. ledermannii* is only species of *Crotalaria* in which the large anthers (as well as the small) are spinulose.

Crotalaria ononoides Benth.
Basally woody herb, 0.3–1.2m; stem suberect; lower branches often procumbent, brownish pilose; leaves 3- or 5-foliolate, oblong-elliptic to oblanceolate, 1–6cm long; inflorescence terminal; flowers yellow, turning orange. Grassland; 1300–1400m.
Distr.: W Africa across to NE Africa and south to southern tropical Africa, and Madagascar [Tropical Africa].
IUCN: LC
Bali Ngemba F.R.: Onana 1881 10/2001; 1907a 10/2001.

Crotalaria orthoclada Welw. ex Baker
Polhill R., *Crotalaria* of Africa: 138 (1982).
Syn. *Crotalaria harmsiana* sensu Hepper
Shrub, to 3m; leaves 3-foliolate; leaflets oblanceolate, 1.5–2.5 × 0.3–0.6cm, densely-hairy on lower surface; inflorescence a raceme; flowers yellow, up to 10mm, standard often tinged red; pod subsessile, cylindrical, inflated, up to 3cm, densely-hairy. Grassland; 1850–2160m.
Distr.: Nigeria across to E Africa and south to Zambia & Angola [Tropical & subtropical Africa].
IUCN: LC
Bali Ngemba F.R.: Biye 85 11/2000; Pollard 654 fl., 10/2001.

Crotalaria recta Steud. ex A.Rich.
Shrub, to 2m; leaves 3-foliolate; leaflets variable in shape, 2–12 × 1.2–5.5cm, sparsely to densely-hairy on the lower surface; inflorescence a raceme; flowers yellow, up to 2.5cm; standard tinged purple or deep red at base; pod shortly stipitate, cylindrical, inflated, becoming slightly broader towards apex, up to 6cm, glabrous. Grassland, forest margins, wetter places at lower altitudes.
Distr.: Nigeria east to Ethiopia, south to South Africa, west to Angola [Tropical Africa].
IUCN: LC
Bali Ngemba F.R.: Ujor 30440 fl., 6/1951.

Crotalaria subcapitata De Wild. subsp. ***oreadum*** (Baker f.) Polhill
Polhill R., *Crotalaria* of Africa: 197 (1982).
Syn. *Crotalaria acervata* sensu Hepper
Annual or perennial, erect or straggling, herb, 0.5–1.3m; leaves 3-foliolate; leaflets very variable; inflorescence a raceme; peduncle shorter than rachis; flowers yellow, darkly veined, 0.5–1cm. Grassland; 1340–2200m.
Distr.: tropical Africa [Afromontane].
IUCN: LC
Bali Ngemba F.R.: Ghogue 1089 11/2000; Pollard 405 fl., fr., 11/2000.

Dalbergia oligophylla Baker ex Hutch. & Dalziel
Liana, to 30m; leaves imparipinnate; leaflets 3–7, alternate, narrowly elliptic to elliptic, 3.5–6 × 1.4–2.8cm, terminal leaflet often larger, underside sparse-moderate pubescent, somewhat appressed, secondary venation, very fine, many pairs; flowers white, 8–10mm; fruit papery, winged, narrowly elliptic-oblong; single central seed, dull. Grassland & forest edge; 2000m.
Distr.: Cameroon & Nigeria [W Cameroon Uplands].
IUCN: EN
Bali Ngemba F.R.: Cheek 10524 11/2000.

***Dalbergia* sp. of Bali Ngemba**
Liane; leaves imparipinnate; leaflets alternate, broadly-elliptic to subrotund (lower leaflet), 2.2–10 × 2.4–6.2cm, discolorous, lower surface puberulous, apex shortly acuminate, base rounded. Forest. 1420m.
Distr.: Bali Ngemba F.R.
Bali Ngemba F.R.: Pollard 417 11/2000.
Uses: none known (*Nkankano Peter* in *Pollard* 417).
Note: it is not possible to identify *Pollard* 417 to species since the specimen is sterile.

Desmodium intortum (Mill.) Urb.
Lock, M. (1989). Legumes of Africa: 246.
Erect herb, to 1m; leaves 3-foliolate; leaflets ovate, 2.5–8.3 × 1.2–3.7cm, hairy on both surfaces, sometimes with a pale band down centre of leaf on upper side; flowers pinkish, blue or purple, 8–11mm; pod compressed, strongly indented along the lower margin, up to 4.5cm long, articles with hooked hairs, readily attaching to clothes. Disturbed forest and roadsides; 1200–1340m.
Distr.: native to neotropics, introduced in Africa.
IUCN: LC
Bali Ngemba F.R.: Onana 1801 10/2001; 1802 10/2001; Pollard 406 fl., fr., 11/2000.

Desmodium repandum (Vahl) DC.
Erect herb, to 1.3m; leaves 3-foliolate; leaflets rhombic-elliptic, 4.2–9.5 × 2.8–7.5cm; flowers orange-red or red, 8–11mm; pod strongly indented along the upper margin, up to 2.5cm. Forest & forest-grassland transition; 1400–1420m.
Distr.: palaeotropical [Montane].
IUCN: LC
Bali Ngemba F.R.: Onana 1917 10/2001; Pollard 412 11/2000.
Local name: Vernacular name (not recorded here) means 'to follow you', referring to the seed pods that adhere to clothing

as you brush past the plant. **Uses:** MEDICINES (*Nkankano Peter* in *Pollard* 412).

Desmodium setigerum (E.Mey.) Benth. ex Harv.

Prostrate or scrambling herb; leaves 3-foliolate; leaflets obovate or broadly obovate, 1.5–3.7 × 1.0–2.5cm; flowers blue, pale purple or pink, 4–6mm; pod indented along both margins, more so along upper margin, up to 1.8cm. Riverine grassland & swampy forest margins; 1400–1560m.
Distr.: widespread in tropical Africa.
IUCN: LC
Bali Ngemba F.R.: Onana 1907b 10/2001; Pollard 616 10/2001.

Desmodium uncinatum (Jacq.) DC.

Lock, M. (1989). Legumes of Africa: 248.
Erect or scrambling herb, to 2m; stems with hooked hairs; leaves 3-foliolate, with stipules and stipels; leaflets ovate or elliptic, 2.2–9 × 0.7–4.8cm, pubescent below; flowers pink, turing pale purple or blue, up to 1.5cm; fruits indented along both margins, up to 3cm, articles up to 3mm wide, covered with hooked hairs rendering fruit 'sticky' - readily attaching to clothes. Naturalised roadside weed; 1900m.
Distr.: native to S America, introduced elsewhere.
IUCN: LC
Bali Ngemba F.R.: Biye 109 11/2000.

Eriosema erici-rosenii R.E.Fr.

Lock, M. (1989). Legumes of Africa: 401.
Erect woody herb, 10–55cm; stem covered with gland-based hairs; leaves 1-foliolate, ovate to almost round, 2–7 × 1–5cm, mucronate, base cordate, multiple-nerved from base, glandular above, sparsely-hairy and glandular below; inflorescence axillary and terminal, raceme to 6cm; flowers yellow; standard glandular, 5–10mm long; sepals and petals with golden, glandular hairs; pod elliptic-oblong, 7–9 × 10–13mm, with dense, golden, occasionally glandular, hairs. Grassland; 2200m.
Distr.: Cameroon east to Tanzania [Tropical Africa].
IUCN: LC
Bali Ngemba F.R.: Ghogue 1092 11/2000.

Eriosema montanum Baker f. var. *montanum*

Subwoody shrub, to 90cm; stem much-branched; leaves 3-foliolate; leaflets elliptic or oblanceolate, 4–10cm long; inflorescence a dense raceme, 6–13cm long; flowers yellow. Grassland, bush, scrub, forest edge; 1340–2100m.
Distr.: Nigeria, Cameroon, Congo (Brazzaville), Ethiopia, Uganda, Kenya, Tanzania, Malawi, Zambia, Zimbabwe, Angola [Afromontane].
IUCN: LC
Bali Ngemba F.R.: Etuge 4707 11/2000; Pollard 402 fl., fr., 11/2000; 478 fl., 11/2000; 687 10/2001.

Eriosema psoraleoides (Lam.) G.Don

Herb or subshrub, to 1.5m; stem ribbed, densely covered with short,brownish hairs, glandular; leaves pinnately 3-foliolate; leaflets oblong to elliptic, retuse or mucronulate, 2.5–9.5 × 1–3.5cm, central leaflet often larger than laterals, secondary venation parallel below and tertiary venation scalariform, densely pubescent below, discolorous; inflorescence a terminal or axillary raceme to 16cm; calyx green, densely pilose; flowers 5–15mm, yellow to orange-yellow; pod ovate to suborbicular, c. 1cm diam., densely covered with golden hairs. Grassland, savanna and forested savanna; 1340–1400m.
Distr.: Gambia to Burundi [Lower Guinea & Congolian].
IUCN: LC
Bali Ngemba F.R.: Biye 94 11/2000; Pollard 404 fl., fr., 11/2000.

Eriosema robustum Baker f.

Lock, M. (1989). Legumes of Africa: 406.
Woody herb or subshrub, to 2.5m; stem subglabrous; leaves 3-foliolate; leaflets ovate to elliptic, 1.5–6.5 × 3.5–13cm, usually acuminate, discolorous, venation prominent below, secondary venation parallel, tertiary venation scalariform, glandular, velutinous below, felty above; petiole to 3cm; inflorescence congested; many foliaceous bracts, densely pubescent; calyx white, pale green or purple, with long silvery hairs; flowers 1–2cm long; corolla yellow with red longitudinal stripes, glandular; pod oblong, 12–17 × 8–12mm. Rocky escarpments and tree-covered savanna; 1340–1400m.
Distr.: Cameroon east to Ethiopia and Kenya [Tropical Africa].
IUCN: LC
Bali Ngemba F.R.: Cheek 10438 11/2000; Pollard 400 fl., 11/2000.

Eriosema scioanum Avetta subsp. *lejeunei* (Staner & Ronse Decr.) Verdc. var. *lejeunei*

Lock, M. (1989). Legumes of Africa: 406.
Herb or subshrub, 20–75cm; leaflets 3, elliptic to round, 0.8–7 × 0.6–3.8cm; racemes axillary, 2–10cm long; flowers with standard yellowish inside, blackish outside, 7–8mm long; pods oval-oblong, 8–12 × 6–9mm. Upland grassland, rocky hillsides; 2100m.
Distr.: Cameroon, Burundi, Rwanda, Uganda, Kenya, Tanzania [Afromontane].
IUCN: LC
Bali Ngemba F.R.: Pollard 468 fl., fr., 11/2000.

Eriosema verdickii De Wild. var. *schoutedenianum* (Staner & Ronse Decr.) Verdc.

Syn. ***Eriosema schoutedenianum*** Staner & De Craene
Creeping or climbing herb, to 45cm; stem slender with white hairs; leaves 1-foliolate, ovate to cordate, 1–2.5 × 3–6cm, acuminate, slightly discolorous, upper and lower surfaces with longish white hairs; inflorescence and axillary and terminal raceme, c. 4.5–11cm long, fairly dense; flowers yellow with red striations, 5–8mm long; pod green with brown hairs, c. 7 × 4mm. Grassland and low vegetation; 2100m.
Distr.: Cameroon east to Sudan and south to Zambia [Tropical & subtropical Africa].
IUCN: LC
Bali Ngemba F.R.: Pollard 690 fl., 10/2001; 695 fl., 10/2001.

Indigofera atriceps Hook.f. subsp. *atriceps*

Syn. ***Indigofera atriceps*** Hook.f. subsp. ***alboglandulosa*** (Engl.) J.B.Gillett
Herb, to 80cm; leaves imparipinnate; leaflets 2–7 pairs plus a terminal leaflet, 8–12 × 3–5mm; inflorescence an axillary

raceme; flowers deep red, 5–7mm; pod narrowly oblong, up to 12mm, covered with glandular-tipped hairs. Grassland; 1800–1900m.
Distr.: tropical Africa [Afromontane].
IUCN: LC
Bali Ngemba F.R.: Pollard 463 11/2000; Tadjouteu 427 11/2000.

Indigofera mimosoides Baker var. *mimosoides*
Shrub or scrambling herb, to 2m; somewhat woody, sparingly glandular; leaves imparipinnate, leaflets elliptic, 6–14 × 3–8mm; flowers red, 4–7mm; pod linear, to 1.6cm. Upland grassland, streambanks, forest margins; 1300m.
Distr.: Cameroon to E & SE Africa & Angola [Tropical Africa].
IUCN: LC
Bali Ngemba F.R.: Onana 1888 10/2001.

Kotschya strigosa (Benth.) Dewit & Duvign.
Robust herb, to 2m; leaves paripinnate, 5–12 pairs; leaflets linear, slightly falcate, 3–10 × 1–2mm; flowers bright blue, with a small yellow blotch at base of standard, 6–9mm; pod resembling a 'caterpillar', up to 8mm. Grassland; 1300–1950m.
Distr.: Cameroon to Mozambique & to Angola, Indian Ocean [Palaeotropics].
IUCN: LC
Bali Ngemba F.R.: Biye 122 11/2000; Onana 1897 10/2001; 2045 fr., 4/2002.
Uses: FOOD – infructescences – fruit edible (*Onana* 2045).

Lotus discolor E.Mey.
Ascending herb; stem of wiry branches arising from woody rootstock; leaflets narrowly oblanceolate, terminal leaflet 9–11mm long; inflorescence a terminal 5–7-flowered head; flowers yellow, c. 10mm long. Fallow; 2100m.
Distr.: Cameroon east to Ethiopia and south to South Africa [Tropical & subtropical Africa].
IUCN: LC
Bali Ngemba F.R.: Pollard 466 11/2000; 688 fl., 10/2001.

Microcharis longicalyx (J.B.Gillett) Schrire
Syn. ***Indigofera longicalyx*** J.B.Gillett
Small herb, to 30cm; stem laxly pubescent; leaves elliptic, 25 × 5mm, mucronate, laxly pubescent; stipules paired, linear, persistent; inflorescence an axillary or terminal raceme to 5.5cm long; single bract at base of pedicel; calyx pubescent, deeply lobed; flowers red, blackish interior, 3mm long; pod linear, 20 × 2mm, pubescent, with raised margin. Grassland on rocky hills; 1300m.
Distr.: Guinea (Conakry) to CAR [Guineo-Congolian].
IUCN: LC
Bali Ngemba F.R.: Onana 1894 10/2001.

Neonotonia wightii (Wight & Arn.) J.A.Lackey subsp. *pseudojavanica* (Taub.) J.A.Lackey
Iselya 2(1): 11 (1981).
Scrambling or climbing herb, to 4m; leaves 3-foliolate; leaflets ovate or elliptic, 1.3–12.5 × 0.7–1cm; flowers white, drying red, 4–7mm; pod linear-oblong, up to 3cm, glabrous. Grassland & old cultivation; 1340m.
Distr.: Sierra Leone to Tanzania [Guineo-Congolian].
IUCN: LC
Bali Ngemba F.R.: Pollard 403 fl., fr., 11/2000.

Pseudarthria hookeri Wight & Arn.
Erect herb or subshrub, to 2m; leaves 3-foliolate; leaflets ovate or narrowly rhomboid to rhomboid, 5.5–13.5 × 2.5–6.5cm; flowers purple, blue, deep pink or white, sometimes pale yellow, 4–8mm; pod narrowly oblong, up to 2.5cm. Grassland, farmbush & grazed grassland; 1300–1400m.
Distr.: widespread in tropical Africa.
IUCN: LC
Bali Ngemba F.R.: Biye 93 11/2000; Kongor 54 10/2001; Onana 1912 10/2001.

Trifolium baccarinii Chiov.
Prostrate (occasionally ascending) herb, sometimes rooting at the nodes; leaves 3-foliolate; leaflets elliptic or obovate, finely toothed, 11–16 × 7–10mm; flowers purple or white, 3–4mm; pod broadly oblong, 2–3 × 1–1.5mm. Grazed grassland; 2100m.
Distr.: N Nigeria to Ethiopia & to Tanzania [Afromontane].
IUCN: LC
Bali Ngemba F.R.: Pollard 682 fl., 10/2001.

Trifolium simense Fresen.
Straggling perennial, to 45cm; leaves 3-foliolate; leaflets linear or linear-oblanceolate, finely toothed, 15–55 × 2.4mm; inflorescence spherical, up to 1.5cm diam.; calyx-nerves 17–20; flowers purple (rarely white); pod oblong, c. 4 × 2mm. Grassland; 2100m.
Distr.: Bioko & Cameroon to Zambia & E Africa [Afromontane].
IUCN: LC
Bali Ngemba F.R.: Pollard 694 fl., 10/2001.

Trifolium usambarense Taub.
Straggling herb, to 1m; leaves 3-foliolate; leaflets oblanceolate, finely toothed, 6–13 × 3–7mm; calyx-nerves 10–12; flowers purple, occasionally white, 4–6mm; pod broadly oblong, c. 3 × 2mm. Marshy places & clearings in forest; 2160m.
Distr.: Bioko & Cameroon, Congo (Brazzaville), Ethiopia to E Africa, Zambia & Rwanda [Afromontane].
IUCN: LC
Bali Ngemba F.R.: Pollard 655 fl., 10/2001.

Vigna gracilis (Guill. & Perr.) Hook.f. var. *gracilis*
Slender, twining or semi-prostrate, herb; leaves 3-foliolate; leaflets ovate, broadly elliptic or rhombic, 1–4.5 × 0.8–2.1cm; flowers pink or bluish, turning yellow; 9–16mm; pod linear, deflexed, up to 4cm. Wooded grassland, grassland & roadsides; 2100m.
Distr.: widespread in W & WC Africa [Guineo-Congolian].
IUCN: LC
Bali Ngemba F.R.: Pollard 472 11/2000.

Vigna vel. sp. aff. *multiflora* Hook.f.
Delicate climbing herb; leaves 3-foliolate; leaflets broadly ovate, 4.4–8.6 × 3.5–5.5cm, base of lateral leaflets asymmetric, hairy on both surfaces; inflorescence a raceme; flowers blue (fide *Ghogue*), not present; fruit compressed, linear, 3.2cm (immature), sparsely-hairy. Forest understorey; 1500m.

Distr.: Bali Ngemba F.R.
Bali Ngemba F.R.: Ghogue 1017 11/2000.

Vigna parkeri Baker subsp. *maranguensis* (Taub.) Verdc.

Opera Bot. 68: 133 (1983).
Syn. *Vigna maranguensis* (Taub.) Harms
Slender, twining or prostrate herb, to 2m; mature stem pale brown, pubescent; leaves pinnately 3-foliolate; leaflets ovate, mucronate-acuminate, 10–40 × 8–30mm, 3-nerved from the base, laxly covered with appressed, strigose hairs; petiole 1.5–3.5(–8.5)cm; stipules persistent, lanceolate; inflorescence an axillary raceme or panicle; peduncle to 13cm long with few flowers clustered at apex; calyx sparsely pubescent; flowers purple/blue, 7–10mm; pod flat, linear-oblong, 20 × 4mm, green with short, whitish, appressed hairs. Degraded secondary forest and savanna (grazed); 2100m.
Distr.: Nigeria east to Burundi and Rwanda, south to Uganda, Kenya and Zambia [Tropical Africa].
IUCN: LC
Bali Ngemba F.R.: Pollard 689 fl., fr., 10/2001.

Vigna vexillata (L.) A.Rich.

Climbing or trailing herb, to 5m; leaves 3-foliolate; leaflets 2.5–16.5 × 0.4–8.3cm, often ovate to lanceolate, but very variable in shape; flowers pink or purple, sometimes yellowish, occasionally with pale spot at base of standard; pod linear, cylindrical, up to 14cm. Grassland; 2200m.
Distr.: pantropical.
IUCN: LC
Bali Ngemba F.R.: Ghogue 1090 11/2000.

LENTIBULARIACEAE

M. Cheek (K)

Utricularia pubescens Sm.

Terrestrial herb; leaves peltate, orbicular, c. 4mm wide; scape usually pubescent, few-flowered; flowers c. 3mm long, white or mauve. Wet places; 1500m.
Distr.: W Africa across to E Africa and south to Angola & Zimbabwe [Tropical & subtropical Africa].
IUCN: LC
Bali Ngemba F.R.: Pollard 707 fl., 10/2001.

Utricularia scandens Benj.

Syn. *Utricularia scandens* Benj. subsp. *schweinfurthii* (Baker ex Stapf) P.Tayl.
Herb, erect or twining, to about 15cm; corolla yellow, 1cm long; spur slightly curved, slender, acute. Basalt pavement and swamps; 1300–1500m.
Distr.: W Africa to Ethiopia [Tropical Africa].
IUCN: LC
Bali Ngemba F.R.: Onana 1884 10/2001; Pollard 526 11/2000; 706 10/2001.

Utricularia striatula Sm.

Epiphytic annual herb, 1–20cm, stoloniferous; leaves petiolate, blade transversely elliptic, to 1–5mm wide; inflorescence non-twining; bracts medifixed, oblong, to 1.5mm; upper calyx obcordate, c. 2mm, lower $^{1}/_{5}$ as large; corolla 0.5–1cm, white, lower lobe flat, 5-lobed, upper inconspicuous; capsule with short,ventral slit. Rocky grassland; 1450m.
Distr.: Guinea (Conakry) to Tanzania, India to New Guinea [Palaeotropics].
IUCN: LC
Bali Ngemba F.R.: Pollard 527 11/2000.

LOGANIACEAE

Y.B. Harvey (K) & M. Cheek (K)

Fl. Cameroun 12 (1972).

Anthocleista vogelii Planch.

Tree, to 25m; sparsely branched, stem and trunk always spiny; leaves below inflorescence, petiolate, oblanceolate, c. 28 × 10cm, apex rounded, base cuneate-decurrent, nerves 10–12 pairs; flowers fragrant, flower buds blunt and rounded at apex, green-orange; outer sepals 1 × 1.2cm. Secondary forest, farmbush; 1350m.
Distr.: Sierra Leone to Uganda [Guineo-Congolian].
IUCN: LC
Bali Ngemba F.R.: Etuge 5450 fl., 4/2004.
Note: field determination.

Strychnos tricalysioides Hutch. & M.B.Moss

Fl. Cameroun 12: 119 (1972).
Climber; stems terete; leaves papery, drying dark green above, brown-green below, ovate-elliptic, 14–28 × 7–11cm, acumen 1.5cm, base rounded-obtuse, palmately 3-nerved, marginal nerves, tertiary nerves 5–10, prominent; petiole 0.5–1cm; tendrils paired; inflorescence axillary, dense, 1–1.5cm; flowers white, 6mm; fruit globose-ellipsoid, 2–4cm, orange, hard, 1-seeded; seed with deep pit on one side. Forest; 1310–1900m.
Distr.: Nigeria to Congo (Brazzaville) [Lower Guinea].
IUCN: NT
Bali Ngemba F.R.: Onana 1928 10/2001; 1996 fl., 4/2002.

LORANTHACEAE

R.M. Polhill (K) & B.J. Pollard (K)

Fl. Cameroun 23 (1982).

Globimetula oreophila (Oliv.) Tiegh.

Parasitic shrub; twigs compressed; leaves lanceolate, ovate, 8–13 × 2.5–6cm, with 6–12 pairs of well-spaced curved-ascending nerves; umbels 1–4, in axils, 8–21-flowered; peduncle 0.5–3.5cm; corolla red or red-purple, darkening apically as bud ripens, or red with a green or cream top, 2.5–3.5cm; basal swelling 5-shouldered. Forest-grassland transition; 2100m.
Distr.: SE Nigeria & Cameroon [W Cameroon Uplands].
IUCN: NT
Bali Ngemba F.R.: Pollard 660 fl., 10/2001.
Note: possibly threatened by forest loss in the Bamenda Highlands.

Phragmanthera capitata (Spreng.) Balle

Fl. Cameroun 23: 29 (1982).
Syn. ***Phragmanthera incana*** (Schum.) Balle
Syn. ***Phragmanthera lapathifolia*** (Engl. & K.Krause) Balle
Parasitic shrub, to 2m; branchlets and lower leaf surface with reddish dendritic hairs to 1mm; leaves glossy above with hairs beneath, ovate-lanceolate to ovate, elliptic or round, acuminate, 6–17 × 3–14cm, with 4–8 pairs of looped nerves; umbels 2–4-flowered; peduncle 0–3mm; corolla 4–5.5(–6)cm, yellow to orange; petals erect. Forest; 1310m.
Distr.: Guinea (Conakry) to Congo (Kinshasa), Angola [Guineo-Congolian (montane)].
IUCN: LC
Bali Ngemba F.R.: Zapfack 2027 fl., 4/2002.

Tapinanthus globiferus (A.Rich.) Tiegh.

Parasitic shrub, to ± 1m, mostly glabrous; leaves linear-lanceolate to oblong-elliptic, ovate or somewhat oblanceolate, 4–8(–17) × 0.5–4(–12)cm, with 4–6 pairs of curved-ascending nerves; umbels mostly in axils, (4–)6–8-flowered; corolla-tube (2–)2.5–3.5(–4)cm, pink to red, spotted; basal swelling conspicuous; petals reflexing at anthesis; stigma capitate. Forest; 2100m.
Distr.: Sudano-Sahelian & subtropical Africa, Arabia.
IUCN: LC
Bali Ngemba F.R.: Pollard 968 fl., 4/2002.

MALVACEAE

M. Cheek (K)

Hibiscus noldeae Baker f.

Subshrub, to 2m, spiny, very sparingly simple-hairy; leaves orbicular in outline, c. 6cm diam., palmately 5-lobed almost to base, lobes elliptic-oblong, serrate, stipules caducous; flowers axillary, subsessile; epicalyx 8–10; bracts linear, 14mm, apex bifurcate; corolla 4cm, yellow. Disturbed forest; 1400m.
Distr.: Nigeria, Cameroon, Uganda, Congo (Kinshasa), Tanzania, Angola [Afromontane].
IUCN: LC
Bali Ngemba F.R.: Biye 95 11/2000.
Local name: Kakachong (Bali) (*Biye* 95). **Uses:** MEDICINE (*Biye* 95).

Kosteletzkya adoensis (Hochst. ex A.Rich.) Mast.

Shrub, 0.3–0.9m, or straggling in trees to 6m; subscabrid; leaves ovate or shallowly 3-lobed, to 5.5 × 3cm, acute, cordate, crenate; petiole 1cm; pedicel 1cm; epicalyx of c. 8 filiform bracts, 4mm; calyx 4mm; corolla pink, centre-purple, 1.5cm; fruit strongly 5-ridged-winged. Forest edge; 1600–1900m.
Distr.: tropical Africa [Afromontane].
IUCN: LC
Bali Ngemba F.R.: Biye 106 11/2000; Onana 2028 fr., 4/2002.

Pavonia urens Cav. var. ***urens***

Subshrub, to 2m, densely persistent-pubescent on stems and leaves; leaves circular in outline, c. 15cm, more or less 5-lobed; fascicles axillary, 5–10-flowered; corolla pink, 1cm; mericarps 5; awns long-exserted with retrorse spines. Forest edge; 1710–2100m.
Distr.: Guinea (Conakry) to Madagascar [Tropical Africa & Madagascar].
IUCN: LC
Bali Ngemba F.R.: Biye 104 11/2000; Pollard 686 fl., 10/2001.

Sida **cf.** ***acuta*** Burm.f. subsp. ***carpinifolia*** (L.f.) Borss.Waalk.

Blumea 14:188 (for *S. acuta* subsp. *carpinifolia*) (1966).
Subshrub, 1.5m, moderately densely stellate puberulent on stems, petiole and nerves; leaves elliptic, c. 7.5 × 2.5cm, acute, serrate; peduncles c. 0.5cm; 3-flowered; pedicels 1cm; corolla pale orange; mericarps not seen. Farmbush; 1790m.
Distr.: West Africa (*S. acuta* subsp. *carpinifolia*).
Bali Ngemba F.R.: Cheek 10500 11/2000.
Uses: MATERIALS – Other materials – ropes – bark once used to make rope to tie young cows (*Cheek* 10500).

Sida javanensis Cav.

Blumea 14: 178–184 (1966).
Syn. ***Sida veronicifolia*** Lam.
Syn. ***Sida pilosa*** Retz.
Prostrate herb, rooting at nodes; stem puberulent; leaves ovate, 4cm, acute, cordate, crenate; petiole 3cm; pedicel 1cm; corolla orange, c. 0.7cm. Forest edge; 1500m.
Distr.: pantropical.
IUCN: LC
Bali Ngemba F.R.: Etuge 4295r 10/2001.

Urena lobata L.

Subshrub, 0.6m, stellate-hairy; leaves elliptic, c. 6cm, slightly 3-lobed or entire, base rounded, teeth glandular, densely grey-hairy below; petiole 2cm; flowers subsessile; epicalyx 5-lobed in upper half, 7mm; corolla pink, centre-purple, 1cm; mericarps 5, spines with grapnel ends. Farmbush; 1560m.
Distr.: pantropical.
IUCN: LC
Bali Ngemba F.R.: Pollard 599 fl., 10/2001.

MELASTOMATACEAE

M. Cheek (K) & E.M. Woodgyer (K)

Fl. Cameroun 24 (1983).

Amphiblemma mildbraedii Gilg ex Engl.

Erect, robust, terestrial herb, 1–2m; stems square, to 1cm diam.; leaves ovate, to 29 × 18.5cm, subacuminate, cordate, 7-nerved, subscabrid above; petiole to 10cm; stipules persistent, c. 1 × 1cm; panicle terminal, c. 18 × 15cm; peduncle 7cm; flowers 3cm, 5-petaled, purple; fruit dry, 5-angled, 0.8cm. Forest edge; 2100m.
Distr.: SE Nigeria, Bioko, Cameroon [W Cameroon Uplands].
IUCN: NT
Bali Ngemba F.R.: Pollard 658 10/2001.

Antherotoma naudinii Hook.f.

Erect, annual, herb, c. 15cm tall; stems hirsute; leaves ovate to oblong-lanceolate; blade c. 2 × 1cm; inflorescence capitulate; flowers 4-merous; petals pink; anthers truncate;

fruit a capsule. Short grassland, usually in damp places; 2100m.
Distr.: Guinea (Conakry) to Cameroon, Ethiopia to Angola & South Africa; Madagascar [Afromontane].
IUCN: LC
Bali Ngemba F.R.: Pollard 664 10/2001.

Dissotis bamendae Brenan & Keay
Syn. ***Dissotis princeps*** (Kunth) Triana var. ***princeps***
Erect herb, 0.5m, setose; leaves 3 per node, ovate, c. 9 × 4.5cm, acute, base rounded, felted below; petiole 1.5cm; flowers c. 20; peduncle 10cm; hypanthium 10 × 7mm, densely white-simple-hairy, emergences nil; sepals caducous; stamens dimorphic; appendages 2, globose. Grassland; 1950–2200m.
Distr.: Cameroon [Uplands of Western Cameroon].
IUCN: VU
Bali Ngemba F.R.: Biye 118 11/2000; Ghogue 1093 11/2000; 1094 11/2000.
Note: sunk into *D. princeps* (South Africa to Ethiopia) by Jacques-Félix but obviously distinct in e.g. 3 leaves per node.

Dissotis brazzae Cogn.
Erect herb, c. 1.5m tall; stems 4-winged; leaves c. 7.5 × 3cm, 9-nerved, ovate, apex acuminate; petiole c. 4mm long; inflorescence a terminal panicle; flowers 5-merous; petals pink to violet; fruit a capsule. Grassland, roadsides; 1300–1500m.
Distr.: Guinea (Conakry) to Ethiopia & to Zambia [Tropical Africa].
IUCN: LC
Bali Ngemba F.R.: Biye 76 11/2000; Ghogue 1062 11/2000; Kongor 53 10/2001; Onana 1903 10/2001.

Dissotis irvingiana Hook.
Annual, erect herb, 60cm; stems red, 4-angled, with patent white hairs, 3mm; leaves lanceolate, c. 4 × 1.1cm, acute, obtuse; petiole 1mm; capitula loose; flowers 2–4, pink, 0.8cm; hypanthium c. 5 × 4mm, upper part with several short emergences; sepals 5, not persistent; anthers homomorphic; Open forest; 1300m.
Distr.: Nigeria to Zambia [Tropical Africa].
IUCN: LC
Bali Ngemba F.R.: Onana 1893 10/2001.
Note: this taxon is treated as a synonym of *D. senegambensis* at K.

Dissotis longisetosa Gilg & Ledermann ex Engl.
Ascending herb, 0.5m, setose; leaves opposite, ovate, c. 9 × 4.5cm, acute, truncate-rounded, upper surface with hairs sigmoid, lower part dilate, adnate to lamina; petiole 1cm; peduncle c. 20cm, robust, with 1–2 pairs reduced leaves, 1–2cm; capitula 2–3-flowered; hypanthia c. 2 × 1.5cm, emergences very short and densely-hairy. Grassland; 1950–2100m.
Distr.: Cameroon [Uplands of Western Cameroon].
IUCN: VU
Bali Ngemba F.R.: Biye 117 11/2000; Pollard 662 fl., 10/2001.

Guyonia ciliata Hook.f.
Prostrate herb, sparsely ciliate on stems, leaves and fruit; leaves ovate, 1.5cm, finely serrate; petiole to 2cm; flowers single, pink, 1cm diam.; stamens equal; fruit 5mm, with leafy calyx. Forest edge; 1400m.
Distr.: Sierra Leone to Tanzania [Guineo-Congolian (montane)].
IUCN: LC
Bali Ngemba F.R.: Cheek 10442 11/2000.

Memecylon dasyanthum Gilg & Ledermann ex Engl.
Tree, 8m; stems terete; leaves coriaceous, elliptic, 5–8 × 3–3.5cm, acumen acute, 0.5cm, acute-decurrent, lateral nerves c. 6 pairs, inconspicuous; petiole 3mm; compound umbels axillary, 2.5cm, c. 20-flowered; peduncle 1.2cm; hypanthium 0.5 × 1mm, entire; petals not seen; fruit oblate, 3mm. Forest; 1310–2000m.
Distr.: Cameroon [W Cameroon Uplands].
IUCN: VU
Bali Ngemba F.R.: Etuge 4266 10/2001; 4711 fr., 11/2000; Onana 1827 10/2001; 1989 fr., 4/2002; Ujor 30371 fr., 2/1951; Zapfack 2029 fr., 4/2002; 2060 fl., 4/2002.
Local Name: Oken. **Uses:** MATERIALS – Personal items – used as a walking stick (*Etuge* 4711).

Warneckea cinnamomoides (G.Don) Jacq.-Fél.
Fl. Cameroun 24: 164 (1983).
Syn. ***Memecylon cinnamomoides*** G.Don
Small tree, 5m; stems 4-angled; leaves papery, elliptic, to 11 × 4cm, acumen 1–2cm, acute, 3-nerved with c. 15 transverse nerves; petiole 3mm; pedicel 8mm; fruits on old wood, blue, ellipsoid, c. 6 × 4.5mm. Forest; 1700m.
Distr.: Guinea (Conakry) to Congo (Kinshasa) [Guineo-Congolian].
IUCN: LC
Bali Ngemba F.R.: Cheek 10517 11/2000; 10529 11/2000.

MELIACEAE

M. Cheek (K)

Carapa grandiflora Sprague
Tree, 6–20m, glabrous; leaves to 1.2m, paripinnate, 4–7-jugate; petiole c. 15cm; leaflets oblong to oblong-obovate, c. 18 × 7cm, rounded, acute; petiolules c. 1cm; inflorescence a terminal panicle, c. 30cm; flowers white, c. 8mm; sepals and petals greenish; staminal tube white; disk orange; stigma white; fruit 5-valved, subglobose, c. 10cm, warty; seeds c. 3cm. Forest; 1400–1700m.
Distr.: Nigeria to Uganda [Lower Guinea & Congolian (montane)].
IUCN: NT
Bali Ngemba F.R.: Etuge 4806 11/2000; Zapfack 2062 fl., fr., 4/2002.
Uses: MATERIALS – wood – timber (*Etuge* 4806).

Entandrophragma angolense (Welw.) C.DC.
Forest tree, to 55m; 3m girth; bole long and straight; crown open; buttresses blunt, broad, low; bark smooth, pale grey-brown to orange-brown with papery scales, scales flaking high up tree; slash dark red and pink; leaves paripinnate, clustered at ends of branches, 7–10-jugate; rachis 25–45cm; petioles 12–16cm, not winged, glabrous or puberulous; leaflets oblong-elliptic, 7–28 × 3–10.5cm, rounded and often

mucronate, base rounded to obtuse; lateral nerves pubescent, 9–12 pairs; flowers yellowish; fruit 14–22 × 3.5–5cm; valves 2.5–3cm wide, 2.5–4mm thick; seeds winged. Forest and farmbush; 1750m.
Distr.: Guinea (Conakry) to Uganda & Sudan [Guineo-Congolian].
IUCN: VU
Bali Ngemba F.R.: Etuge 5369 fr., 4/2004.
Uses: MATERIALS – wood – timber (*Etuge* 5369).
Note: field determination.

Trichilia dregeana Sond.
Meded. Land. Wag. 68(2): 28 (1968).
Tree, 6–18m, puberulent; leaves imparipinnate, 30–60cm, 5-jugate; petioles 12cm; leaflets drying dark brown, oblanceolate, 20 × 6cm, acumen 1cm, acute to obtuse, lateral nerves 16 pairs; petiolule 5mm; panicles 30cm, axillary; flowers white, 1cm. Forest; 1400–1500m.
Distr.: Guinea to South Africa [Afromontane].
IUCN: LC
Bali Ngemba F.R.: Cheek 10472 11/2000; 10730 10/2001; 10733 10/2001; Etuge 4812 11/2000.

Trichilia gilgiana Harms
Tree, 12m, 1m girth, slash red; densely puberulous; leaves imparipinnate, c. 5-jugate, rachis 14cm; petiole 10cm; leaflets with dots and dashes in transmitted light, oblanceolate to oblong, c. 13 × 3.5cm, acuminate, acute, lateral nerves, c. 12 pairs; fruits globose, 2cm, yellow-brown puberulent.
Distr.: Nigeria to Congo (Kinshasa) [Lower Guinea & Congolian].
IUCN: LC
Bali Ngemba F.R.: Tiku 30421 fl. 6/1951.

Turraea vogelii Hook.f. ex Benth.
Woody climber, to 12m, rarely a shrub to 2m, glabrous; leaves simple, elliptic, c. 11 × 5cm, subacuminate, obtuse; petiole 0.5cm; inflorescence subumbellate, axillary; peduncle 3.5cm; pedicels 1cm; flowers white, 3–10; staminal tube 1.5cm; fruit globose, 2.5cm. Forest; 1310m.
Distr.: Ghana to Uganda [Guineo-Congolian].
IUCN: LC
Bali Ngemba F.R.: Onana 1980a fl., 4/2002.

Turraeanthus africanus (Welw. ex C.DC.) Pellegr.
Tree, c. 20–35m; 60cm dbh; slash yellow; leaves imparipinnate, c. 12-jugate, rachis 4–60cm; petioles 5–10cm, terete, densely chocolate-hairy when young; leaflets oblong, c. 16 × 4.5cm, rounded submucronate, base acute, lateral nerves c. 20 pairs; infructescence c. 19cm, stout; partial peduncles c. 1cm, 1–5-fruited; fruits leathery, globose, c. 5cm, dehiscent; seeds 4, orange-segment-shaped, coat yellow, pulpy. Forest; 1310–1700m.
Distr.: Sierra Leone to Uganda [Guineo-Congolian].
IUCN: LC
Bali Ngemba F.R.: Cheek 10451 11/2000; 10732 10/2001; Etuge 4697 11/2000; Zapfack 2028 fl., 4/2002; 2061 fr., 4/2002.
Uses: MATERIALS – wood – timber (*Etuge* 4697).

MELIANTHACEAE

Y.B. Harvey (K)

Bersama abyssinica Fresen. sens. lat.
Syn. ***Bersama maxima*** Baker
Syn. ***Bersama acutidens*** Welw. ex Hiern
Tree or shrub, 2–8m, glabrous; leaves alternate, c. 30cm, variable, imparipinnate; leaflets 5–6 pairs, densely pubescent or glabrous, glossy, oblong-elliptic, c. 15 × 5.5cm, apex acute, base obliquely obtuse, lateral nerves c. 10 pairs, sometimes serrate in upper half; petiolule 0.5cm, rachis ± winged; petiole c. 12cm; stipules c. 1cm, intrapetiolar; inflorescence a terminal raceme to c. 40cm; rachis c. 7cm; pedicels 1cm; flowers white, 1cm; fruit magenta-red, dehiscent, ovoid, 2cm; seeds 1cm, arillate. Forest; 1730–2100m.
Distr.: tropical Africa [Afromontane].
IUCN: LC
Bali Ngemba F.R.: Pollard 673 fr., 10/2001; Tiku 30413 fl., 6/1951.
Note: at least three varieties might be recognised from our material.

MENISPERMACEAE

B.J. Pollard (K)

Stephania abyssinica (Quart.-Dill. & A.Rich.) Walp. var. *abyssinica*
Slender, glabrous, liana, to 10m; leaves ovate to orbicular-ovate, 5–10 × 4–13cm, entire, dark green above, glaucous beneath; petiole 4–12cm; inflorescences 4–7cm diam., to 40cm long; pseudo-umbel on a single peduncle to 10cm; rachis fleshy, red; flowers green or purple. Forest; 1310–1500m.
Distr.: tropical Africa [Afromontane].
IUCN: LC
Bali Ngemba F.R.: Cheek 10504 11/2000; Ghogue 1050 11/2000; Zapfack 2005 fr., 4/2002.

Stephania abyssinica (Quart.-Dill. & A.Rich.) Walp. var. *tomentella* (Oliv.) Diels
As for var. *abyssinica*, but densely pubescent, with fine, brown indumentum. Forest.
Distr.: widespread in central and East Africa [Afromontane].
IUCN: LC
Bali Ngemba F.R.: Ujor 30322 fl., 5/1951.

Tinospora bakis (A.Rich.) Miers
Slender, glabrous, twiner; older branches lenticellate; leaves broadly ovate, widely cordate, shortly acuminate, 5–8 × 5–8cm, membranaceous, digitately 7-nerved at base; petiole 3–8cm, slender, twisting to assist climbing habit; ♂ inflorescences axillary, racemose, slender, to 9cm; flowers c. 5mm. Forest, farmbush; 1500m.
Distr.: tropical Africa in dry areas [Tropical Africa].
IUCN: LC
Bali Ngemba F.R.: Ghogue 1023 11/2000.
Note: this sterile specimen (Ghogue 1023) closely resembles others referred to this taxon at K, but differs slightly, in: leaf apex long-acuminate (not shortly acuminate, occasionally

mucronulate), leaf margin bearing 1–2 minute denticulations on either side (not truly entire). With collection of fertile material this determination can be confirmed or rejected.

MONIMIACEAE

Y.B. Harvey (K)

Fl. Cameroun 18 (1974).

Xymalos monospora (Harv.) Baill. ex Warb.
Shrub or small tree, 3–8(–25)m; leaves opposite, leathery, elliptic, c. 10 × 4cm, acute, serrate; inflorescences c. 4cm, below leaves; fruit elliptic, 1cm with apical knob. Forest & forest-grassland transition; 1400–1500m.
Distr.: SE Nigeria to E & South Africa [Afromontane].
IUCN: LC
Bali Ngemba F.R.: Etuge 4272r 10/2001; Ghogue 1016 11/2000; Pollard 408 fr., 11/2000; Ujor 30354 5/1951.

MORACEAE

R. Becker (RNG) & N. Rønsted (K)

Fl. Cameroun 28 (1985).

Dorstenia barteri Bureau var. ***barteri***
Fl. Cameroun 28: 66 (1985).
Herb to 90cm; stem ascending, sometimes branched; leaves in spirals or subdistichous; lamina variable, lanceolate to elliptic, broadest above, narrowed below, (3–)6–22 × (1.5–)3–8.5cm, usually entire, but sometimes with 1–2(–5) blunt teeth in the upper third; receptacle with c. 5 or 6 primary appendages, to 3cm, secondary to 0.7cm; fringe 1–5(–10)mm. Forest.
Distr.: SE Nigeria, Bioko, W Cameroon & Congo (Kinshasa) [Lower Guinea & Congolian].
IUCN: LC
Bali Ngemba F.R.: Ujor 30374 fl., 5/1951.

Dorstenia barteri Bureau var. ***subtriangularis*** (Engl.) Hijman & C.C.Berg
Fl. Cameroun 28: 70 (1985).
Syn. ***Dorstenia subtriangularis*** Engl.
Herb; leaves lanceolate to elliptic, 4.5–11 × 4.5–6.3cm, with 6–10 pairs of secondary veins; stipules persistent; inflorescences pedunculate, 1.5–7cm long; receptacle submultitriangular with large margin, 3–8mm. Forest understorey and stream edges; 1400m.
Distr.: SE Nigeria and Cameroon [Lower Guinea].
IUCN: LC
Bali Ngemba F.R.: Etuge 4794 11/2000.

Ficus ardisioides Warb. subsp. ***camptoneura*** (Mildbr.) C.C.Berg
Fl. Cameroun 28: 238 (1985); Kew Bull. 43: 77 (1988).
Syn. ***Ficus camptoneura*** Mildbr.
Epiphytic shrub, to 6m, occasionally a tree to 12m; epidermis flaking, glabrous; leaves obovate, elliptic, or elliptic-oblong, 9–22 × 4–9cm, acuminate, obtuse, lateral nerves c. 4 pairs, basal-most arising 0.5cm or more above leaf-base, quaternary nerves conspicuous; stipules subpersistent, 9mm; figs axillary, sessile, globose 0.7–1cm diam., often verrucose, subrostrate, basal bracts 2, 4mm. Forest; 1790m.
Distr.: Ivory Coast, Nigeria to E Congo (Kinshasa) & N Zambia [Guineo-Congolian].
IUCN: LC
Bali Ngemba F.R.: Cheek 10499 11/2000; Rønsted 222 fr., 4/2004.

Ficus cyathistipula Warb. subsp. ***pringsheimiana*** (Braun & K.Schum.) C.C.Berg
Fl. Cameroun 28: 240 (1985); Kew Bull. 43: 83 (1988).
Syn. ***Ficus pringsheimiana*** Braun & K.Schum.
Epiphyte; leaves spirally arranged, obovate-oblong to elliptic, 7–16(–27) × 3.5–6.5(–8)cm, glabrous; figs sessile to subsessile, solitary, 1–2cm diam. in dry state. Forest, edges of rivers and seasonally flooded areas; 1500m.
Distr.: Sierra Leone to Congo (Kinshasa) [Guineo-Congolian].
IUCN: LC
Bali Ngemba F.R.: Ghogue 1057 11/2000.

Ficus exasperata Vahl
Fl. Cameroun 28: 121 (1985).
Tree, c. 15m tall, beginning as a strangler; leaves elliptic (lobed and longer when juvenile), 10 × 6.5cm, obtuse at base and apex, obscurely crenate-dentate, scabrid above and below, lower 2 nerve pairs with waxy axils; petiole c. 2.5cm; stipules not amplexicaul; figs c. 1.5cm, on leafy branches, single; bracts 3, scattered on pedicel; ostiole circular. Farmbush and secondary, deciduous forest; 1700m.
Distr.: tropical Africa to S India & Sri Lanka [Palaeotropics].
IUCN: LC
Bali Ngemba F.R.: Rønsted 226 4/2004; Zapfack 2045 fr., 4/2002.
Note: distinctive in the combination of non-amplexicaul stipules, leaf indumentum and fig position.
Uses: MATERIALS – Abrasives - leaves important as pot scourers (*Zapfack* 2045).

Ficus glumosa Delile
Fl. Cameroun 28: 170 (1985).
Tree, pubescent; leaves oblong-elliptic to ovate, 6–14.5 × 3.5–11.5cm, leathery, raised venation on the lower surface; petiole 1.5–5cm; figs sessile or short-pedunculate, mostly solitary, 0.5–1cm diam. in dry state. Gallery forest and savanna; 1450m.
Distr.: Senegal to Ethiopia, south to Namibia and South Africa and also Yemen [Tropical Africa & Arabian Peninsula].
IUCN: LC
Bali Ngemba F.R.: Pollard 948 fr., 4/2002.

Ficus ingens (Miq.) Miq.
Fl. Cameroun 28: 149 (1985).
Tree, to 18m; leaves ovate-elliptic, 5–14 × 2.4–7cm, apex acuminate to acute, glabrous, margin entire; figs in pairs, subsessile, 0.5–1cm diam., wrinkled in dry state. Savanna and wooded grassland; 1400m.
Distr.: Senegal to Cameroon, Somalia & Ethiopia to South Africa, Yemen [Africa and Arabian Peninsula].
IUCN: LC
Bali Ngemba F.R.: Cheek 10446 11/2000.

Ficus oreodryadum Mildbr.

Fl. Cameroun 28: 200 (1985).
Tree, to 30m, or epiphytic shrub, glabrous; leaves coriaceous, drying dark brown below, oblanceolate, c. 20 × 5cm, obtuse, subacuminate, base acute; secondary nerves c. 13 pairs; petiole 5.5cm, exfoliating, stout; stipules caducous; figs axillary, sparse, subglobular, 1.3cm, slightly warty, glabrous, sessile, stoutly beaked; basal bracts 3, 2mm. Forest; 2050m.
Distr.: SW Cameroon to Burundi, Rwanda & Uganda [Lower Guinea & Congolian (montane)].
IUCN: LC
Bali Ngemba F.R.: Rønsted 227 4/2004; 228 fr., 4/2004.

Ficus ovata Vahl

Epiphytic shrub, glabrous; leaves thickly coriaceous, ovate, to 28 × 17cm, acuminate, obtuse, entire; lateral nerves c. 10 pairs; petiole c. 6cm; stipules caducous; figs axillary, single, sessile, elliptic, 3.8 × 2.5cm, rarely globular, dark yellow-puberulent; basal bracts forming a sinuate-margined puberulent cup, 1cm diam. Farmbush; 1380–1400m.
Distr.: Senegal to Mozambique [Tropical Africa].
IUCN: LC
Bali Ngemba F.R.: Rønsted 225 4/2004; 229 4/2004.
Note: *Rønsted* 229 has been tentatively placed in this taxon, it is a free standing tree to 5m with leaves in whorls.

Ficus sansibarica Warb. subsp. *macrosperma* (Mildbr. & Burret) C.C.Berg

Fl. Cameroun 28: 220 (1985); Kew Bull. 43: 94 (1988); Kirkia 13(2): 276 (1990).
Syn. ***Ficus macrosperma*** Mildbr. & Burret
Tree, up to 10m; leaves oblong to lanceolate, 4.5–11 × 2.5–3.6cm, glabrous; fig-bearing spurs up to 10(–15)cm long. Rainforest, edges of lakes and rivers.
Distr.: Sierra Leone to Angola, Uganda and into N Zambia [Tropical & subtropical Africa].
IUCN: LC
Bali Ngemba F.R.: Ujor 30345 fr., 5/1951.

Ficus sur Forssk.

Fl. Cameroun 28: 135 (1985).
Syn. ***Ficus capensis*** Thunb.
Tree, 5–20(–30)m, but fruiting at only 5m; stem to 60cm dbh; leaves elliptic-oblong, c. 14 × 7cm, shortly acuminate, rounded to obtuse, margin with c. 5 well-marked serrations on each side, subglabrous; petiole c. 4cm; stipules caducous; figs on branches c. 15cm long on main branches or trunk apex, (lowest c. 6m from ground), c. 2cm diam.; peduncle 7mm; bracts 3, c. 1.5mm, whorled. Farmbush and secondary forest; 1000–1500m.
Distr.: Senegal to S Africa [Tropical Africa].
IUCN: LC
Bali Ngemba F.R.: Ghogue 1031 11/2000; Onana 1923 10/2001; Ujor 30433 fr., 7/1951.

Ficus thonningii Blume

Fl. Cameroun 28: 175 (1985).
Syn. ***Ficus dekdekena*** (Miq.) A.Rich.
Syn. ***Ficus iteophylla*** Miq.
Epiphytic shrub, glabrous; leaves elliptic, elliptic-oblong or oblanceolate-elliptic, 7.5–13 × 3.5–5.5cm, shortly-acuminate, obtuse, entire; lateral nerves c. 12 pairs, fine, quaternary nerves conspicuous; petiole 2.3cm; stipules caducous; figs axillary, amongst and below the leaves, dense, sessile, globose, c. 0.8cm; basal bracts 2, fused to form a bilobed, brown-puberulent, saucer-shaped structure, 0.7cm diam. Forest; 1400m.
Distr.: tropical and S Africa.
IUCN: LC
Bali Ngemba F.R.: Rønsted 223 fr., 4/2004; .224 4/2004.

Ficus vallis-choudae Delile

Fl. Cameroun 28: 142 (1985).
Tree, 7–20m, densely-puberulent; bole short; crown spreading; leaves broadly ovate, sometimes slightly 3-lobed, c. 20 × 16cm, obtuse, gradually acuminate, base rounded, margin slightly serrate; secondary nerves 5 pairs, the basal-most pair ascending in the upper half, smooth above, almost scabrid below; petiole 5–9cm; stipules 4cm, lanceolate, subpersistent; figs axillary, globose, 1.5cm, solitary; peduncle 0.8cm; bracts triangular, 3mm. Riverine forest, lakesides, ground-water forest; 1400m.
Distr.: Guinea (Conakry) to Ethiopia, E Africa, Mozambique, Zimbabwe [Tropical Africa].
IUCN: LC
Bali Ngemba F.R.: Darbyshire 418 fr., 4/2004.

Trilepisium madagascariense DC.

Fl. Cameroun 28: 103 (1985); Bull. Jard. Bot. Natl. Belg. 47: 299 (1977).
Syn. ***Bosqueia angolensis*** Ficalho
Tree, 20m, glabrous; leaves elliptic, c. 10 × 4.5cm, acuminate, obtuse, entire; lateral nerves c. 5 pairs, basal pair acute; petiole 1cm; stipules caducous; inflorescence axillary at leafless nodes, ellipsoid, 1.5 × 0.8cm, glabrous; peduncle 1.5cm; flowers emerging from apical aperture in inflorescence, in a cluster c. 0.5 × 0.5cm. Forest, farmland; 1390m.
Distr.: Guinea (Conakry) to Congo (Kinshasa) [Guineo-Congolian]
IUCN: LC
Bali Ngemba F.R.: Rønsted 230 4/2004.

MYRSINACEAE

Y.B. Harvey (K)

Ardisia staudtii Gilg

Bull. Jard. Bot. Natl. Belg. 49: 112 (1979).
Syn. ***Afrardisia staudtii*** (Gilg) Mez
Syn. ***Afrardisia cymosa*** (Baker) Mez
Shrub, (0.5–)1.5–4(–5)m tall, glabrous; leaves elliptic to ovate, 90–180 × 30–70mm, glandular dots present on lower surface, very shallowly crenate; petioles 7–15mm; flowers in axillary fascicles; peduncles 2–5mm; 6–12 flowers per fascicle; pedicel 6–10mm; calyx c. 2.5mm wide, fimbriate margin; flowers white or pink, to 4mm with glandular spots/streaks; fruits globose, 3–6.5mm, red with red gland-dots. Forest; 1400–1800m.
Distr.: Nigeria to Congo (Kinshasa) & CAR [Guineo-Congolian (montane)].
IUCN: LC

Bali Ngemba F.R.: Biye 135 11/2000; Ghogue 1042 11/2000; Onana 1832 10/2001; Pollard 921 fl., 4/2002; Tadjouteu 405 11/2000; Zapfack 2048 fl., 4/2002.

Maesa lanceolata Forssk.

Tree or shrub, 6–8m tall; stem glabrous, dark brown with paler lenticels; leaves elliptic, 9–16 × 3.5–6cm, serrulate, glabrous; petioles 2–2.5cm, glabrous; inflorescence many-branched, 5–7mm, profusely covered with minute hairs (<0.5mm); flowers pale green, to 1.5mm, subsessile; fruits globose, 4–5mm; pedicel to 3.5mm. Forest-grassland transition, forest; 1400–1450m.
Distr.: Guinea (Conakry) to Madagascar [Afromontane].
IUCN: LC
Bali Ngemba F.R.: Cheek 10475 11/2000; Tadjouteu 408 11/2000.

MYRTACEAE

M. Cheek (K)

***Eucalyptus* sp.**

Tree, to 12m, bole pink-brown; leaves narrowly lanceolate, slightly falcate, c. 12 × 2.5cm, apex long-attenuate, base unequally obtuse, midrib bright white below, lateral nerves infinite, inconspicuous; petiole 1.5cm; inflorescences 5–7cm; partial-peduncles flattened, 7mm; 3-flowered; flowers white, 5mm diam.; calyx 6 × 3mm, including slightly narrower stipe, 1.5mm. Cultivated; 1550–1900m.
Distr.: cosmopolitan.
IUCN: LC
Bali Ngemba F.R.: Etuge 5357 4/2004.
Note: a great many species of *Eucalyptus* have been introduced to Africa and elsewhere from E Malesia (few) and Australia.

Eugenia gilgii Engl. & Brehmer

Shrub or small tree, to 8m; leaves elliptic, 6–10cm long, margin revolute basally; inflorescence a very short raceme, central bracteate axis, to 4mm long; flowers pink, borne below leafy parts of shoots. Forest; 1700m.
Distr.: Cameroon [Cameroon Endemic].
IUCN: CR
Bali Ngemba F.R.: Cheek 10530 11/2000.

Syzygium staudtii (Engl.) Mildbr.

Tree, 8–20m; bole white, usually with 2–3-numerous laterally-flattened root buttresses, arising up to 60cm above ground; stems near apex red when young, 4-ridged, glabrous; leaves (fruiting stems) elliptic, 6–7 × 2–3.5cm, acute, secondary nerves numerous; petiole c. 1.2cm; juvenile leaves to 11 × 5cm, sometimes briefly-acuminate; inflorescence terminal, 10cm, 10–30-flowered; flowers white, 0.6cm; fruit subumbellate, obovoid, 1cm. Forest; 1700–2100m.
Distr.: Liberia to Cameroon [Upper & Lower Guinea (montane)].
IUCN: NT
Bali Ngemba F.R.: Cheek 10514 11/2000; Pollard 970 fl., 4/2002.
Note: unjustifiably reduced to a subspecies of *S. guineense* by White. Likely to rate as VU when taxon better delimited.

OCHNACEAE

I. Darbyshire (K)

Ochna membranacea Oliv.

Shrub or small tree; bark red-brown; paired stipules linear, to 1.1cm, fimbriate towards base, straw-coloured, caducous; leaves elliptic-obovate, c. 10.5 × 3cm, base cuneate, apex acuminate, margins serrulate; midrib and secondary nerves prominent yellow-brown below in mature leaves; petiole 1–3mm, red-brown; fruit in axillary clusters. Forest; 1550m.
Distr.: Gambia to Congo (Kinshasa) [Guineo-Congolian].
IUCN: LC
Bali Ngemba F.R.: Darbyshire 400 fr. 4/2004.
Note: field determination.

OLACACEAE

M. Cheek (K)

Fl. Cameroun 15 (1973).

Strombosia scheffleri Engl.

Tree, to 33m; branchlets strongly angled; leaves ovate-elliptic or oblong, 6–20 × 3–13cm, 5–8 pairs of main lateral nerves, venation distinct; petioles 1–3cm long; flowers greenish yellow or white; petals 3–5mm; fruits obconical, c. 2cm. Forest; 1310–1900m.
Distr.: SE Nigeria to Uganda & E Africa [Tropical Africa].
IUCN: LC
Bali Ngemba F.R.: Cheek 10737 10/2001; Onana 1990 fr., 4/2002; Pollard 1039 fr., 4/2002; Zapfack 2019 fr., 4/2002.

***Strombosia* sp. 1**

Tree, 6–20m tall; stems and leaves resembling *S. scheffleri*, but leaves usually cordate; pedicels 2mm; flower buds black, 4mm; petals white, 5mm, inner surface sparsely and very shortly puberulent; fruit 1cm, resembling *Diogoa*: turbinate, with a calyx-derived wing around the equator. Forest; 1400–1430m.
Distr.: Nigeria (Chappal Waddi) & Cameroon (Kupe-Bakossi, Bali Ngemba F.R., Bamboutos Mts) [W Cameroon Uplands].
Bali Ngemba F.R.: Etuge 4703 11/2000.
Note: probably a new and threatened species. More investigation needed. The fruits are quite unlike any other in the genus.

OLEACEAE

P.S. Green (K) & I. Darbyshire (K)

Chionanthus africanus (Knobl.) Stearn

Bot. J. Linn. Soc. 80: 197 (1980).
Syn. *Linociera africana* (Knobl.) Knobl.
Tree, 6–12m; bark pale brown, lenticels concolorous; leaves oblong-oblanceolate, 19.5–22.7 × 7–9.7cm, acumen abrupt, 0.6–1.4cm, base acute, glabrous, midrib raised and bronze below, lateral nerves 11–12 pairs, domatia minute, tufted; petiole 1.2–2.2cm, swollen, flaking; panicles axillary, c. 5 × 1.5cm, appressed-pubescent; calyx lobes ovate, 1mm; corolla lobes subulate, 0.5cm, free almost to base, white; stamens 2;

filaments short; fruit ellipsoid, 1.9 × 1.4cm, grey-brown. Forest & forest edge.
Distr.: Sierra Leone to Tanzania [Guineo-Congolian].
IUCN: LC
Bali Ngemba F.R.: Ujor 30335 fr., 5/1951.

Chionanthus mildbraedii (Gilg & Schellenb.) Stearn

Bot. J. Linn. Soc. 80: 202 (1980).
Shrub or tree, to 8m; bark pale; leaves oblong-oblanceolate, 11–14.2 × 3.9–5.5cm, acumen abrupt 1–2.2cm, base obtuse, lateral nerves 6–7 pairs, inconspicuous, domatia minute, tufted in young leaves; petiole to 0.7cm, swollen and slightly flaking; inflorescence axillary, 3–6-flowered, lax, rachis to 6cm, thin, glabrous; pedicels 1.2cm, fine, single on opposite secondary peduncles, 0–1.5cm; calyx cupular with spreading triangular lobes, glabrous; corolla lobes lanceolate, 0.45 × 0.25cm, free to near base, yellow to red-green; fruit not seen. Forest & forest edge; 1200–1600m.
Distr.: Cameroon Congo (Kinshasa), Ethiopia to Tanzania [Tropical Africa].
IUCN: LC
Bali Ngemba F.R.: Onana 1805 10/2001; 2031 fr., 4/2002; Pollard 509 11/2000.

Jasminum pauciflorum Benth.

Liana, to 6m; rather sparsely branched, subglabrous; leaves with tufts of hairs in the axils of the main nerves beneath, 3–9 × 1.5–5cm, ovate; petiole 2–10mm; few-flowered cymes; peduncles to 20mm; pedicels 20–30mm long; calyx pilose, lobes to 7mm; corolla white; tube to 27mm long; lobes 6–8, to 20mm long. Forest; 1350m.
Distr.: tropical Africa.
IUCN: LC
Bali Ngemba F.R.: Etuge 5447 fl., 4/2004.

ONAGRACEAE

I. Darbyshire (K) & M. Cheek (K)

Fl. Cameroun 5 (1966).

Ludwigia africana (Brenan) Hara

J. Jap. Bot. 28: 291 (1953).
Syn. ***Jussiaea africana*** Brenan
Herb, 0.9m, erect; stems, leaves and inflorescence with thin, spreading, white hairs; leaves elliptic, to c. 6 × 3.5cm, acute; petiole c. 1.5cm; inflorescence with c. 5-flowered peduncles 1.5cm; ovary 10mm; sepals 5mm; fruit not seen. Riverbanks; 1550m.
Distr.: Guinea (Conakry) to Congo (Kinshasa) [Guineo-Congolian].
IUCN: LC
Bali Ngemba F.R.: Darbyshire 402 fl., 4/2004.
Note: field determination.

OXALIDACEAE

I. Darbyshire (K)

Biophytum umbraculum Welw.

Brittonia 33: 451 (1981).
Syn. ***Biophytum petersianum*** Klotzsch
Annual herb, to c. 15cm, glabrous; leaves in terminal rosette, 1–3cm long, with up to 6 leaflets; leaflets ovate, c. 7mm; inflorescence 1–3-flowered; peduncle 1.5cm; flowers c. 0.5cm, orange, yellow in centre. Fields & grassland; 1300m.
Distr.: palaeotropical.
IUCN: LC
Bali Ngemba F.R.: Onana 1889 10/2001.

Oxalis corniculata L.

Prostrate, rooting-stemmed herb; stems to c. 30cm, borne horizontally, densely white-hairy when young; leaflets transversely elliptic, 1cm long, apical sinus 0.5cm, base obtuse, sessile; inflorescence 1-flowered; flower 0.7cm; yellow. Grassland & forest edge; 1920m.
Distr.: cosmopolitan.
IUCN: LC
Bali Ngemba F.R.: Darbyshire 360 fl., 4/2004.
Note: field determination.

Oxalis latifolia Humb., Bonpl. & Kunth var. *latifolia*

Bradea 7(2): 585 (2000).
Acaulous herb; leaves trifoliolate, 5–8cm diam., lobes distinctly obtriangular, emarginate, glabrous; petioles to 23cm; inflorescence umbellate, 6–12-flowered; peduncle to 27cm; pedicels to 1.8cm; sepals lanceolate, 4mm, green with orange tips; petals obovate, 1cm, violet. Farmbush; 1800m.
Distr.: native to S America, cultivated as an ornamental elsewhere [Neotropics].
IUCN: LC
Bali Ngemba F.R.: Pollard 956 fl., 4/2002.
Note: possibly the first record for Cameroon. Abundant in farmbush within the reserve, April 2004 (pers. obs.).

PASSIFLORACEAE

I. Darbyshire (K)

Adenia cf. *cissampeloides* (Planch. ex Benth.) Harms

Climber; stems glabrous, tendrils simple; leaves trilobed, 9 × 9.5cm, lobes rounded, apiculate, base truncate, margin entire to undulate, glabrous, upper surface green, mottled paler, lower surface grey-green; petiole 7cm, paired glands fused at base of leaf. Forest; 1790m.
Distr.: Guinea (Conakry) to Kenya (*A. cissampeloides*) [Tropical Africa (*A. cissampeloides*)].
Bali Ngemba F.R.: Cheek 10498 11/2000.
Note: leaves more strongly lobed than usual for this taxon. Fertile material required to confirm.

Adenia cf. *lobata* (Jacq.) Engl.

Climber, to 4m; stems glabrous; tendrils simple; leaves ovate, 9–13 × 6–7cm, apex shortly acuminate, base cordate, margin

subentire-undulate, glabrous; petiole 4.5cm; paired glands separate at apex of petiole. Forest; 1700m.
Distr.: Senegal to Cameroon (*A. lobata*) [Upper & Lower Guinea (*A. lobata*)].
Bali Ngemba F.R.: Cheek 10516 11/2000.
Note: fertile material required for confirmation.

Passiflora edulis Sims
F.T.E.A. Passifloraceae: 15 (1975).
Climber, to 10m; stems glabrous, tendrils simple; leaves deeply trilobed, c. 11 × 11.5cm, lobes elliptic, acute, base rounded, margins serrate, glabrous, shiny; petiole c. 2cm, with paired glands near apex; fruit globular, 4cm diam., brown, fleshy. Villages; 1430m.
Distr.: native of S America, cultivated in Africa [Neotropics].
IUCN: LC
Bali Ngemba F.R.: Etuge 4708 11/2000.
Uses: FOOD – infructescences – fruit eatable (*Etuge* 4708).

PIPERACEAE

M. Cheek (K)

Peperomia fernandopoiana C.DC.
Epiphytic herb, c. 4m from ground; stems erect, branched, 30cm, drying black, glabrous; leaves alternate, ovate-lanceolate, c. 7 × 3.5cm, acumen long, acute; inflorescences terminal and axillary, 2–3 per peduncle, to 6cm. Forest, secondary forest, beside water; 1310–1500m.
Distr.: Sierra Leone to Kenya [Tropical Africa].
IUCN: LC
Bali Ngemba F.R.: Ghogue 1044 fl., 9/2000; Tadjouteu 396 11/2000; Zapfack 2003 fl., 4/2002.

Peperomia retusa (L.f.) A.Dietr. var. *mannii* (Hook.f.) Düll
Bot. Jahrb. Syst. 93: 90 (1973).
Syn. ***Peperomia mannii*** Hook.f. ex C.DC.
Epiphytic herb, at first mat-forming with prostrate stems and circular leaves, 0.5–1cm; flowering from erect stems with leaves elliptic, slightly reflexed, to 3cm, apex rounded; inflorescence 4(–7)cm. Forest; 1700m.
Distr.: Liberia, Nigeria, Cameroon, Bioko [Upper & Lower Guinea].
IUCN: NT
Bali Ngemba F.R.: Plot voucher BAL58 fl., 4/2002.

Peperomia tetraphylla (G.Forst.) Hook. & Arn.
Bot. Jahrb. Syst. 93: 81 (1973).
Syn. ***Peperomia reflexa*** (L.f.) A.Dietr.
Epiphytic herb, glabrous; leaves obovate-elliptic, whorled in fours, glabrous, shining, 1.5cm long. Forest; 1800m.
Distr.: pantropical [Montane].
IUCN: LC
Bali Ngemba F.R.: Onana 1816 10/2001.

Peperomia thomeana C.DC.
Bot. Jahrb. Syst. 93: 104 (1974).
Syn. ***Peperomia vaccinifolia*** C.DC.
Epiphytic herb, stoloniferous, glabrous, flowering from erect stems; leaves opposite at uppermost node, obovate or elliptic, to 3 × 2cm, apex retuse; inflorescences 1–2 in uppermost axils, to 6cm. Forest; 1700m.
Distr.: Bioko & W Cameroon [Lower Guinea].
IUCN: NT
Bali Ngemba F.R.: Zapfack 2052 fl., 4/2002.
Note: for details on conservation status, see Cheek *et al.* (2000).

Piper capense L.f.
Pithy shrub, c. 1(–5)m; peppery aroma emitted when crushed; leaves opposite, broadly ovate, c. 15 × 10cm, cordate, glabrous except on nerves; inflorescences of leaf-opposed single, erect, white spikes c. 3 × 0.5cm. Forest; 1310–1950m.
Distr.: Guinea (Conakry) to South Africa [Tropical & subtropical Africa].
IUCN: LC
Bali Ngemba F.R.: Ghogue 1000 11/2000; Pollard 416 fr., 11/2000; 608 fl., 10/2001; Ujor 30358 fr., 5/1951; Zapfack 2008 fl., fr., 4/2002.
Local name: Mumbot. **Uses:** SOCIAL USES – ritual – when twins are born, mother wears crown of leaves on her head as birthing nears (*Pollard* 608); MEDICINES (*Nkankano Peter* & *Achu John* in *Pollard* 416).
Note: above 1400m, plants commonly 2.5m tall.

Piper umbellatum L.
Pithy shrub, 1–2m; peppery aroma emitted when crushed; leaves orbicular, c. 24 × 24cm, deeply cordate, white below; inflorescences 2–5, clustered on peduncle, c. 3.5cm. Forest edge; 1400–1490m.
Distr.: Guinea (Conakry) to Cameroon [Upper & Lower Guinea].
IUCN: LC
Bali Ngemba F.R.: Cheek 10452 11/2000; Ghogue 1005 11/2000.
Uses: SOCIAL USES – unspecified parts; MEDICINES (*Cheek* 10452).

PITTOSPORACEAE

Y.B. Harvey (K)

Pittosporum viridiflorum Sims *'mannii'*
Meded. Land. Wag. 82(3): 260 (1982); Kew Bull. 42: 328 (1987).
Syn. ***Pittosporum viridiflorum*** Sims subsp. ***dalzielii*** (Hutch.) Cuf.
Syn. ***Pittosporum mannii*** Hook.f.
Shrub or small tree, to 10m, immature branches and petioles pubescent to glabrescent; leaves usually crowded towards the end of the branches, similar in size to *'ripicolum'*, lanceolate to spathulate, acuminate, cuneate; inflorescences paniculate, with pubescent branches; pedicels to 5mm; sepals to 1.2 × 0.8mm, glabrous or ciliolate, free or basally connate; petals to 5mm; capsule valves to 8mm diam.; seeds mostly 4 on each valve. Forest, bushland and occasionally farmland.
Distr.: Cameroon, Nigeria and Bioko [Guineo-Congolian].
IUCN: NT
Bali Ngemba F.R.: Ujor 30430 fr., 6/1951.

Pittosporum viridiflorum Sims ***'ripicolum'***
Kew Bull. 42: 329 (1987).
Small tree, to 12m; bark grey; branchlets and petioles glabrescent; leaves crowded towards end of branches, obovate, 7–17 × 1.5–4mm, cuneate, acuminate; petioles to 20mm; inflorescences paniculate; pedicels 3–10mm; sepals 1.5–3 × 0.8–1mm; petals narrowly elongate, 4–7 × 1.5–2mm; capsule valves up to 8mm diam.; 4-seeded. Riverine forest, swamp and humid woodland; 1600–1950m.
Distr.: Guinea to Congo (Kinshasa) & Ethiopia [Tropical Africa].
IUCN: LC
Bali Ngemba F.R.: Biye 130 11/2000; Onana 1977 11/2001; 2035 fl., 4/2002; Pollard 1046 fl., fr., 4/2002.

POLYGALACEAE

I. Darbyshire (K)

Polygala persicariifolia DC.
Herb, to 50cm; stem puberulent, few-branched; leaves ovate-lanceolate, to 6.5 × 2cm, sparsely pubescent, subsessile; flowering spikes axillary and terminal, to 4cm long, 10–15-flowered; pedicels 4mm; lateral calyx lobes ovate-orbicular, 7 × 6mm, glabrous, white-pink to magenta; fruit 4.5mm long, margin ciliate, surface glabrous. Grassland; 1400m.
Distr.: tropical Africa & Asia [Palaeotropics].
IUCN: LC
Bali Ngemba F.R.: Onana 1915 10/2001.

Polygala tenuicaulis Hook.f. subsp. ***tayloriana*** Paiva
Fontqueria 50: 207 (1998).
Erect, annual, herb, 30–80cm; stem wiry, pubescent, unbranched in basal half; leaves sessile, linear-lanceolate, c. 30 × 1.5mm, pubescent; main inflorescence a dense terminal raceme, to 12cm, continuous, many-flowered, short,subterminal spikes may also be present; pedicels 2mm, flowers pale pink to purple; lateral sepals obovate, to 6 × 4mm, puberulent towards the base; fruit 4mm long, puberulent; seeds with long hairs towards apex, shorter below. Grassland; 2100m.
Distr.: Nigeria & Cameroon [W Cameroon Uplands].
IUCN: NT
Bali Ngemba F.R.: Pollard 471 fl., fr., 11/2000; 665 fl., 10/2001.

POLYGONACEAE

I. Darbyshire (K)

Fagopyrum snowdenii (Hutch. & Dalziel) S.P.Hong
Grana 27(4): 295 (1988).
Syn. ***Harpagocarpus snowdenii*** Hutch. & Dalziel
Herb, weakly erect or scandent, to 1m; leaves sagittate, lower ones ovate-triangular, up to 8 × 3cm, upper linear-lanceolate; inflorescences in slender terminal and axillary racemes, up to 15cm long; flowers green; fruits with red or purple barbed setae. Forest; 1400m.
Distr.: Cameroon, Congo (Kinshasa), Uganda & Tanzania [Tropical Africa].
IUCN: LC
Bali Ngemba F.R.: Cheek 10431 11/2000.

Rumex abyssinicus Jacq.
Stout herb, to 2m; stems hollow, cylindrical, to 1.3cm diam.; leaves papery, hastate, 10.5–16 × 10–13cm, uppermost becoming linear subhastate, palmately-nerved, glabrous; petiole 3–7cm; panicles many-branched, to 30cm long, leafless, many-flowered; flowers in fascicles; pedicels 3–4cm; perianth 2mm; fruiting valves orbicular, 6–7mm diam., base cordate, tubercles absent, reticulately nervose; nut acutely trigonous, 3.5 × 1.5cm, shining pale brown. Forest edge & grassland; 1450–2000m.
Distr.: tropical Africa & Madagascar [Afromontane].
IUCN: LC
Bali Ngemba F.R.: Biye 102 11/2000; Pollard 462 11/2000.

Rumex crispus L.
F.T.E.A. Polygonaceae: 10 (1958).
Robust erect herb, to 1.2m; stems glabrous, 0.7cm diam. at base; stem leaves brittle when dry, oblong-lanceolate, 16–20 × 2.5cm, margin crisped, glabrous; inflorescence a several-branched, dense panicle, to c. 30 × 8cm, branching at c. 10 degrees to the peduncle; flowers in dense fascicles, narrowly spaced; pedicels filiform, c. 0.8cm, articulated at c. 0.2cm; fruiting valves deltoid, c. 4mm long, apex rounded, base truncate, margin entire, reticulately nerved, tubercles well-developed and approximately equal on each valve, c. 1.5 × 0.5mm, yellow, smooth; nut acutely trigonous, ovoid, c. 2mm long, shining-brown. Grassland & disturbed areas; 2000m.
Distr.: Native of Europe and Asia, widely introduced elsewhere [Temperate & tropics].
IUCN: LC
Bali Ngemba F.R.: Pollard 461 11/2000.
Note: *Pollard* 461 represents probably the first record for Lower Guinea, the only previous West African record known is from Liberia.

Rumex nepalensis Spreng.
Fragm. Florist. Geobot., Suppl. 2(1): 93 (1993).
Syn. ***Rumex bequaertii*** De Wild.
Erect herb, to 0.6(–1.5)m; stems hollow, cylindrical; leaves papery, oblong-lanceolate, to 25 × 5cm, apex and base acute, margin subentire, glabrous; petiole to 0.7cm; spikes to 35cm, rarely-branched; flowers in dense, widely-spaced fascicles, c. 13–15-flowered, reduced leaves present in the axes of the lower fascicles; pedicels 3.5–4mm; perianth c. 2mm long; fruiting valves deltoid, 3.5–4mm long, apex acute, margins with 5–7 hooked bristles, c. 1mm; nut trigonous, 2 × 1mm, shining-brown. Forest edge & disturbed areas; 1920m.
Distr.: tropical Africa, S Europe & Asia-Minor to China.
IUCN: LC
Bali Ngemba F.R.: Darbyshire 359 fr., 4/2004.
Note: field determination.

PROTEACEAE

B.J. Pollard (K)

Protea madiensis Oliv. subsp. ***madiensis***
Fire adapted shrub, to 2m; inflorescence a capitulum; bracts yellow, tinged pink at tips, densely tomentose; flowers

numerous, whitish yellow or pink; flower 'limb' glabrous. Upland savanna, grassland; 1410–2000m.
Distr.: S Nigeria, W Cameroon, from Ethiopia south to Angola, Mozambique [Afromontane].
IUCN: LC
Bali Ngemba F.R.: Nkeng 190 fl., 10/2001; Pollard 480 fr., 11/2000; 515 fl., 11/2000; 516 fl., 11/2000; 517 fl., 11/2000; 714 10/2001; 715 10/2001.
Note: pink-flowered populations occur here and elsewhere in the Bamenda Highlands, a feature not known elsewhere in this subspecies. There may well be other differences and further investigation is required for clarification. It may be necessary to recognise a new infraspecific taxon, at varietal or formal rank.

RANUNCULACEAE

I. Darbyshire (K) & M. Cheek (K)

Clematis simensis Fresen.

Tall, woody climber, to 20m, pubescent on young growth, becoming glabrous; leaves imparipinnate, with 5-leaflets (reduced in association with the inflorescence); leaflets ovate to ovate-lanceolate, occasionally with 1–2 lobes, subglabrous below; inflorescence many-flowered; pedicels 1–5cm; sepals 7–18mm, cream or white. Forest edge & farmbush; 1700–2250m.
Distr.: Cameroon to South Africa [Afromontane].
IUCN: LC
Bali Ngemba F.R.: Ghogue 1098 11/2000; Tadjouteu 415 11/2000.

Clematis villosa DC. subsp. *oliveri* (Hutch.) Brummitt

Kew Bull. 55: 104 (2000).
Erect, perennial herb, with stout rhizome; stems 0.7–1.5m, strongly striate, 1–5-flowered; leaves pinnate to bipinnate or trifoliate, lobes very irregular in shape; flowers solitary, 3.5–5cm diam., white, pink or mauve; achenes up to 10cm diam. Savanna; 1450m.
Distr.: Nigeria, W Cameroon, Congo (Kinshasa), Rwanda, Burundi, Sudan, E Africa [Afromontane].
IUCN: LC
Bali Ngemba F.R.: Pollard 942 fl., 4/2002.

Delphinium dasycaulon Fresen.

Erect, perennial herb, to 1m high with yellowish pubescent stems and inflorescences; leaves often persisting after inflorescences have died, alternate, palmisect, to 8cm diam., segments acute; petiolules of lower leaves up to 15cm; inflorescence c. 30cm; flowers numerous, 2–2.5(–3)cm wide; sepals usually deep blue, rarely violet, pinkish or white. Grassland; 1850m.
Distr.: Cameroon east to Ethiopia and south to Zambia & Mozambique [Tropical & subtropical Africa].
IUCN: LC
Bali Ngemba F.R.: Biye 114 11/2000.

Thalictrum rhynchocarpum Quart.-Dill. & A.Rich. subsp. *rhynchocarpum*

Lidia 4(3): 89 (1998).
Herb c. 60cm, glabrous; leaves tripinnate, c. 20cm; leaflets orbicular-elliptic, c. 1.2 × 1.2cm, entire or 3–5-lobed, petiolules capillary; inflorescence c. 30cm; flowers numerous, c. 0.4cm, green. Forest edge; 1700m.
Distr.: Bioko to Tanzania & South Africa [Afromontane].
IUCN: LC
Bali Ngemba F.R.: Pollard 1027 4/2002; Ujor 30361 fl., 5/1951.

RHAMNACEAE

I. Darbyshire (K) & M. Cheek (K)

Fl. Cameroun 33 (1991).

Gouania longispicata Engl.

Climber, tendrillate; leaves alternate, ovate, c. 4.5 × 3cm, base truncate, lateral nerves 6–7 pairs, densely red-hairy, serrulate; petiole 1.5cm; racemes axillary, erect, c. 10cm long; flowers white, 3mm diam. Grassland & forest edge; 1700m.
Distr.: S Nigeria to Mozambique [Afromontane].
IUCN: LC
Bali Ngemba F.R.: Tadjouteu 411 11/2000.

Maesopsis eminii Engl.

Tree, stinking when wounded, 35m, glabrous; leaves subopposite or alternate, papery, elliptic-oblong, c. 11 × 5cm, subacuminate, obtuse, serrate, the teeth often partly covered towards the petiole, lateral nerves c. 6; petiole 2cm; inflorescences axillary, cymose, few-flowered; flowers white c. 2mm; fruit obovoid, 3cm, fleshy, orange, 1-seeded. Forest & grassland; 1450m.
Distr.: Liberia to Uganda [Guineo-Congolian].
IUCN: LC
Bali Ngemba F.R.: Nkeng 198 fr., 10/2001.
Uses: MATERIALS – wood – timber (*Etuge*, pers. comm.).

RHIZOPHORACEAE

Y.B. Harvey (K)

Cassipourea gummiflua J.Lewis var. *verticillata* (N.E.Br.) J.Lewis

Tree, slash yellow, to 1m diam.; sapwood light yellow; leaves simple, opposite; leaf blades 5–14(–21) × 3–5(–10)cm, elliptic, obtuse to acuminate, base cuneate to truncate, margins shallowly serrate above the middle, rarely entire; pedicels (1.5–)2–3(–4)mm; calyx lobes (4–)5–6; stamens (8–)10–12; capsule puberulous above. Forest.
Distr.: Cameroon, Tanzania, Mozambique, Malawi, South Africa, Mascarene Is, Seychelles [Palaeotropics].
IUCN: LC
Bali Ngemba F.R.: Ujor 30415 fr., 6/1951.

Cassipourea malosana Alston

Kew Bull. 1925: 258 (1925).
Tree, 3m; leaves opposite, elliptic, 9 × 3.5cm, including 1.5cm acumen, base cuneate, serrate in upper ⅔, lateral

nerves c. 6 pairs; petiole c.1cm; interpetiolar stipule triangular, sericeous, 6 × 3mm, rounded; flowers 1–3, axillary-fasciculate, c. 4 fertile nodes per stem; pedicel 7mm; flower 8mm. Forest; 2100m.
Distr.: Cameroon to Kenya [Afromontane].
IUCN: LC
Bali Ngemba F.R.: Pollard 957 fl., 4/2002.
Note: further research may well show that the Cameroon montane population merits specific distinction from the E African populations. Likely to be confused with Rubiaceae but for the toothed leaves and superior ovary.

ROSACEAE

B.J. Pollard (K)

Fl. Cameroun 20 (1978).

Alchemilla kiwuensis Engl.

Herb with rosettes and stolons; leaves 7-palmatilobed to palmatipartite, median lobe with 13–23 teeth; flowers with 5–12 carpels. Forest; 2000m.
Distr.: Nigeria to Ethiopia, E Africa & south to Zambia and Zimbabwe [Afromontane].
IUCN: LC
Bali Ngemba F.R.: Pollard 452 fl., fr., 11/2000; 453 11/2000.

Prunus africana (Hook.f.) Kalkman

Fl. Cameroun 20: 209 (1978).
Syn. ***Pygeum africanum*** Hook.f.
Tree, to c. 20m; leaves alternate, lanceolate, 3–6 × 6–15cm, serrate; petiole 2cm long, bearing 2 glands near apex, or at base of lamina; inflorescence a dense panicle; flowers white, 5mm diam.; fruit a drupe, succulent, red, c. 1cm diam. Forest-grassland transition; 1400m.
Distr.: tropical and subtropical Africa [Afromontane].
IUCN: NT
Bali Ngemba F.R.: Etuge 4710 11/2000.

Rubus pinnatus Willd.

Syn. ***Rubus pinnatus*** Willd. var. ***afrotropicus*** (Engl.) Gust.
Scandent, prickly shrub, to 5m; leaves less than 2.5 times as long as broad, glabrous, not glandular below; inflorescences terminal or less often axillary, many-flowered, rachis densely-appressed with short, silver, velvety hairs; petals inconspicuous or caducous; infructescence with many fewer than 100 drupelets. Forest-grassland transition; 1750m.
Distr.: tropical and southern Africa [Afromontane].
IUCN: LC
Bali Ngemba F.R.: Biye 83 fl., 11/2000.

Rubus rosifolius Sm.

Prickly shrub, to 1m; leaves more than 2.5 times as long as broad, glabrous, glandular below; inflorescence axillary, few-flowered, sparsely pilose; petals white, c. 1.5cm long, showy; infructescence of about 100 drupelets. Forest-grassland transition; 1420–1500m.
Distr.: introduced from Asia [Palaeotropics].
IUCN: LC
Bali Ngemba F.R.: Ghogue 1049 fr., 11/2000; Pollard 415 fl., fr., 11/2000; 919a fl., fr., 4/2002.
Local name: colloquially referred to as 'Bush Allowance', along with any other foodstuffs growing naturally by the wayside, or in the forest, that can be harvested and eaten along the way (observations by *B.J. Pollard* in *Pollard* 415); Tele-Tele (*Nkankano Peter* in *Pollard* 919). **Uses:** FOOD – infructescences – fruit is eaten raw, in the wild (observations by *B.J. Pollard* in *Pollard* 415); fruits eaten as 'bush allowance'. (*Peter Nkankano* in *Pollard* 919).

RUBIACEAE

M. Cheek (K), S.E. Dawson: herbaceous taxa (K) & B. Sonké: *Rothmannia* & *Oxyanthus* (University of Yaoundé I)

Chassalia laikomensis Cheek

Kew Bull. 55(4): 884 (2000).
Shrub, 2–3(–8)m; leaves narrowly elliptic, 4–12 × 1.5–4cm, acuminate, lateral nerves 7–10 pairs; stipules 4mm, with conspicuous yellow raphides; panicles terminal, 5 × 5cm, loosely branched; flowers white, 6–10mm long; fruits black, ovoid, 6–9mm long. Forest; 1310–1830m.
Distr.: S Nigeria & W Cameroon [W Cameroon Uplands].
IUCN: CR
Bali Ngemba F.R.: Onana 1997 fl., 4/2002; Pollard 1033 fl., 4/2002.

Coffea liberica Bull. ex Hiern

Evergreen tree, to 10m, glabrous; leaves broadly elliptic, 7–13 × 3.5–6(–10)cm, acuminate, base acute to cuneate, lateral nerves 8–12, domatia on nerves, pit-like; petiole 0.8–2cm; stipule 2–4.5mm, apiculate, midrib inconspicuous; flowers several per axil, subsessile; corolla tube 5–6mm, lobes 6, 15 × 6mm, white; fruit red, ellipsoid, 14–18 × 8–10mm. Forest; 1950m.
Distr.: Guinea Bissau to Uganda [Guineo-Congolian].
IUCN: LC
Bali Ngemba F.R.: Biye 116 11/2000.

Cremaspora triflora (Thonn.) K.Schum. subsp. *triflora*

Shrub or climber, 1–8m; stem shortly pubescent; leaves papery, elliptic, c. 8.5 × 4.5cm, acuminate, rounded, nerves 3–5; petiole 0.5–1cm; flowers white, subsessile, 1cm; fruits red, axillary, fasciculate, ellipsoid, 13 × 6mm; pedicel 1mm. Evergreen forest; 1400–1900m.
Distr.: tropical Africa [Tropical Africa].
IUCN: LC
Bali Ngemba F.R.: Cheek 10468 11/2000; Ghogue 1014 11/2000; Onana 1926 10/2001; Pollard 925 fl., 4/2002; Ujor 30343 fl., 5/1951; Zapfack 2054 fl., 4/2002.

Cuviera longiflora Hiern

Shrub or tree, to 8m, glabrous; stems with ants; leaves papery, lanceolate-oblong, c. 27 × 10cm, acuminate, base subcordate or rounded, nerves 9–10; petiole 1cm; stipule sheathing, 5mm; flowers 10–20, in axillary panicles; peduncle 3–8cm; bracts and calyx lobes leafy; corolla green and white; tube c. 1cm; lobes c. 0.2cm; fruit ellipsoid, 9 × 3cm, brown, fleshy; pyrenes 5, c. 3 × 1cm. Evergreen forest; 1450–1830m.
Distr.: Cameroon to Angola [Lower Guinea].

IUCN: NT
Bali Ngemba F.R.: Cheek 10496 11/2000; 10513 11/2000; Pollard 1030 fl., 4/2002.
Uses: FOOD – infructescences – fruits harvested when brown by the Kom (fide *Clement Toh*), seeds are sucked for the flesh around them, which tastes extremely good (*Cheek* 10496).

Diodia sarmentosa Sw.

F.T.E.A. Rubiaceae: 336 (1976).
Syn. ***Diodia scandens*** sensu Hepper
Straggling herb to 5m, scabrid; leaves ovate or ovate-lanceolate, 2.5–5 × 1–2.5cm, pubescent below, scabrid above; petioles to 7mm; stipule cupular 1mm, with 5 aristae, 3mm; axillary fascicles sessile; flowers white, 4-merous; fruit ellipsoid, 3–4mm, didymous, dry. Forest; 1560m.
Distr.: pantropical.
IUCN: LC
Bali Ngemba F.R.: Etuge 4262r 10/2001.

Euclinia longiflora Salisb.

Deciduous tree, 5–8m with *Terminalia* branching; leaves clustered, elliptic, 12 × 5cm, acuminate, stipules chaffy, persistent; flower terminal, 20 × 6cm, yellowish white, marked purple; berry green, ellipsoid, 4 × 3cm, seeds numerous. Evergreen forest; 1400–1700m.
Distr.: Guinea Bissau to Uganda [Guineo-Congolian].
IUCN: LC
Bali Ngemba F.R.: Cheek 10463 11/2000; 10739 10/2001; Etuge 4793 fl., 11/2000; 4819 fr., 11/2000; Ghogue 1028 11/2000; Kongor 40 10/2001; Onana 1828 10/2001; 2041 fl., fr., 4/2002; 2042 fr., 4/2002; Ujor 30357 fr., 5/1951; Zapfack 2056 fr., 4/2002; 2120 fr., 4/2002.

Galium simense Fresen.

Straggling herb, clinging by minute hooked hairs; leaves 8, in a whorl, linear-oblanceolate, 10–40 × 2–4mm; flowers solitary, axillary; pedicels to 1.5cm; fruits globose, 4mm, black. Forest-grassland transition; 1900–2000m.
Distr.: Bioko, W Cameroon & Ethiopia [Afromontane].
IUCN: LC
Bali Ngemba F.R.: Onana 1846 11/2001; Pollard 454 fl., fr., 11/2000.

Ixora bauchiensis Hutch. & Dalziel

Shrub or tree, 2–8m, glabrous; leaves elliptic, drying grey above, golden-brown below, thickly papery, 19 × 8cm, acumen 0.5cm, base cuneate-decurrent, lateral nerves 15 pairs; petiole 2cm, articulated at base; inflorescence terminal, 10 × 15cm; peduncle 1.5cm; secondary branches c. 3 pairs, basal ones 10-flowered; partial peduncles 4cm; pedicels 1mm; corolla yellow; tube 8mm; lobes 7mm; fruit globose, orange, 5mm. Forest; 1310–1900m.
Distr.: Nigeria & Cameroon [W Cameroon Uplands].
IUCN: NT
Bali Ngemba F.R.: Biye 133 11/2000; 134 11/2000; Cheek 10449 11/2000; Ghogue 1001 11/2000; Onana 1852 10/2001; 1985 fr., 4/2002; Pollard 617 10/2001; 920 fr., 4/2002; Zapfack 2026 fr., 4/2002.
Note: likely to prove Vulnerable when data on habitat loss in Nigeria are available.

Ixora foliosa Hiern

Tree, (2–)7–12m; leaves coriaceous, elliptic-oblong, c. 10 × 5cm, shortly acuminate, acute, nerves c. 12; petiole 1cm; inflorescence with reduced leaves 4cm; peduncle erect, c. 9cm; flowers subcapitate, sessile; corolla tube red, 14mm; lobes white. Forest; 1900m.
Distr.: SE Nigeria & Cameroon [W Cameroon Uplands].
IUCN: VU
Bali Ngemba F.R.: Pollard 1040 fl., 4/2002.

Mussaenda erythrophylla Schum. & Thonn.

Climber, densely puberulent; leaves elliptic, to 1.5 × 7cm, acuminate, obtuse, lateral nerves 9–13 pairs; petiole to 2cm; flowers 10–20, the calyx with one leaf-sized, bract-like, red lobe; corolla cream to orange; tube 2cm; lobes 0.5cm. Evergreen forest; 1600m.
Distr.: Guinea (Conakry) to S Africa [Tropical Africa].
IUCN: LC
Bali Ngemba F.R.: Etuge 4809 11/2000; Onana 2046 fl., 4/2002.
Note: *Etuge* 4809 (alt. 1600m) is tentatively placed in this taxon, flowers are needed for confirmation.

Oldenlandia herbacea (L.) Roxb.

Erect or spreading herb, 7–60cm tall, glabrous, 4-ribbed; leaf blades linear-lanceolate, 0.6–5.5cm × 1–3.5mm, acute, cuneate, sessile; stipule sheath 0.5mm, with a few setae; flowers solitary or paired at each node; pedicels 0.8–3cm; calyx 4-toothed; corolla white or mauve; tube 0.2–1.1cm; capsule subglobose, 2–5mm. Grassland; 1340m.
Distr.: tropical & South Africa, Asia [Palaeotropics].
IUCN: LC
Bali Ngemba F.R.: Pollard 399 11/2000.

Oldenlandia rosulata K.Schum.

F.T.E.A. Rubiaceae: 290 (1976).
Erect, slender, annual herb to 35cm, glabrous; basal leaves rosulate, blades spathulate to elliptic, 3–7 × 1.3–3mm, obtuse; stem leaves filiform, 0.3–3cm × 0.3–2mm; stipule sheaths with 2 teeth, each 0.5mm; inflorescence paniculate; peduncles to 2.5cm; pedicels to 3.5cm; calyx 4-toothed; corolla white to pink; tube 1.2–3.6mm; lobes 4; capsule subglobose, 1–1.8mm. Grassland; 2100m.
Distr.: Cameroon to Natal [Tropical Africa].
IUCN: LC
Bali Ngemba F.R.: Pollard 676 fl., 10/2001.

Otomeria cameronica (Bremek.) Hepper

Herb; stems 0.3–1m long, usually prostrate; leaves ovate-elliptic or ovate-lanceolate, 2–8 × 0.4–4cm; flowers white; corolla-tube 3–5mm long; fruits ovoid. Forest and grassland; 2100m.
Distr.: Sierra Leone to Bioko and Cameroon [Guineo-Congolian (montane)].
IUCN: LC
Bali Ngemba F.R.: Pollard 693 fl., 10/2001.

Oxyanthus formosus Hook.f. ex Planch.

Shrub or tree, 3–10m; leaves coriaceous, oblong, to 25 × 12cm, short-acuminate, unequally obtuse; petiole 1cm; stipule lanceolate, 2–3 × 1–1.5cm; corolla tube 12 × 0.1cm; lobes 2.5 × 0.1cm; fruit ellipsoid, 5 × 2.5cm; pedicel 2.5cm. Evergreen forest; 1700m.

Distr.: Mali to Uganda [Guineo-Congolian].
IUCN: LC
Bali Ngemba F.R.: Zapfack 2066 fr., 4/2002.

Pauridiantha paucinervis (Hiern) Bremek.

Shrub, 2–4m; stems puberulent; leaves elliptic-oblong, c. 10 × 3cm, base acute, nerves 8, domatia usually absent; petiole 7mm; stipule subulate, 7 × 1mm; flowers 5–10, on 1–2 peduncles to 1.5cm, white; corolla tube 3mm; fruit 5mm, red or black. Evergreen forest; 1400–1700m.
Distr.: Bioko & Cameroon [Lower Guinea].
IUCN: NT
Bali Ngemba F.R.: Etuge 4694 fr., 8/2000; Zapfack 2057 fl., fr., 4/2002.
Note: 15 sites known apart from those cited here.

Pavetta hookeriana Hiern var. *hookeriana*

Shrub, 2–3m, subglabrous; flowering twigs 15cm; leaves papery, elliptic, to 13 × 6cm, acumen to 1cm, cuneate, lateral nerves 10 pairs, domatia arched, hairy, nodules not seen, tertiary venation inconspicuous; petiole 2cm; inflorescence to 10cm across, flowers to 100; calyx lobes rotund, 2mm; corolla white; tube 2–5mm; lobes 4–8mm. Forest-grassland transition; 1400–1800m.
Distr.: Bioko & W Cameroon [W Cameroon Uplands].
IUCN: VU
Bali Ngemba F.R.: Cheek 10450 11/2000; Etuge 4273r 10/2001; Ghogue 1015 11/2000; Onana 1823 10/2001.

Pentas ledermannii K.Krause

Kew Bull. 31: 185 (1976).
Syn. ***Pentas pubiflora*** S.Moore subsp. ***bamendensis*** Verdc.
Weak shrub, 0.6–1m, densely pubescent; leaves elliptic, to 10 × 5cm, subacuminate, acute, lateral nerves 10 pairs; petiole 1cm; stipules divided into 7–10 aristae, 4mm; corymbs terminal, 5cm wide, 10–20-flowered; peduncle 3cm; calyx lobes 3mm; corolla white; tube 5mm; lobes 5. Grassland, forest edge; 1790–1900m.
Distr.: Mt Kupe to Mt Oku [W Cameroon Uplands].
IUCN: VU
Bali Ngemba F.R.: Cheek 10493 11/2000; Onana 1847 11/2001.

Pentas schimperiana Vatke subsp. *occidentalis* (Hook.f.) Verdc.

Herb or subshrub, 1.5m; leaves ovate-elliptic, 9 × 4cm, pubescent below; petiole 1cm; flowers yellow-white; corolla tube 17mm long. Forest edge; 2100m.
Distr.: Cameroon, Bioko, São Tomé & Ituri, Congo (Kinshasa) [Lower Guinea & Congolian (montane)].
IUCN: LC
Bali Ngemba F.R.: Pollard 663 fl., 10/2001.

Psychotria sp. A aff. *calva* Hiern

Shrub, 3m; stems chalky-green; leaves papery, drying black above, grey below, bacterial nodules linear, along midrib; blade oblong-elliptic, 20 × 9cm, acumen 1.5cm, base obtuse-rounded, lateral nerves 10–12 pairs; petiole to 3cm; stipule ovate, aristate, bifid, to 8mm; inflorescence c. 4cm, 10–20-flowered; peduncle 2cm; infructescence 4–8cm, fruit subglobose, red, 6mm. Forest; 1310–1700m.
Distr.: Kupe-Bakossi & Bali Ngemba F.R. [W Cameroon Uplands].
Bali Ngemba F.R.: Cheek 10501 11/2000; Zapfack 2031 fl., fr., 4/2002; 2121 fl., 4/2002.
Note: easily confused with *P. martinetugei*. Distinguished by the slender fruiting pedicels.

Psychotria camptopus Verdc.

Kew Bull. 30: 259 (1975).
Syn. ***Cephaelis mannii*** (Hook.f.) Hiern
Tree, 3–5m, glabrous; leaves leathery, obovate, 26 × 13cm, acumen obtuse, 1cm, acute, lateral nerves 14 pairs; petiole 6cm; stipules 2 × 1.5cm; peduncle 1–3m, pendulous, red; flowers white, 1cm, 10–20, enveloped in fleshy bracts; involucre 3 × 5cm, glabrous. Forest; 1400–1800m.
Distr.: SE Nigeria, Bioko, SW Cameroon & Congo (Kinshasa) [Guineo-Congolian (montane)].
IUCN: NT
Bali Ngemba F.R.: Cheek 10458 11/2000; Onana 1834 10/2001.

Psychotria latistipula Benth.

Shrub 0.3–2m, glabrous; leaves thinly papery, drying dark brown below, elliptic, 16 × 7cm, subacuminate, decurrent, 20-nerved; petiole 2cm; stipule ovate, 1.2cm, bifurcate to half its length, glabrous; inflorescence diffuse, 10–20cm; bracts linear, 1cm, patent; flowers white, 4mm; infructescence pendulous; fruit globose, 4mm, red, ridged; pedicel not fleshy. Forest; 1400m.
Distr.: Nigeria, Bioko, Cameroon & Gabon [Lower Guinea (montane)].
IUCN: LC
Bali Ngemba F.R.: Cheek 10448 11/2000.

Psychotria peduncularis (Salisb.) Steyerm. var. *peduncularis*

Kew Bull. 30: 257 (1975).
Syn. ***Cephaelis peduncularis*** Salisb. var. **A** sensu Keay
Syn. ***Cephaelis peduncularis*** Salisb. var. **B** sensu Keay
Shrub, 1–5m, glabrous; leaves elliptic, to 15 × 8cm, acumen 0.5cm, acute, lateral nerves 12–15 pairs; petiole 2–3cm; stipule translucent, bifurcate, 1 × 0.8cm; inflorescence capitate; peduncle 2–4cm, nodding in accrescence, puberulent; involucral bracts fleshy; flowers 10–15, white, 5mm; infructescence umbellate; bracts absent; pedicels white, 1.5cm; berries blue, 7mm. Forest; 1400–1900m.
Distr.: tropical Africa (Afromontane).
IUCN: LC
Bali Ngemba F.R.: Cheek 10435 11/2000; Ghogue 1034 11/2000; Onana 1841 10/2001; Tadjouteu 395 11/2000.

Psychotria psychotrioides (DC.) Roberty

Shrub, 2m, glabrous; leaves elliptic, to 14 × 7cm, subacuminate, cuneate, lateral nerves 12 pairs; petiole 2cm; stipule obovate, sheathing, 1cm; inflorescence sessile, capitate, 2cm diam.; flowers 10–15, sessile, white, each 8mm; fruit ellipsoid, 1.5cm, calyx foliose. Forest; 1440–1900m.
Distr.: tropical Africa [Tropical Africa].
IUCN: LC
Bali Ngemba F.R.: Biye 75 11/2000; Onana 1839 10/2001; Plot voucher BAL49 fl. 4/2002.

Psychotria schweinfurthii Hiern

Bull. Jard. Bot. État. Bruxelles 34: 146 (1964).
Syn. ***Psychotria obscura*** Benth.
Shrub, c. 1m, resembling *P. latistipula*, but infructescence 3–4cm. Forest; 1900m.
Distr.: Ivory Coast to Uganda [Guineo-Congolian].
IUCN: LC
Bali Ngemba F.R.: Onana 1930 10/2001.

Psychotria succulenta (Hiern) Petit

Shrub, 1–3m, drying dark brown, matt; leaves leathery, elliptic-oblong, 15 × 7cm, acumen 0.5cm, acute-obtuse, lateral nerves 12 pairs; petiole 1cm; stipule broadly elliptic, 1.5cm, entire; inflorescence loosely capitate, 3 × 3cm; peduncle 7cm; flowers white, 3mm; fruit ovoid, 5mm. Forest; 1450–1560m.
Distr.: Nigeria to Zimbabwe [Afromontane].
IUCN: LC
Bali Ngemba F.R.: Ghogue 1029 11/2000; Pollard 618 10/2001; 918a 4/2002.
Note: although sterile, *Pollard* 918a is almost certain to be this taxon.

Psychotria sp. A of Bali Ngemba Checklist

Shrub, 2m, glabrous; leaves elliptic, thickly papery, drying grey-green above, pale below, bacterial nodules conspicuous, punctate, in clusters, 10 × 6cm, acumen 1cm, acute-decurrent, lateral nerves 9 pairs, brown; petiole 1.5cm; stipule ovate, 6mm, apical third bifid; panicle 2 × 2cm; peduncle 1cm, 10–20-flowered; corolla 3mm, white; infructescence 5cm; peduncle 2-winged; fruits 10–20, ellipsoid, red, 14 × 6mm, smooth. Forest; 1400–1750m.
Distr.: only known from the Bamenda Highlands [Cameroon Endemic].
Bali Ngemba F.R.: Cheek 10485 11/2000; Etuge 4795 11/2000.

Psychotria sp. B of Bali Ngemba Checklist

Shrub, 0.5–1.5m; stems, stipules and inflorescences densely brown puberulent; leaves papery, drying dark green above, lacking bacterial nodules, elliptic, 13 × 6cm, acute to subacuminate, base obtuse, lateral nerves 9–10 pairs, arched, white above, lacking domatia; petioles 1.5cm; stipules to 1.4 × 0.7cm, upper quarter bifid; inflorescence densely capitate, 1.5 × 1.2cm; peduncle 0.5cm; corolla tube 2mm, white; infructescence erect, to 4.5 × 4cm; peduncle 1.5cm; fruits numerous, subglobose, 5mm. Forest; 1310–1560m.
Distr.: known only from Bali Ngemba fr., [Cameroon Endemic].
Bali Ngemba F.R.: Cheek 10467 11/2000; Etuge 4263 10/2001; Onana 1998 fl., fr., 4/2002; Zapfack 2023 fl., fr., 4/2002.

Psydrax dunlapii (Hutch. & Dalziel) Bridson

Kew Bull. 40: 699 (1985).
Syn. ***Canthium dunlapii*** Hutch. & Dalziel
Medium-sized tree; branches horizontal; leaves oblong or oblong-elliptic, 15–20 × 6cm; flowers white, in axillary 'corymbs'; fruit 2-lobed. Forest; 1400–1900m.
Distr.: Nigeria, Bioko and Cameroon. [Lower Guinea (montane)].
IUCN: VU
Bali Ngemba F.R.: Cheek 10477 11/2000; Etuge 4274r 10/2001; 4701 11/2000; Ghogue 1055 11/2000; Onana 1929 10/2001.

Rothmannia urcelliformis (Hiern) Bullock ex Robyns

Shrub or tree, to 20m, stems glabrescent; leaves elliptic, c. 13 × 5cm, acuminate, white-tuft domatia, puberulent below, 7–8 pairs nerves; flowers white, blotched purple; calyx tube 0.3cm, limb 2cm, teeth 1.5cm; corolla with basal tube 2.5 × 0.4cm, upper tube 5 × 4cm; lobes 3 × 1.5cm. Evergreen forest; 1400–1800m.
Distr.: Guinea (Conakry) to Mozambique [Tropical Africa].
IUCN: LC
Bali Ngemba F.R.: Cheek 10459 11/2000; 10520 11/2000; Onana 1824 10/2001.

Rytigynia laurentii (De Wild.) Robyns

Shrub, 6m, mostly appressed, dense white-pubescent on young stem, petioles, midribs, upper and lower blade sparsely so; leaf blade membranous, elliptic, to 5 × 2.5cm, acumen 1cm, broad, base obtuse to rounded, lateral nerves 4 pairs; petiole 0.5cm; stipule sheathing 1mm; arista 1mm; fascicles axillary, 3–5-flowered; peduncle 2mm; pedicels 4mm; corolla tube 7mm, yellowish green. Forest; 1310m.
Distr.: Nigeria, Cameroon & Congo (Kinshasa) [Afromontane].
IUCN: NT
Bali Ngemba F.R.: Zapfack 2015 fl., fr., 4/2002.
Note: morphological differences suggest west-east vicariants. If the Cameroonian population proves taxonomically distinct, it will merit threatened status.

Rytigynia sp.

Shrub, 2m, inconspicuouly puberulent; leaves papery, drying green, oblanceolate, to 9 × 3.5cm, acumen 0.75cm, acute, lateral nerves 7 pairs; petiole 5mm; stipule narrowly triangular, 4mm; umbels axillary, 1cm, 2–5-flowered; peduncle 2mm; bracts 2–3, aculeate; corolla white, fallen; style 6mm, apex subglobose. Forest; 1700m.
Distr.: known only from Bali Ngemba [Narrow endemic].
Bali Ngemba F.R.: Zapfack 2055 fl., 4/2002.
Note: not determined to genus with certainty (more material required to resolve).

Sabicea calycina Benth.

Climber, puberulous; leaves membranous, oblong, c. 9 × 4cm, acuminate, cordate, whitish green below, nerves 9; petiole to 3.5cm; flowers 10–15; peduncle c. 6cm; bracts ovate, 1.2cm; calyx lobes elliptic, purple, 1.2cm; corolla white, tube 2cm. Forest; 1400m.
Distr.: Sierra Leone to Congo (Kinshasa) [Guineo-Congolian].
IUCN: LC
Bali Ngemba F.R.: Etuge 5421 4/2004.
Note: *Etuge* 5421 is tentatively placed in this taxon.

Spermacoce pusilla Wall.

F.T.E.A. Rubiaceae: 356 (1976).
Syn. ***Borreria pusilla*** (Wall.) DC.
Erect annual herb, 10–20cm; stems wiry, terete; leaves linear, 2.5 × 0.2cm; corolla pink; lobes 4. Roadsides & villages; 1300–1500m.

Distr.: tropical Africa & Asia [Palaeotropics].
IUCN: LC
Bali Ngemba F.R.: Onana 1882 10/2001; Pollard 703 10/2001.

Tricalysia sp. B aff. *ferorum* E.Robbrecht
Tree, 7–13m, densely white-appressed-hairy; leaves thinly coriaceous, oblong-acuminate, acumen abrupt, 7mm, base acute-decurrent, nerves 9; petiole 1.2cm; inflorescence on naked stem, 3-flowered, both bracts and bracteoles united, disk like; calyx 5-lobed; corolla tube 5mm; lobes 5; fruit green, globose, 1cm, disc 5mm diam. Forest; 1310–1700m.
Distr.: Mt Kupe & Bali Ngemba [Narrow endemic].
Bali Ngemba F.R.: Etuge 4690 11/2000; Plot voucher BAL35 fl. 4/2002; BAL63 fl. 4/2002; Tadjouteu 399 11/2000; Zapfack 2016 fr., 4/2002; 2046 fr., 4/2002.
Note: a new species, apparently endemic to Mt Kupe and Bali Ngemba.

Virectaria major (K.Schum.) Verdc. var. *major*
Weak-stemmed shrub, 60cm, terete, puberulent; leaves ovate-elliptic, to 12 × 6cm, acumen 1cm, base obtuse, abruptly decurrent, lateral nerves 10–12 pairs, sparsely softly-hairy on both surfaces; petiole 2cm; stipule 5mm, entire; flowers in erect terminal clusters; calyx lobes linear, 1cm; corolla pale purple, 2 × 1.5cm; stamens 10, exserted 2cm. Forest, grassland edges; 1560m.
Distr.: Nigeria to Zimbabwe [Afromontane].
IUCN: LC
Bali Ngemba F.R.: Pollard 594 fl., 10/2001.

RUTACEAE

M. Cheek (K)

Fl. Cameroun 1 (1963).

Clausena anisata (Willd.) Hook.f. ex Benth.
Shrub or tree, 2.5m, non-spiny, puberulent, strongly aromatic; leaves imparipinnate, 15cm, 4–9-jugate; leaflets alternate, lanceolate-oblique, c. 6 × 2.5cm, acuminate, obtuse, lateral nerves c. 10 pairs; petiolules 1mm; panicle c. 12cm, slender; flowers white, 5mm; fruit indehiscent. Forest; 1975m.
Distr.: Guinea (Conakry) to Malawi [Tropical Africa].
IUCN: LC
Bali Ngemba F.R.: Darbyshire 414 fl., 4/2004.
Note: field determination.

Vepris sp. A
Tree, to 20m; grey scurfy-hairy; leaves alternate, trifoliolate, to 42cm; leaflets obovate, 29cm, acumen 2cm, acute, lateral nerves 12 pairs; petiolule 0.5cm; petiole 11cm; male panicle terminal, 12cm, shaggy-hairy; male flowers white, 3mm, A4; pedicel c. 1mm; female flowers 4mm, G4 apocarpus; pedicels 4mm; fruits lacking. Forest; 1310–1500m.
Distr.: Cameroon [Uplands of Western Cameroon].
Bali Ngemba F.R.: Cheek 10437 7/2000; Darbyshire 385 fl., 4/2004; Ghogue 1048 11/2000; Pollard 930 fl., 4/2002; Zapfack 2039 fl., 4/2002.
Note: in FDC, keys out to *V. trifoliolatum* but occurs at a higher altitude, shaggier indumentum, much larger and wider leaves, longer indumentum, longer acumen, shorter male pedicels, and lacks persistent bracts. The specimen cited is male. A female and sterile specimen are known from Baba II. Not known elsewhere. Probably a new oricioid species. Fruit required.

Vepris sp. B
Tree, 90cm girth, minutely and inconspicuously densely brown puberulent; leaves opposite, trifoliolate, 15cm; leaflets elliptic, 12 × 5.5cm, acumen 0.5cm, blunt, acute-decurrent, lateral nerves 6 pairs, inconspicuous; panicles terminal, 10cm, few-flowered, few-branched; pedicels 1.5mm; male flowers white, 3mm, A8, G2; female flowers and fruits required. Forest; 1700m.
Distr.: Cameroon [Uplands of Western Cameroon].
Bali Ngemba F.R.: Ujor 30422 fl., 6/1951.
Note: noted as *Vepris* sp. nov.? in FWTA, as identical to *Toddaliopsis ebolowensis* Engl. by Letouzey 1961, on specimen, and as det. as *V.* cf. *heterophylla* by Mziray. Probably a new species. The opposite leaves have been overlooked by previous research efforts.

Zanthoxylum rubescens Hook.f.
F.T.E.A. Rutaceae: 44 (1982).
Syn. ***Fagara rubescens*** (Hook.f.) Engl.
Shrub or tree, 2.5–20m; leaves to c. 0.6m, 4(–6)-jugate, rachis smooth; leaflets drying pale green below, oblong, to 18 × 7cm, lateral nerves c. 10; petiolule 12mm; inflorescence a terminal panicle, 20cm; flowers numerous, 5mm, greenish or whitish; fruit globose, pink, 1cm, dehiscent, seed glossy. Farmbush; 1400m.
Distr.: Guinea Bissau to Angola [Guineo-Congolian].
IUCN: LC
Bali Ngemba F.R.: Cheek 10469 11/2000.

SAPINDACEAE

M. Cheek (K)

Fl. Cameroun 16 (1973).

Allophylus bullatus Radlk.
Tree, 15–18m; leaves trifoliolate; leaflets drying blackish green above, brown below, secondary nerves 10–12 pairs, domatia conspicuous, white tufted, along midrib and secondary nerves, elliptic, c. 19 × 8cm, long-acuminate, cuneate, margin serrate; petiole c. 8cm; inflorescence in the leaf axils, 10–21cm, branches 6–12 in the upper half, to 10cm long; flowers white, 2mm. Forest; 1830m.
Distr.: Nigeria, Cameroon, Príncipe & São Tomé [Western Cameroon Uplands (montane)].
IUCN: VU
Bali Ngemba F.R.: Pollard 1031 fl., 4/2002.
Note: *Pollard* 1031 is tentatively placed in this taxon.

Allophylus conraui Gilg ex Radlk.
Tree, 3–4m; bole 2m; spines to 2cm; stems, petiole and midrid with grey patent hairs, 1mm; leaves trifoliolate, 18cm; leaflets obovate, to 10 × 6cm, obtuse, cuneate, coarsely serrate, lateral nerves 7 pairs; petiolule 5mm; petiole 7cm; panicles axillary, 10cm, with 4 branches, each to 2cm; fruit red to 5mm. Forest; 1400m.
Distr.: Cameroon [Uplands of Western Cameroon].
IUCN: EN

Bali Ngemba F.R.: Cheek 10432 11/2000; Ujor 30334 fr., 5/1951.

Allophylus sp. A

Spiny small tree, 3m, inconspicuously densely patent-pubescent, glabrescent; trifoliolate, 2cm; lateral leaflets elliptic, to 9cm, central obovate, to 12 × 6.5cm, subacuminate, cuneate, inconspicuously serrate, lateral nerves 8 pairs; petiolule 3mm; petiole to 7cm, very sparsely and finely-hairy; panicles 7cm, axillary, lateral branches 2, 2cm; peduncle to 3.5cm. Forest; 1310m.
Distr.: Cameroon [Uplands of Western Cameroon].
Bali Ngemba F.R.: Onana 1984 fl., 4/2002.
Note: not matched, more material needed to identify, possibly new to science. Only two other spiny species are known in Cameroon: *A. conraui* and *A. hamatus*. Also unusual inflorescence being as long as the petiole.

Deinbollia sp. 1

Monopodial shrub, 0.8–3(–5)m; stems brown with bright white raised lenticels; leaves 25–63cm, (2–)3–4(–5)-jugate; leaflets pale green, nerves yellow, oblong-elliptic, c. 15–24 × 5.5–9cm, acuminate; petiole 9–16.5cm; panicle terminal, 8–20 × 5–20cm; flowers white, 3–4mm, glabrous; fruit orange, all but 1 carpel aborting, globose, 3cm diam. Forest; 1200–1800m.
Distr.: Cameroon & SE Nigeria [W Cameroon Uplands].
Bali Ngemba F.R.: Etuge 4717 fl., fr., 9/2000; Ghogue 1040 11/2000; 1054 11/2000; Onana 1804 10/2001; 1830 10/2001; Pollard 413 fl., fr., 11/2000; 923 fr., 4/2002; Tadjouteu 398 11/2000; Ujor 30355 fr., 5/1951; Zapfack 2047 fr., 4/2002.
Uses: FOOD – infructescences – fruits eaten, tasting sweetly (*Nkankano Peter* in *Pollard* 413).

Deinbollia sp. 2

Tree 4m; leaves 60cm, 10-jugate; leaflets narrowly elliptic, c. 14–15cm, acumen 1cm, lateral nerves 16 pairs, brochidodromous; petiole 20cm, terete, white, upper edge black scurfy puberulent; panicle 18cm, terminal, densely red-pubescent; flowers 4mm. Forest; 1310m.
Distr.: Cameroon [W Cameroon Uplands].
Bali Ngemba F.R.: Onana 1981 fr., 4/2002.
Note: specimen needs more investigation. Not matched.

Eriocoelum macrocarpum Gilg

Tree, 20m, glabrescent; leaves punctate, 2–3-jugate, paripinnate, 30cm; uppermost leaflets largest, obovate, to 30 × 14.5cm, basal leaflets 1cm, resembling stipules, shortly acuminate, soon glabrous below; inflorescence spike-like, 10–20cm; flowers c. 3mm; fruit hard, woody, orange, smooth, depressed globose, to 3 × 4cm; valves 3, hairy inside. Forest; 1400m.
Distr.: SE Nigeria to Congo (Kinshasa) [Lower Guinea & Congolian].
IUCN: LC
Bali Ngemba F.R.: Cheek 10476 11/2000; Ujor 30403 5/1951.
Note: *Ujor* FHI 30403 has been tentatively placed in this taxon.

Lychnodiscus grandifolius Radlk.

Fl. Cameroun 16: 165 (1973).
Tree, 7m; stems stout, densely pubescent; leaves glossy, leathery, c. 1m, c. 5-jugate, paripinnate; leaflets elliptic, c. 30 × 12cm, subacuminate, cuneate, lateral nerves c. 10 pairs, petiolule 5mm; petiole 30cm; inflorescence 30cm or more, dark brown puberulent, few-branched; flowers 5mm, white; capsules 3-valved, 3 × 2.5cm; sepals persistent, reflexed; seeds brilliant orange-red. Forest; 1700–1950m.
Distr.: Cameroon & Gabon [Lower Guinea].
IUCN: NT
Bali Ngemba F.R.: Cheek 10503 11/2000; Ghogue 1080 11/2000.

Paullinia pinnata L.

Woody liana, to 25m; leaves c. 12cm, 2-jugate, rachis winged; leaflets elliptic, to 11 × 6cm, obscurely toothed, apex rounded; petiole c. 10cm; inflorescence tendriliform, as long as leaves, spicate; flowers white, 3mm; fruit red, 3-lobed, obovoid, 4 × 1cm, stipitate; seed white and red, 0.5cm. Forest & farmbush.
Distr.: tropical Africa & America [Amphi-Atlantic].
IUCN: LC
Bali Ngemba F.R.: Pollard 518 11/2000.
Uses: MEDICINES (*Nkankano Peter* in *Pollard* 518).

SAPOTACEAE

Y.B. Harvey (K)

Fl. Cameroun 2 (1964).

Chrysophyllum gorungosanum Engl.

Tree, to 40m with long, straight, fluted-bole; immature growth with chestnut to golden-brown indumentum; leaf blades elliptic, oblanceolate to oblong-elliptic, 7–25 × 2.5–8cm, acuminate, cuneate, lower surface with chestnut coloured hairs; petiole 1–4cm; flowers clustered in current leaf axils; pedicels 1–2mm long; calyx with chestnut hairs; sepals to 3.5 × 3mm; corolla cream; tube to 2mm; lobes to 2mm; fruits to 4 × 3cm with chestnut coloured hairs, patchy; seeds obliquely ellipsoid, to 2.8 × 1.2cm, dark brown or blackish. Forest; 1700m.
Distr.: SW Cameroon, Angola and E Africa [Afromontane].
IUCN: LC
Bali Ngemba F.R.: Cheek 10528 11/2000.

Manilkara obovata (Sabine & G.Don) J.H.Hemsl

Syn. ***Manilkara multinervis*** **(Baker) Dubard**
Tree, to 35m; with slightly fluted bole and buttressed base; bark pale brown-grey; young growth deep purplish brown with pale lenticels; leaves clustered towards the apices of branches; leaf blades obovate, 3–17 × 1.5–7.2cm, apex rounded or emarginate, cuneate, drying brownish grey; petiole 5–18mm long; flowers fascicled in axils of older or fallen leaves; pedicels 4–10mm; calyx lobes to 6 × 3.5mm; corolla white; tube to 1.5mm; lobes up to 5mm; fruits yellow, obovoid, up to 2.5cm long. Rainforest, riverine and swamp forest; 1900m.
Distr.: Sierra Leone east to Zambia, Angola and E Africa [Tropical Africa].
IUCN: LC

Bali Ngemba F.R.: Pollard 508 11/2000; Ujor 30408 fl., fr., 5/1951.

Pouteria altissima (A.Chev.) Baehni

Pennington T. (1991). Gen. of Sapot.: 203.
Syn. ***Aningeria altissima*** (A.Chev.) Aubrév. & Pellegr.
Tree, to 50m; with straight, cylindrical bole and slightly buttressed base; bark pale greyish; young growth finely pubescent; leaf blades elliptic to oblong-elliptic, 5–16 × 3–17cm, obtuse, emarginate or acuminate, broadly cuneate; petioles to 1.5cm long; flowers clustered in axils of current leaves; pedicels 3–6mm long; sepals 3.5–5.5 × 2.5–4mm; corolla greenish cream to pale yellow; tube to 3.5mm; lobes to 2mm; fruit red, obovoid to subglobose, to 2cm diam; seeds obovoid, to 1.5cm long. Rainforest & riverine forest; 1450–2000m.
Distr.: Guinea (Conakry) to Sudan, SW Ethiopia & E Africa [Tropical Africa].
IUCN: LC
Bali Ngemba F.R.: Cheek 10510 11/2000; 10526 11/2000.
Uses: MATERIALS – wood – timber – when a tree is felled, the whole area stinks of bad meat or 'fart' (*Cheek*10510)

Synsepalum brevipes (Baker) T.D.Penn.

Pennington T. (1991). Gen. of Sapot.: 249.
Syn. ***Pachystela brevipes*** (Baker) Baill. ex Engl.
Medium tree with dense crown, to 35m, bole often fluted; young growth with dense, short-appressed hairs; leaf blades oblanceolate to obovate, 9–26 × 3.5–10cm, acuminate, obtuse or emarginate, cuneate, lower surface with greyish pubescence; petioles to 1cm long; flowers in dense clusters in older leaf axils; pedicels to 2mm long; sepals to 4 × 3mm; corolla yellowish green or cream; tube to 2mm; lobes to 4.5 × 2.5mm; fruits (orange-)yellow, ellipsoid, beaked, to 2.5cm long; seeds ellipsoid, to 2cm, shiny-brown. Forest; 1500–1790m.
Distr.: Guinea (Conakry) to Sudan, E Africa and south to Mozambique & Angola [Tropical Africa].
IUCN: LC
Bali Ngemba F.R.: Cheek 10494 11/2000; 10519 11/2000; Ghogue 1019 11/2000.
Uses: MATERIALS – wood –timber; ANIMAL FOOD – fruit eaten by animals, not by men (*Cheek*10519).

Synsepalum msolo (Engl.) T.D.Penn.

Pennington T. (1991). Gen. of Sapot.: 249.
Syn. ***Pachystela msolo*** (Engl.) Engl.
Tree, 5–25m tall, 15–30cm dbh; bole smooth, grey and green, slightly twisted with pairs of smooth knobs at intervals of 30–60cm, slash pink-red and white to pale brown inside, with white latex; branching *Terminalia*-style; leaves in whorls at the ends of branches, oblanceolate, 14.5–51 × 5–14cm, glabrous, apex rounded with acuminate tip, towards base narrowly cuneate (base is truncate), upper surface pale green, lower surface pale chestnut and iridescent, secondary nerves prominent below, 14–22 pairs of nerves; petiole 9–25mm long, inflated, to 5mm wide and fissured in dried state; stipules triangular, 3–5mm long; inflorescence in dense fascicles on branches; pedicels 4–6mm long; sepals 5, 6.5–7mm long; corollas with 5 lobes, 7.5mm long; tube 3–4mm long, irregularly sized staminodes; fruit ovoid, orange-brown, 3 × 2.5cm in diam.; single-seeded; seeds ovate. Evergreen forest; 1310–1800m.
Distr.: Ghana to Tanzania [Guineo-Congolian].
IUCN: LC
Bali Ngemba F.R.: Cheek 10518 11/2000; Onana 1833 10/2001; 1999 4/2002; Pollard 950 fl., fr., 4/2002; Ujor 30348 fr., 5/1951; Zapfack 2040 fr., 4/2002.
Local name: Bangbali. **Uses:** FOOD – infructescences – fruits are eaten: the exocarp is removed and the whole seed and gelatinous endocarp is swallowed whole, tastes sweet (*Eric Nucam* in *Pollard* 950); ANIMAL FOOD – infructescences – fruits eaten by monkeys (*Onana* 1999).

SCROPHULARIACEAE

B.J. Pollard (K) & M. Cheek (K)

Alectra sessiliflora (Vahl) Kuntze var. *senegalensis* (Benth.) Hepper

Slender, erect, roughly pilose herb, c. 30cm, usually drying black; leaves opposite, alternate on floriferous stem, sessile, ovate, serrate, 10–50 × 10–20mm; flowers yellow, 5mm. Grassland; 2100–2200m.
Distr.: Senegal to Mozambique [Tropical Africa].
IUCN: LC
Bali Ngemba F.R.: Ghogue 1083 11/2000; Pollard 691 fl., 10/2001.

Buchnera welwitschii Engl.

Herb, to 12–45(–85)cm, erect, drying jet-black; stems rarely branched; leaves: basal 17–40 × 10–25mm, elliptic-oblong to broadly ovate or suborbicular, entire to crenate, scabrid above, 3–5-nerved beneath; cauline opposite to subopposite, 2–3(–5) pairs, 3–5-nerved; inflorescence spicate, bracteolate, 8–30 × c. 8mm; ± quadrangular; flowers crowded, with some borne singly along the stem; bracteoles 3.5–4.5 × c. 0.5mm, linear to lanceolate; flowers blue; calyx c. 5mm; corolla tube 4.5–5.5(–8)mm, pilose outside; capsule 4–5 × 1.5–2.5mm. Grassland, open woodland; 1300–1600m.
Distr.: W Cameroon, Tanzania, Angola, Zambia [Afromontane].
IUCN: LC
Bali Ngemba F.R.: Etuge 4811 fl., fr., 11/2000; Onana 1891 fr., 10/2001.

Cycnium adonense E.Mey. ex Benth. subsp. *camporum* (Engl.) O.J.Hansen

Dansk. Bot. Ark. 32(3): 53 (1978).
Syn. ***Cycnium camporum*** Engl.
Syn. ***Cycnium petunioides*** Hutch.
Perennial herb, 30–90cm, drying black; many-branched, pilose, scabrid; leaves opposite or subopposite, slightly petiolate, 3–8 × 1–3cm, unevenly serrate; flowers showy, white, solitary, supra-axillary; pedicels 1.5–2cm; calyx 1.5–2cm; corolla tube 3.5–4.5cm; limb 3–4cm across. Savanna, rocky grassland; 1400–1680m.
Distr.: Guinea to Cameroon, east to Sudan, E Africa, south to Angola, Mozambique [Tropical & subtropical Africa].
IUCN: LC
Bali Ngemba F.R.: Onana 2048 fl., 4/2002; Pollard 972 fl., 4/2002; Ujor 30330 fl., fr., 5/1951.

Lindernia abyssinica Engl.

Small perennial, erect or ascending, 5–10cm; leaves elliptic-oblong, 3–5 × 7–13mm, ciliate, rather fleshy and congested; calyx-lobes lanceolate; flowers white, the lobes suffused purple; fruits c. 1cm, shortly beaked. Grassland.
Distr.: W Cameroon, Sudan, Ethiopia, Uganda [Afromontane].
IUCN: LC
Bali Ngemba F.R.: Ujor 30325 fl., 5/1951.

Rhabdotosperma densifolia (Hook.f.) Hartl

Syn. ***Celsia densifolia*** Hook.f.
Robust, erect, herb, to 90cm; stems pithy, woody at base, tomentose; leaves tomentose beneath, lanceolate, closely serrate, 2–7 × 0.7–2cm; inflorescence a terminal raceme; pedicels 1.5cm in fruit; flowers yellow, c. 2cm diam.; fruits 6–8mm. Grassland; 2100m.
Distr.: Bioko, W Cameroon [W Cameroon Uplands].
IUCN: NT
Bali Ngemba F.R.: Tadjouteu 431 fl., fr., 11/2000.

Sopubia mannii Skan var. *mannii*

Erect, rigid, branched undershrub, 30–150cm, drying dark greenish brown; rootstock woody; internodes short; leaves very many, densely congested on stem, 0.75–1mm; inflorescences terminal, racemose, to 10cm; flowers actinomorphic, purplish or magenta, numerous, often densely arranged; calyx 3.5–5.5mm, subglabrous without, tomentose on margin; corolla limb 10–12(13.5)mm diam. Grassland; 1450–2200m.
Distr.: tropical Africa [Afromontane].
IUCN: LC
Bali Ngemba F.R.: Ghogue 1097 fl., 11/2000; Pollard 705 fl., 10/2001; 946 fl., 4/2002.

Veronica abyssinica Fresen.

Prostrate, creeping herb, usually drying dark brown; stem branched from the base, pilose; leaves opposite, petiolate, ovate, serrate except towards base, 2–4 × 1–2cm; inflorescence a slender axillary peduncle; flowers blue or pinkish, paired or a few together, 8–10mm diam.; fruit bilobed, pubescent. Grassland, forest-grassland transition, roadsides; 1900–2200m.
Distr.: Nigeria to Zimbabwe [Afromontane].
IUCN: LC
Bali Ngemba F.R.: Ghogue 1091 fl., 11/2000; Tadjouteu 429 11/2000.

SIMAROUBACEAE

M. Cheek (K)

Quassia sanguinea Cheek & Jongkind ined.

Syn. ***Hannoa ferruginea*** Engl.
Tree, (1.5–)2–4(–6)m, glabrous; 3(–4)-jugate, rachis and midrib violet; leaflets usually drying green below, c. 13 × 5cm, apex subacuminate. Forest; 1400–1800m.
Distr.: Nigeria & Cameroon [W Cameroon (uplands)].
IUCN: VU
Bali Ngemba F.R.: Cheek 10491 11/2000; 10731 10/2001; Onana 1836 10/2001.

SOLANACEAE

B.J. Pollard (K)

Capsicum annuum L.

Annual or biennial herb; leaves broadly lanceolate to ovate, apex acutely acuminate, 5–8 × 2–5cm; inflorescences axillary, 3–8-flowered; pedicels to c. 1.5cm, nodding at maturity, dilated distally; calyx obscurely 5-toothed, 10-ribbed; corolla rotate-campanulate, deeply 5-lobed, usually straight, white or greenish; fruits single, a ± elongated berry, 1–severalcm. Cultivated in gardens and farms, sometimes escaping into forest understorey; 1450m.
Distr.: widely dispersed throughout the tropics, cultivated and sometimes naturalised [Pantropical].
IUCN: LC
Bali Ngemba F.R.: Pollard 922 fl., fr., 4/2002.

Discopodium penninervium Hochst.

Small tree or shrub, 5–7m; leaves elliptic to oblong-elliptic, mostly glabrous, lateral nerves in 10–12 pairs, 10–25 × 3–10cm; flowers white or yellowish, fading to brown, fascisculate, axillary; corolla cylindrical; lobes reflexed or spreading, c. 8mm; berry globose, 6–8mm diam. Forest, forest edges; 1850m.
Distr.: Bioko, Cameroon, Congo (Kinshasa), Ethiopia, E Africa [Afromontane].
IUCN: LC
Bali Ngemba F.R.: Biye 92 fl., 11/2000.

Physalis peruviana L.

Erect or straggling perennial, to 1m, densely-hairy; from creeping rootstock; leaves rhomboid to deltoid, entire or with a few large teeth, 8–10 × 6–7.5cm; flowers yellow with purple centre, 15mm; fruiting calyx large, to 4 × 3cm, villous. Fallow, rocky grassland; 1450–1850m.
Distr.: tropical America, naturalized in W Africa [Pantropical].
IUCN: LC
Bali Ngemba F.R.: Biye 113 fl., 11/2000; Pollard 523 fl., 11/2000.
Uses: FOOD – infructescences (*Biye* 113).

Solanum

A.E.Gonçalves' treatment of *Solanum* spp. (ined.) for Flora Zambesiaca is followed here. This is particularly relevant with regard to the taxonomy of the difficult *Solanum anguivi* Lam. complex.

Solanum aculeastrum Dunal var. *albifolium* (C.H.Wright) Bitter

Tree or shrub, to 7m; unarmed; white tomentum on all parts except surface of leaves; spines to 15mm long, compressed, base to 5mm long, sharply recurved; inflorescence lateral, axillary. Riverine forest; 1400m.
Distr.: Cameroon, tropical E Africa to Angola [Tropical & subtropical Africa].
IUCN: LC
Bali Ngemba F.R.: Biye 96 fl., 11/2000; 137 fl., fr., 11/2000.
Local name: Tamkekan. **Uses:** MEDICINES (*Biye* 96).

Solanum aculeatissimum Jacq.
Undershrub, 30–60cm armed, with almost straight spines; leaves distinctly petiolate, pinnately lobed, pubescent; flowers white, subsolitary, c. 1cm diam.; fruits brownish yellow. Forest.
Distr.: tropical and southern Africa [Afromontane].
IUCN: LC
Bali Ngemba F.R.: Ujor 30333 fr., 5/1951.

Solanum anguivi Lam.
Fl. Rwanda: Spermatophytes 3: 376 (1985).
Syn. ***Solanum distichum*** Schum. & Thonn.
Syn. ***Solanum indicum*** *sensu* auctt et collectt. afric. plur., non L., 1753
Syn. ***Solanum indicum*** subsp. ***distichum*** (Schum. & Thonn.) Bitter
Syn. ***Solanum indicum*** subsp. ***distichum*** var. ***grandemunitum*** Bitter
Syn. ***Solanum indicum*** subsp. ***distichum*** var. ***modicearmatum*** Bitter
Coarse tomentose undershrub, to 2m, spiny or not, with stellate hairs; leaves elliptic, very shortly pubescent above, subtomentose beneath, to 10–16cm; inflorescence a racemose-like cyme; flowers white, c. 5mm; fruits erect, globose, red, 1–1.5cm across. Forest; 1850–1900m.
Distr.: tropical Africa, Madagascar & Mascarene Is., Arabia.
IUCN: LC
Bali Ngemba F.R.: Biye 87 fl., fr., 11/2000; Pollard 506 11/2000.

Solanum betaceum Cav.
Taxon 44: 584 (1995).
Syn. ***Cyphomandra betacea*** (Cav.) Sendtn.
Tree, to 6m; leaves alternate, ovate, to c. 20 × 15cm, base cordate; petiole c. 10cm long; inflorescence few-flowered, axillary, pendulous; flowers campanulate, c. 1.5 cm × 0.7cm wide; petals pink; fruit ellipsoid, c. 7 × 4cm, orange, edible. Forest; 1200–1500m.
Distr.: Native to Peru, but naturalised in many parts of the tropics [Pantropical].
IUCN: LC
Bali Ngemba F.R.: Biye 79 fr., 11/2000; Ghogue 1041 fr., 11/2000; Onana 1803 fr., 10/2001; Pollard 924 fr., 4/2002.
Local name: Njanga Wankah. **Uses:** FOOD – infructescences – suck raw fruits (*Florence Azamah* in *Pollard* 924); fruits edible (*Ghogue* 1041).

Solanum giganteum Jacq.
Tree or shrub, to 8m; whole plant white-tomentose, except upper leaf surface; stems armed; leaves to c. 30 × 10cm, upper leaf surface glabrous; inflorescence cymose, c. 5–6cm across; flowers pale violet. Forest; 1700–1850m.
Distr.: tropical and subtropical Africa, India, Sri Lanka [Palaeotropics].
IUCN: LC
Bali Ngemba F.R.: Onana 1924 10/2001; Pollard 1035 fl., fr., 4/2002.

Solanum seretii De Wild.
Miss. E. Laurent, i: 439 (1907).
Shrub, shortly tomentose; stems with spines; leaves ovate, to 15 × 5cm, upper surface subglabrous, lower shortly tomentose, margins slightly serrate at apex; petiole to 2.5cm; inflorescence cymose, to 3cm diam. in flower, extending to 8cm in fruit; flowers light blue, densely tomentose; fruits globose, to 7cm diam. On top of hill.
Distr.: tropical Africa.
IUCN: LC
Bali Ngemba F.R.: Ujor 30356 fl., fr., 5/1951.

Solanum terminale Forssk.
Syn. ***Solanum terminale*** Forssk. subsp. ***inconstans*** (C.H.Wright) Heine
Syn. ***Solanum terminale*** Forssk. subsp. ***sanaganum*** (Bitter) Heine
Woody climber, to c. 10–15m; leaves elliptic, c. 12 × 5cm, acuminate, glabrous; petiole 1–2cm; inflorescence terminal or lateral, paniculate or cymose, very rarely spicate, 10–20cm; flowers c. 5 × 8mm; petals purple; staminal tube yellow. Forest; 1310–1800m.
Distr.: tropical Africa.
IUCN: LC
Bali Ngemba F.R.: Onana 1982 fl., 4/2002; Pollard 953 fl., 4/2002.

Solanum torvum Sw.
Shrub to 3m; stems occasionally armed, densely stellate-hairy; leaves large, ± elliptic, lobate-sinuate to subentire, to 10–16 × 4–12cm, subscabrid; inflorescence of corymbose cymes, 2–5(–14)-flowered; corolla white (rarely purple), to 2.5cm; fruit c. 1cm, globose, dirty-brown, occasionally drying black. A common weed in farmbush or forest; 1450–1700m.
Distr.: pantropical.
IUCN: LC
Bali Ngemba F.R.: Onana 1925 fl., 10/2001; Tadjouteu 409 fl., 11/2000.
Uses: SOCIAL USES – religious uses – magic – 'les fruits sont utilisés chez les Bulu pour se protéger contre les sorciers' (*Tadjouteu* 409)

STERCULIACEAE

M. Cheek (K)

Cola anomala K.Schum.
Tree, 15–20m; crown dense; stems bright white from waxy cuticle; leaves in whorls of 3(–4), simple, entire, elliptic, to 17 × 7.5cm, subacuminate, base rounded to obtuse, lateral nerves c. 7 pairs; petiole c. 2cm; stipules caducous; panicles 3cm in leaf axils; flowers yellow, without red markings, 1.5cm; fruit follicles to 12cm, 2-seeded, with knobs and ridges. Forest; 1440m.
Distr.: Cameroon & Nigeria [W Cameroon Uplands].
IUCN: NT
Bali Ngemba F.R.: Nana, F 134 fl., 4/2004.
Note: field determination.

Dombeya ledermannii Engl.
Tree, 3.5–15m; leaf-blade suborbicular to ovate, slightly 5-lobed, 11–19 × 5.3–15cm; cordate; petiole 4–7cm long; inflorescence an axillary cyme; peduncle 3–7cm long; petals white, 0.8–1.3cm long. Forest edge; 1700m.
Distr.: Bamenda Highlands [Uplands of Western Cameroon].
IUCN: CR
Bali Ngemba F.R.: Etuge 4815 11/2000.

Leptonychia sp. 1 of Bali Ngemba checklist

Tree to 17m, 30 dbh; leaves elliptic or rhombic-oblong, to 23 × 8cm, acumen slender, to 2cm, obtuse, 3-nerved; petiole 1.5cm; panicles subfasciculate, axillary, 1–5-flowered, 3cm; flowers 1cm, dull-white; anthers in 5 bundles; fruit globose, 4cm, minutely rugose, densely brown-puberulent; seed 3cm. Forest; 1300–2000m.
Distr.: Bamenda Highlands [Uplands of Western Cameroon].
Bali Ngemba F.R.: Cheek 10521 11/2000; Kongor 59 10/2001; Onana 1826 10/2001; 1919 10/2001; 1986 fr., 4/2002; Tadjouteu 423 11/2000.
Note: possibly a new species, more analysis required, specimens not seen from elsewhere.

Pterygota mildbraedii Engl.

Tree, c. 50m; leaves alternate, ovate, c 15 × 15cm, base cordate; domatia conspicuous; petiole 10cm. Forest; 1400m.
Distr.: Bamenda Highlands and Albertine Rift [Afromontane].
IUCN: NT
Bali Ngemba F.R.: Etuge 4693 11/2000.
Note: the population of this species at Bali Ngemba is the largest known W of the Congo Basin. May prove to be globally threatened when better data on Albertine Rift populations are available.

THYMELAEACEAE

M. Cheek (K)

Fl. Cameroun 5 (1966).

Gnidia glauca (Fresen) Gilg

Syn. ***Lasiosiphon glaucus*** Fresen.
Tree, to 15m; trunk much-branched; leaves oblanceolate, very acute, glabrous, 5–8cm long; flowerheads numerous, subsessile, c. 5cm diam.; petals spathulate, surrounded by large, ovate, glabrescent bracts. Grassland; 2100m.
Distr.: Cameroon, Ethiopia, Zambia and E Africa [Afromontane].
IUCN: LC
Bali Ngemba F.R.: sight record by Pollard, in notes to Pollard 968 4/2002 [Host plant for *Tapinanthus globiferus*].

TILIACEAE

M. Cheek (K)

Triumfetta annua L.

F.T.E.A. Tiliaceae: 79 (2001).
Herb, 0.5m, glabrous apart from a line of hairs on stem and petiole; basal leaves ovate, c. 7 × 4cm, acuminate, obtuse, serrate; inflorescences at uppermost nodes, c. 1cm long; flowers yellow, 3mm; fruit indehiscent, glabrous, globose, 8mm diam., including hooked bristles 3mm. Grassland; 1850m.
Distr.: Cameroon to China [Palaeotropics].
IUCN: LC
Bali Ngemba F.R.: Biye 86 11/2000.

ULMACEAE

I. Darbyshire (K)

Fl. Cameroun 8 (1968).

Trema orientalis (L.) Blume

Syn. ***Trema guineensis*** (Schum. & Thonn.) Ficalho
Tree, to 8m; young stems densely pubescent; leaves variable, distichous, ovate-lanceolate, 6.5–13.5 × 2.8–5.3cm, apex acuminate, base truncate, margin serrulate, lateral nerves 4(–6) pairs, alternate above basal pair, upper surface scabrid, lower surface scabrid or sparsely pubescent to densely pubescent; cymes axillary, c. 10–20-flowered; peduncle 0–0.5cm; flowers white, c. 2mm; sepals broadly elliptic, obtuse, puberulent; fruit globose, 2–3mm diam., green; styles and sepals persistent. Forest & farmbush; 1400–1560m.
Distr.: widespread in tropical Africa & Asia [Palaeotropics].
IUCN: LC
Bali Ngemba F.R.: Etuge 4265 10/2001; Ghogue 1056 11/2000; Pollard 438 fr., 11/2000.

UMBELLIFERAE

I. Darbyshire (K)

Fl. Cameroun 10 (1970).

Agrocharis melanantha Hochst.

Fl. Cameroun 10: 52 (1970).
Syn. ***Caucalis melanantha*** (Hochst.) Hiern
Suberect, perennial, herb, to 60cm; rootstock woody; robust stems rounded, finely ridged; leaves bi-pinnate on rachis 7–15cm, c. 6 pairs of pinnae, each divided into c. 3 pinnules, serrate to approx. half their width, blades finely pubescent; petioles sheathing, membranous at node, densely pubescent; umbel dense, globular, terminal, c. 1.5–2cm diam.; peduncle to 20cm, pubescent; involucral bracts numerous, lanceolate, ciliate, 3–4mm; c. 12 subsessile flowers per umbel, each with a subtending involucel of lanceolate bracts; corolla white; fruits green, ellipsoid, c. 5 × 3mm, ridged, with reflexed, barbed bristles along ridges, ciliate hairs between ridges. Grassland; 2000m.
Distr.: Cameroon, Bioko, Congo (Kinshasa), E Africa [Afromontane].
IUCN: LC
Bali Ngemba F.R.: Pollard 456 fr., 11/2000.
Note: recognised by some authors (e.g. Lee (2002): Israel J. Pl. Sc. 50(3): 211) as a synonym of *A. gracilis* Hook.f.

Centella asiatica (L.) Urb.

Creeping, perennial herb; long, glabrous, internodal stolons to 10cm, nodal rooting; petioles 5–20cm, pubescent particularly when young, sheathing, with subtending, leafy, lanceolate-stipule c. 5mm long; blade reniform with regular, crenate margin, non-lobed, glabrous, c. 2–3cm diam.; umbel 3–5mm; peduncle 1–1.5cm, pubescent; 1–5 umbels per node, subtended by leafy bract; 3–4 subsessile flowers per umbel, petals pink-purple, subtended by 2 pubescent bracts 1–2mm; fruits ellipsoid, truncate at apex, 2(–4) × 1(–3)mm, with reticulate sculpturing. Damp grassland.
Distr.: pantropical [Montane].

IUCN: LC
Bali Ngemba F.R.: sight record 4/2004 (*Darbyshire*, pers. comm.).
Note: common around Mantum.

Eryngium foetidum L.

Rank smelling, upright, perennial herb, to 60cm; rootstock fleshy; stems robust, green, glabrous, furrowed; basal rosette of oblanceolate leaves, c. 17 × 3cm, finely dentate, spiny, glabrous, on membranous, sheathing petioles; cauline leaves sessile, ± lobed, spiny; 2-bifurcating cymes above cauline leaves, each inflorescence a dense, rounded, green spike c. 0.5–1 × 0.5cm, subtended by involucre of 4–6 spiny, lanceolate bracts, c. 2–3 × 0.3–0.5cm, with spiny teeth. Streambanks in forest; 1400m.
Distr.: originating in the Americas, introduced in W Africa.
IUCN: LC
Bali Ngemba F.R.: Darbyshire 420 fl., 4/2004.
Note: field determination.

Hydrocotyle hirta R.Br. ex A.Rich.

Fl. Cameroun 10: 34 (1970).
Syn. ***Hydrocotyle mannii*** Hook.f.
Creeping herb; stems ± glabrous with nodal rooting; leaves reniform to suborbicular, c. 2.7 × 3cm, 7-lobed, indented to 2mm, with further sublobing, distinctive pilose hairs, to 1mm, on both surfaces; petioles 8–12cm, pubescent, particularly towards leaf-bases; umbel axillary, c. 10 subsessile flowers on peduncle, c. 1cm, pubescent. Open, damp habitats.
Distr.: tropical Africa, Madagascar, Mascarenes, Australia. [Palaeotropics].
IUCN: LC
Bali Ngemba F.R.: Ujor 30331 fl., fr., 5/1951.

Peucedanum angustisectum (Engl.) Norman

Erect herb, c. 1m; stems glabrous; leaves pinnately ternate at base to (bi)ternate on stem leaves; leaflets lanceolate, to 6.5–9 × 0.6–1.2cm, serrate, glabrous, sessile; petiole 5.5–6cm; petiolules 1–1.5cm; umbels compound, c. 6–8 ray florets per umbel; involucre absent; ray florets c. 1.2cm; pedicels 4mm, 7–11 per ray; involucel of lanceolate bracts, 2mm; flowers <1mm, green; mericarps flattened, elliptic, c. 4.5mm long, wings narrow, stylopodium conical. Grassland; 2000m.
Distr.: SE Nigeria & W Cameroon [W Cameroon Uplands].
IUCN: LR
Bali Ngemba F.R.: Cheek 10525 11/2000.

Sanicula elata Buch.-Ham.

Upright herb, 0.5(–1)m; short stolon at base; upright stems c. 2mm diam., glabrous, ridged; basal rosette of 2–4 leaves on petioles, to 15cm, 3(–5)-lobed almost to base, with irregular sublobing and dentate-mucronate margins, c. 6 × 8cm, glabrous; cauline leaves smaller on petioles, to 5cm; inflorescence cymose, bifurcating 3–4 times with a central cyme on peduncle, 1–1.5cm long; lateral primary peduncles 5–6cm, secondary peduncles c. 1–2cm; each floret of 2–3 sessile flowers; florets and peduncles subtended by lanceolate bracts, 6–12mm; flowers 1–2mm; corolla white; fruits ellipsoid, 3 × 2mm, covered in hooked bristles, green. Forest including stream edges; 1300–1400m.
Distr.: tropical & South Africa, Madagascar, Comores, temperate Asia [Montane].
IUCN: LC
Bali Ngemba F.R.: Kongor 56 fl., 10/2001; Onana 1916 10/2001; Pollard 436 fl., fr., 11/2000.

URTICACEAE

I. Darbyshire (K)

Fl. Cameroun 8 (1968).

Boehmeria macrophylla Hornem.

F.T.E.A. Urticaceae: 44 (1989).
Syn. ***Boehmeria platyphylla*** D.Don
Shrub, to 2(–3)m; branches glabrous, except when young; leaves opposite, anisophyllous, ovate, 10–13.5 × 5.5–9cm, acuminate, base acute to rounded, margin serrate, basal lateral nerves prominent, upper surface sparsely pubescent, cystoliths punctiform, lower surface glabrescent; petiole to 6cm; spikes axillary, 7–50cm, whip-like, with glomerules of flowers spaced 1–10mm apart; male glomerules 1–2mm; female 2–3mm. Forest & forest edge; 1310–1900m.
Distr.: tropical Africa & Madagascar, tropical Asia to SW China [Palaeotropics].
IUCN: LC
Bali Ngemba F.R.: Cheek 10466 11/2000; Onana 1829 10/2001; Zapfack 2022 fl., 4/2002.

Elatostema paivaeanum Wedd.

Herb, to 50cm, rarely branched, stipules conspicuous, lanceolate, 7–10mm; leaves drying green-black, large, 7.5–16 × 3–5.5cm, highly asymmetric with distal base subcordate, proximal base cuneate, marginal teeth 12–18 distally, 9–14 proximally, cystoliths dense, conspicuous; inflorescence sessile, c. 13cm wide; bracts broadly ovate (male) to lanceolate (female), c. 5mm, ciliate; bracteoles pilose, clearly so in female inflorescence. Forest; 1400–1500m.
Distr.: Guinea (Conakry) to E Africa & Malawi [Tropical Africa].
IUCN: LC
Bali Ngemba F.R.: Cheek 10453 11/2000; Etuge 4270r fl., 10/2001; Ghogue 1011 11/2000; Ujor 30328 5/1951.

Girardinia diversifolia (Link) Friis

Kew Bull. 36: 145 (1981).
Syn. ***Girardinia condensata*** (Steud.) Wedd.
Erect herb, 0.5(–2)m, few-branched, densely setose throughout, bristles to 5mm, brown; leaves alternate, c. 12 × 10cm, deeply 5–7-lobed, lobing subpalmate, lobes acuminate, margin coarsely serrate, blade pilose and setose; petiole 3.5cm; stipules oblong-lanceolate, pilose, 1.5cm, caducous; male inflorescence a few-branched panicle to 10cm; female inflorescence a dense, many-branched panicle, to 3cm, densely-hairy, extending in fruit. Clearings in forest; 1950m.
Distr.: tropical & South Africa, Madagascar, Yemen, India to Indonesia [Palaeotropics].
IUCN: LC
Bali Ngemba F.R.: Biye 124 11/2000.

Laportea alatipes Hook.f.

Erect herb, to 1m, stinging-setulose throughout; stipules lanceolate, to 1.2cm; leaves alternate, ovate, 8–18 × 4–10cm, acuminate, base rounded, margin serrate, cystoliths punctiform, lateral nerves c. 6 pairs; petiole to 6.5cm; female

panicles towards apex of shoots, to 13cm with c. 8 branches, on peduncles to 11cm, puberulent and stinging-hairy; male panicles shorter, peduncle to 1cm, found lower on the shoots; pedicels 1–2mm, membranously 2-winged in fruit; achene flattened-globose, 1.5–2mm, centre rugose, pale brown, margins smooth, dark brown. Forest; 1490m.
Distr.: Cameroon to E Africa & South Africa. (Afromontane).
IUCN: LC
Bali Ngemba F.R.: Ghogue 1010 11/2000.

Laportea ovalifolia (Schumach.) Chew
Fl. Cameroun 8: 131 (1968).
Syn. ***Fleurya ovalifolia*** (Schumach.) Dandy
Stoloniferous herb, to 60cm, erect to prostrate, except for male inflorescence; stems stinging-hairy; stipules 7mm, lanceolate; leaves alternate, ovate, 3–9 × 2–5cm, acuminate, base obtuse, margin serrate, cystoliths punctiform; male inflorescence axillary or from stolons, erect, paniculate, c. 13cm on peduncle, to 22cm; branches short, flower clusters 0.5–1cm; female inflorescence geocarpic or rarely in axils of upper leaves, densely racemose, 1.5(–5)cm; peduncles to 14cm, densely stinging-hairy; achene flattened, ovoid with a membranous margin and warted centre surrounded by a ridge. Roadside & forest; 1550m.
Distr.: Sierra Leone to Zimbabwe [Tropical Africa].
IUCN: LC
Bali Ngemba F.R.: Darbyshire 399 fl., 4/2004.
Note: field determination.

Parietaria debilis G.Forst.
Syn. ***Parietaria laxiflora*** Engl.
Decumbent or erect, non-stinging herb, to 50cm tall; stems puberulent; stipules absent; leaves alternate, ovate, 3.5–6 × 1.4–2.7cm, long-acuminate, base obliquely subcordate, margin entire, ciliate; petiole to 3cm; cymes axillary, dense, 0.4–1cm diam., with mixed male and female flowers; bracts pubescent. Farmbush, forest edge; 1560–1730m.
Distr.: pantropical [Montane].
IUCN: LC
Bali Ngemba F.R.: Pollard 598 10/2001; 653 fl., fr., 10/2001.

Pilea angolensis (Hiern) Rendle subsp. *angolensis*
Succulent herb, to 20cm, few-branched; stems glabrous; stipule <1mm, inconspicuous; leaves opposite, ovate, 2.5–5 × 2–3.5cm, acuminate, base truncate or rounded, margin deeply serrate-apiculate, surfaces sparsely pubescent or glabrescent, cystoliths punctiform; petiole 1.5–5cm; panicles axillary, few- or unbranched, 0.4–1.5cm; flowers in cymose clusters, 3mm diam. Rocks in forest; 1400–1560m.
Distr.: Guinea (Conakry) to Sudan & Tanzania [Guineo-Congolian].
IUCN: LC
Bali Ngemba F.R.: Cheek 10454 11/2000; Pollard 597 fl., 10/2001.

Pilea tetraphylla (Steud.) Blume
Herb, to 20cm; internodes decreasing and leaf size increasing towards apex; leaf pairs decussate, blade as *P. angolensis*, but to 3 × 2.2cm, apex scarcely acuminate, cystoliths linear, sparsely long-hairy, uppermost leaves subsessile; inflorescence appearing terminal, a corymb of 4 uneven parts, 2–3cm broad, dense. Forest, villages; 1490–2000m.
Distr.: tropical Africa [Afromontane].
IUCN: LC
Bali Ngemba F.R.: Ghogue 1009 11/2000; Pollard 458 fl., 11/2000.

Urera cordifolia Engl.
Robust climber or scrambler, to 8m; stems to 1cm diam., with numerous epidermal appendages; leaves alternate, ovate-orbicular, 16–27 × 11.5–21cm, apex short-acuminate, base shallowly cordate, margin shallowly crenate, cystoliths linear, glabrous; petiole 7–23cm; panicles axillary, many-branched, 10–17cm; flowers numerous, <1mm with short, stinging hairs; pedicel 1mm. Forest; 1440m.
Distr.: Nigeria, Cameroon, Gabon & CAR [Lower Guinea].
IUCN: LC
Bali Ngemba F.R.: Darbyshire 424 fl., 4/2004.
Note: field determination.

Urera cf. *gravenreuthii* Engl.
Climber, to 5m; stems lacking epidermal appendages; leaves ovate, to 13 × 8.7cm, apex short-acuminate, base subcordate-rounded, margin crenate, glabrous, cystoliths shortly linear; petiole to 9cm; inflorescence an axillary panicle to 11cm, much branched; peduncle 5cm; flowers in cymose clusters; pedicels <1cm; perianth with short, stinging hairs. Forest, by stream; 1450m.
Distr.: Nigeria, Cameroon, Angola [Lower Guinea].
Bali Ngemba F.R.: Pollard 917a fl., 4/2002.

Urera trinervis (Hochst.) Friis & Immelman
F.T.E.A. Urticaceae: 6 (1989).
Syn. ***Urera cameroonensis*** Wedd.
Robust, climber, glabrous; stems cylindrical; leaves ovate-elliptic, 10.5–14 × 5.5–7.5cm, acumen to 1.7cm, base acute to rounded, margin entire, basal lateral nerve pair prominent, cystoliths inconspicuous; petiole 2–4.5cm; male panicles to 6cm; peduncles 1cm; pedicels 1.5mm; female panicles denser, c. 4cm, with clusters of stinging hairs; flowers sessile. Forest; 1500m.
Distr.: Ghana to E & S Africa, Madagascar [Tropical Africa & Madagascar].
IUCN: LC
Bali Ngemba F.R.: Darbyshire 352 fl., fr., 4/2004.
Note: field determination.

VERBENACEAE

B.J. Pollard (K)

Note: here follows the treatment of Verbenaceae *sensu stricto*, without *Clerodendrum* L. and *Vitex* L. which are now treated in Labiatae *sensu lato*. See reference under that account.

Lippia rugosa A.Chev.
Robust, woody perennial, to 3m; stems distinctly appressed-pubescent; leaves ternate, to c. 10 × 3cm, oblong-lanceolate, upper surface scabrid; venation prominently rugose-reticulate; inflorescences spreading, much-branched cymes; flowers small, whitish. Grassland, savanna; 1400m.
Distr.: Guinea, Nigeria, Cameroon [Upper & Lower Guinea].

IUCN: LC
Bali Ngemba F.R.: Onana 1914 10/2001.

Stachytarpheta cayennensis (Rich.) Vahl

Shrubby herb, to 2m; stems glabrous; leaves ovate or elliptic, 1.8–8 × 0.5–4cm, attenuate into petiole, 1–1.5cm; inflorescence a slender spike, up to 20–25(– 34)cm, with some pubescence; bracts linear to triangular-subulate, 4–5mm, acuminate; calyx 4–5mm, with 4 equal teeth; corolla white or mostly pale blue; tube 4–5mm, scarcely exceeding the calyx. Roadsides, fallow; 1340–1500m.
Distr.: Sierra Leone to Cameroon, Uganda, Mozambique, Zimbabwe, widespread in tropical America, naturalised throughout the tropics [Pantropical].
IUCN: LC
Bali Ngemba F.R.: Pollard 401 fl., 11/2000; 701 10/2001.

VIOLACEAE

G. Achoundong (YA)

Rinorea preussii Engl.

Tree, 2–12m; stems glabrous, pale; leaves (oblong-)elliptic, 14–21 × 6–11cm, acumen 1.5cm, base acute to obtuse, margin serrulate, lateral nerves prominent, 11–12 pairs, glabrous, eglandular; petioles 2.5–5cm; panicles terminal, elongate, to 20cm, somewhat lax, finely puberulent, many-flowered; sepals rounded, 2.5mm; petals ovate, 4.5mm, white to pale yellow; fruit glabrous. Forest and forest edge; 1310–1490m.
Distr.: Liberia, SW Cameroon [Upper and lower Guinea].
IUCN: NT
Bali Ngemba F.R.: Onana 1918 10/2001; Tadjouteu 421 11/2000; Ujor 30354 fl., fr., 5/1951; Zapfack 2001 fr., 4/2002.
Note: specimens from Congo (Kinshasa) previously identified as this species are, in fact, *R. mildbraedii*.

VISCACEAE

R.M. Polhill (K) & B.J. Pollard (K)

Viscum congolense De Wild.

Globose dioecious parasitic shrub, < 0.5m; nodes often dilated; leaves highly variable, elliptic-ovate to oblong, 4–6(–10) × 2–4cm, conspicuously triplinerved above and below; base cuneate; margin sometimes crisped; ♂ and ♀ flowers occurring in triads (occasionally up to 6), <3mm; berries subsessile, 6–9mm, greenish white, translucent. Forest; 1440m.
Distr.: Ivory Coast to Congo (Kinshasa), Rwanda, Burundi, Ethiopia, Angola [Guineo-Congolian].
IUCN: LC
Bali Ngemba F.R.: Darbyshire 426 fl., fr., 4/2004.
Note: field determination.

VITACEAE

I. Darbyshire (K)

Fl. Cameroun 13 (1972).

Cayratia gracilis (Guill. & Perr.) Suesseng.

Fl. Cameroun 13: 20 (1972).
Syn. ***Cissus gracilis*** Guill. & Perr.
Herbaceous climber, to 6m; stems cylindrical, fine, glabrous; tendrils bifid; leaves papery, pedately 5-foliolate; leaflets ovate, central 5.2–8.5 × 2.1–4.2cm, apex acute or acuminate, base obtuse-rounded, margin serrate, sparsely pubescent or glabrous below; petiole 4–6cm; compound cyme c. 5cm long; flowers 2mm long; calyx cupular; corolla buds truncate; fruit globose, 0.8cm diam., glabrous; seeds cordiform, 0.4cm long with 2 depressions on the ventral side and a dorsal furrow. Forest & thicket.
Distr.: widespread in (sub)tropical Africa, also Yemen [Tropical, subtropical Africa & Arabian Peninsula].
IUCN: LC
Bali Ngemba F.R.: Ujor 30391 fl., 5/1951.

Cissus diffusiflora (Baker) Planch.

Herbaceous climber, to 7m; stems cylindrical, pubescent; tendrils few, bifid; leaves papery, ovate-lanceolate, 9–16 × 3.3–6.8cm, apex attenuate-acuminate, base truncate to shallowly cordate, margin finely dentate, principal veins of lower surface pilose; petiole 0.4–4.5cm, pubescent; inflorescence to 1.7cm long; flowers in umbellate clusters; peduncles 1–5mm, pubescent; flowers to 1.5mm long; calyx cupular, buds rounded at apex; pedicel to 2.5mm, glabrous; fruit pyriform or globose, 0.6–0.8 × 0.5cm, glabrous, red-black; seeds c. 0.45cm long, flattened, ovoid with a prominent dorsal ridge. Forest and thicket; 1400–1560m.
Distr.: Guinea Bissau to Uganda [Guineo-Congolian].
IUCN: LC
Bali Ngemba F.R.: Cheek 10436 fr., 7/2000; Pollard 607 fr., 10/2001.

Cissus oreophila Gilg & Brandt

Climber, to 6m; stems cylindrical, red-brown-pubescent when young, becoming glabrescent; leaves broadly ovate, 7.5–11 × 6.5–8.5cm, apex acuminate, base deeply cordate, margin denticulate, undersurface finely white-puberulent throughout and with reddish, medifixed, V-shaped hairs mainly along the veins; petiole 3–5cm; panicles with numerous umbellate clusters; peduncle to 8.5cm, partial peduncles and pedicels pubescent; calyx cupular; buds conical, c. 1.5mm long, minutely puberulent; petals white; fruit subglobose, c. 5mm; seeds subreniform, highly sculptured. Forest & forest edges; 1440m.
Distr.: Liberia to Gabon [Upper & Lower Guinea].
IUCN: LC
Bali Ngemba F.R.: Darbyshire 421 fr., 4/2004.
Note: field determination.

Cissus producta Afzel.

Herbaceous climber, to 20m; stems robust, glabrous; tendrils simple; leaves variable, ovate to oblong, 5–11.7(–13) × 2.2–6(–10.2)cm, acumen 0.8cm, base truncate to shallowly cordate, margins toothed, prominent when young, glabrous, usually drying brown-green; inflorescence to 3cm, with 3–4

dense umbellate cymes; peduncle c. 1cm; flowers c. 3mm; calyx cupular; corolla buds acute at apex; pedicels 0.5–1cm, puberulent; fruit obovoid, 1(–1.6) × 0.8(–1.2)cm, glabrous, black when ripe; seeds 0.8–1.4cm long, ovoid, ventral side with 2 cavities, dorsal ridge c. 1mm broad. Farmbush and open forest; 1400m.
Distr.: Senegal to Zambia [Tropical Africa].
IUCN: LC
Bali Ngemba F.R.: Etuge 5430 fl., 4/2004.
Note: field determination.

Cyphostemma mannii (Baker) Desc.
Fl. Cameroun 13: 68 (1972).
Syn. ***Cissus mannii*** (Baker) Planch.
Herbaceous climber; stems cylindrical, pubescent; tendrils bifid; leaves palmately 5-foliolate, c. 11 × 12cm; leaflets obovate-elliptic, central 8–9.5 × 3.8–4.6cm, apex shortly acuminate, base acute, margin dentate, puberulent along veins; petiole 3.5–4.5cm, pubescent; compound cyme c. 10 × 14cm, subcorymbiform, puberulent throughout; flowers 0.4cm; calyx cupular; corolla pinched in centre, apex truncate; fruit globose, 6mm diam., glabrous; seeds ellipsoid, 4mm, ridged. Forest and thicket; 1850m.
Distr.: Bioko & Cameroon [W Cameroon Uplands].
IUCN: LC
Bali Ngemba F.R.: Biye 89 11/2000.

Cyphostemma rubrosetosum (Gilg & Brandt) Desc.
Fl. Cameroun 13: 56 (1972).
Syn. ***Cissus rubrosetosa*** Gilg & Brandt
Herbaceous climber; stems cylindrical, densely glandular-hairy, hairs to 5mm, red, tendrils bifid; leaves papery, palmately 5-foliolate, 12.5 × 14cm; leaflets obovate, central 12 × 4.8cm, apex shortly acuminate, base cuneate, margin crenate-dentate, sparsely pilose along veins; petiole 7–10cm, pilose; compound cyme, c. 11 × 17cm; peduncles pubescent; flower 0.4cm; calyx cupular; corolla pinched at the centre, apex truncate, glabrous; pedicels 0.3cm, pubescent; fruit subglobose, 0.5cm diam., glabrous; seeds subglobose, c. 4mm, striate. Farmbush & open forest; 1700–1800m.
Distr.: Guinea (Conakry) to CAR [Upper & Lower Guinea].
IUCN: LC
Bali Ngemba F.R.: Onana 1822 10/2001; Tadjouteu 412 11/2000.

MONOCOTYLEDONAE

AMARYLLIDACEAE

I. Darbyshire (K)

Fl. Cameroun 30 (1987).

Crinum zeylanicum (L.) L.
Fl. Cameroun 30: 20–21 (1987).
Bulb globose, to 15cm diam.; leaves 6–10 per bulb, thin, to 100 × 10cm, margin slightly scabrous; peduncle 15–65cm, tinged red; flowers 10–20 per umbel, fragrant; perianth-tube curved, to 14cm long; lobes oblong, acute, as long as tube, white with red stripe. Farmbush; 1390m.
Distr.: tropical Africa and Asia.
IUCN: LC
Bali Ngemba F.R.: Etuge 5308 fl., 4/2004.
Note: field determination.

Scadoxus multiflorus (Martyn) Raf.
Fl. Cameroun 30: 8 (1987).
Syn. ***Haemanthus multiflorus*** Martyn
Bulbous herb, 25–80cm; bulb cylindrical, c. 2 × 1.5cm; leaves expanding after flowering, ovate-lanceolate, to 25 × 8cm, base attenuate; inflorescence lateral, 7–25cm, globose, many-flowered; flowers scarlet; pedicels 1.5–3.5cm. Forest edge & semi-deciduous forest; 1500m.
Distr.: Senegal to Somalia & to South Africa, also Yemen [Tropical Africa & Arabian Peninsula].
IUCN: LC
Bali Ngemba F.R.: Etuge 5318 fl., fr., 4/2004.
Note: field determination.

ANTHERICACEAE

I. Nordal (O) & Y.B. Harvey (K)

Chlorophytum sparsiflorum Baker
Herb, 25–60cm, drying light green, sometimes viviparous; leaves oblanceolate or oblanceolate-ligulate, c. 25 × 6cm, acute-mucronate, base tapering into a variably defined petiole; inflorescence about as long as leaves, or longer. Forest; 1500–1700m.
Distr.: Sierra Leone to Kenya [Afromontane].
IUCN: LC
Bali Ngemba F.R.: Tadjouteu 401 11/2000; Zapfack 2063 fl., fr., 4/2002.
Note: this taxon is soon to be reduced to a variety of *Chlorophytum comosum* (Thunb.) Jacques.

ARACEAE

Y.B. Harvey (K) & P.C. Boyce

Fl. Cameroun 31 (1988).

Amorphophallus staudtii (Engl.) N.E.Br.
Herb; leaf like a tattered umbrella; leaflets acute, not fishtail-shaped; petiole to 1.2m, smooth; inflorescence pale dirty-cream-white; peduncle very short; spadix base not swollen. Forest; 1310–1700m.
Distr.: Cameroon, Equatorial Guinea [Lower Guinea].
IUCN: NT
Bali Ngemba F.R.: Cheek 10502 11/2000; Zapfack 2013 fr., 4/2002.
Note: known from only 10 sites in Cameroon, where it is likely threatened in the Bamenda Highlands by agricultural encroachment into existing forest patches. Also recorded from Equatorial Guinea, but poorly documented there.

Anchomanes difformis (Blume) Engl.

Syn. ***Anchomanes difformis*** (Blume) Engl. var. ***pallidus*** (Hook.) Hepper
Syn. ***Anchomanes welwitschii*** Rendle
Herb; leaf like a tattered umbrella; leaflets fishtail-shaped; petiole spiny (prickles); spathe green-tinged purple; styles curved, scabrid. Forest & forest margins; 1400m.
Distr.: Sierra Leone to Congo (Kinshasa), Angola and Sudan [Guineo-Congolian].
IUCN: LC
Bali Ngemba F.R.: Cheek 10461 11/2000.
Uses: MEDICINES (*Cheek* 10461).
Note: *A. hookeri* was until recently treated as a synonym of *A. difformis* but the two are separable on differences in the styles.

Culcasia insulana N.E.Br.

Slender climber; stems smooth; lamina lanceolate, asymmetric, drying pale, without transparent lines; petiole as long as lamina. Forest; 1310m.
Distr.: Bioko, Cameroon, Congo (Kinshasa) & possibly CAR [Lower Guinea & Congolian (montane)].
IUCN: NT
Bali Ngemba F.R.: Zapfack 2041 fr., 4/2002.
Note: rare throughout its range, only c. 10 locations known, poorly recorded in Congo (Kinshasa). This taxon's submontane habitat remains largely unthreatened.

Culcasia sapinii De Wild.

Syn. ***Culcasia seretii*** De Wild.
Herbaceous climber; innovations (new shoots) copper-orange; leaves with translucent dots; petiole more than half the length, but always shorter than, lamina. Forest; 1500m.
Distr.: W Africa to Congo (Kinshasa) [Guineo-Congolian].
IUCN: LC
Bali Ngemba F.R.: Zapfack 2119 fr., 4/2002.
Note: uncommon throughout much of its range; only 5 previous collections at K, but 16 locations recorded in Fl. Cameroun. Submontane forest habitat largely unthreatened.

Culcasia sp.

Slender climber; stems smooth; lamina ovate, 17–18.5 × 8.5–9cm, asymmetric, base rounded to slightly cordate, without transparent lines; petiole ⅔ length of lamina; spathe 35–40mm long. Forest; 1790m.
Distr.: Bali Ngemba F.R.
Bali Ngemba F.R.: Cheek 10490 11/2000.

Nephthytis poissonii (Engl.) N.E.Br.

Syn. ***Nephthytis gravenreuthii*** (Engl.) Engl.
Syn. ***Nephthytis constricta*** N.E.Br.
Syn. ***Nephthytis poissonii*** (Engl.) N.E.Br. var. ***constricta*** (N.E.Br.) Ntépé-Nyamé
Terrestrial creeping herb; leaves triangular, posterior lobes considerably more developed than the anterior lobes; fruits orange-red, subtended by a spreading, persistent, green spathe. Forest; 1310–1430m.
Distr.: Sierra Leone to Gabon [Upper & Lower Guinea].
IUCN: LC
Bali Ngemba F.R.: Etuge 4724 11/2000; Zapfack 2014 fl., fr., 4/2002.

COLCHICACEAE

B.J. Pollard (K)

Wurmbea tenuis (Hook.f.) Baker subsp. *tenuis*

Cormous herb, to 15cm above ground; corm ovoid, c. 1cm, tunicate, bulb-like; stem with membranous basal sheaths, truncate, 1–3cm; basal leaf 1, erect, linear, 5–17cm × 1–3mm; cauline leaves 2–3, shorter than spike, or lower leaf somewhat longer; spike 2–6-flowered; perianth 5.5–7.5mm, connate, $^1/_6{}^{th}$ to $^1/_{10}{}^{th}$, white or lilac-pink to lilac, often with two purple blotches per segment. Grassland; 2100m.
Distr.: Bioko, W Cameroon [Uplands of Western Cameroon].
IUCN: NT
Bali Ngemba F.R.: Pollard 959 fl., 4/2002.

COMMELINACEAE

R.B. Faden (US), M. Cheek (K) & Y.B. Harvey (K)

Aneilema dispermum Brenan

Weak erect herb, to c. 1m; leaves elliptic, to 14 × 4cm, acuminate, sessile or shortly petiolate, margin ciliate; inflorescence terminal, dense, c. 4.5 × 3cm; flowers 40–50, white; capsules broader than long, 1-seeded. Forest edge, forest-grassland transition; 1440m.
Distr.: Bioko, SW Cameroon, Malawi & Tanzania [Afromontane].
IUCN: NT
Bali Ngemba F.R.: Biye 84 11/2000.

Aneilema umbrosum (Vahl) Kunth subsp. *ovato-oblongum* (P.Beauv.) J.K.Morton

Straggling herb, 10–30cm high, resembling *A. umbrosum* subsp. *umbrosum*, but rarer, smaller and more slender; leaves to 8 × 3cm; sheaths lacking rusty hairs; inflorescence with 2–8 branches. Lowland forest gaps; 1400–1560m.
Distr.: tropical Africa & tropical America [Montane].
IUCN: LC
Bali Ngemba F.R.: Cheek 10445 11/2000; Ghogue 1006 11/2000; Pollard 610 fl., 10/2001.

Aneilema umbrosum (Vahl) Kunth subsp. *umbrosum*

Straggling herb, to 1m; leaves elliptic or lanceolate, up to 13 × 4cm, acuminate, petiolate; sheath with rusty hairs at apex and sometimes on surface; inflorescence terminal, lax, up to 12 × 7cm, with 8–30 branches; flowers white. Lowland farmbush, forest; 1500m.
Distr.: Sierra Leone to Congo (Kinshasa) [Guineo-Congolian (montane)].
IUCN: LC
Bali Ngemba F.R.: Ghogue 1026 11/2000; Ujor 30359 5/1951.

Commelina africana L. var. ***krebsiana*** (Kunth) C.B.Clarke
Lebrun, J.-P. & Stork, A. (1995). E.P.F.A.T. 3: 23.
Syn. ***Commelina africana*** L. var. ***villosior*** (C.B.Clarke) Brenan
Scrambling herb, to 1m, softly-hairy throughout; leaves elliptic, to 5 × 2cm, greenish white below, deep green above, subsessile; sheath with rusty hairs at mouth; spathes from 2–5cm long; flowers yellow; capsule 1-seeded. Grassland; 1450m.
Distr.: upper Guinea to southern tropical Africa [Tropical Africa].
IUCN: LC
Bali Ngemba F.R.: Nkeng 197 fl., 10/2001.

Commelina africana L. var. ***mannii*** (C.B.Clarke) Brenan
Scrambling herb, to 30cm, sparingly-hairy throughout; stems pale pink; leaves elliptic, to 3 × 2cm; spathes c. 1cm long, ovate; flowers yellow; capsule small, 3-seeded. Grassland; 2100m.
Distr.: SW Cameroon and Ethiopia [Afromontane].
IUCN: NT
Bali Ngemba F.R.: Pollard 681 fl., 10/2001.
Note: although rare in Cameroon, assumed secure in Ethiopia, but needs verification.

Commelina benghalensis L. var. ***hirsuta*** C.B.Clarke
Erect herb, c. 30(–150)cm tall; leaves ovate, to 6 × 3.5cm, subacuminate, truncate, petiolate; sheath with conspicuous rusty hairs all over outside; spathe c. 2 × 1cm; flowers bright blue, open 08:30am to 12:00pm. Farmbush; 1560m.
Distr.: Guinea (Conakry) to Malawi [Afromontane].
IUCN: LC
Bali Ngemba F.R.: Pollard 627 10/2001.

Commelina cameroonensis J.K.Morton
Erect herb, 0.5(–1)m tall, gregarious; leaves elliptic, c. 12 × 4.5cm, acumen well-defined, base strongly oblique; petiole 0.5cm; spathes 1–3, 1.5–3 × 0.5–1cm, margin brown-hairy, base drying pale yellow, subsessile; flowers white; capsules 2-loculate, 1 smooth seed per locule. Forest, forest-grassland transition; 1790m.
Distr.: SE Nigeria, Bioko & W Cameroon [W Cameroon Uplands].
IUCN: NT
Bali Ngemba F.R.: Cheek 10488 11/2000.
Note: local in the W Cameroon Uplands, where it is possibly threatened in E Nigeria and NW Province, Cameroon by forest destruction.

Commelina* sp. B of FWTA, cf. *schweinfurthii C.B. Clarke
Creeping herb; stems ascending, to 0.3m, violet with a line of short, patent hairs; internodes 10cm; leaves lanceolate, 6–8 × 2cm, sessile, confluent with 4mm wide, purple-veined, 3cm sheath; spathe ovate, 2cm, purple-veined; corolla violet. Grassland and forest edge; 1300–1950m.
Distr.: SE Nigeria and Cameroon [W Cameroon Uplands].
Bali Ngemba F.R.: Biye 131 11/2000; Kongor 48 10/2001.

Cyanotis barbata D.Don
Erect herb, 10–30cm; with underground rootstock; leaves linear-lanceolate, to 12 × 1cm, white-pubescent, sessile; spathes 4–5, c. 1 × 0.5cm, pedunculate; flowers blue, actinomorphic; filaments bearded. Grassland; 1400–2200m.
Distr.: tropical Africa & Asia [Palaeotropics (montane)].
IUCN: LC
Bali Ngemba F.R.: Ghogue 1086 11/2000; Onana 1908 10/2001.

Palisota mannii C.B.Clarke
Herb 20–80cm; lacking aerial stem; leaves forming a basal rosette, lanceolate or lanceolate-obovate, 25–40 × 5–9cm, apex acuminate, base cuneate, margin hairy, lower surface white, glabrous; inflorescence cylindrical, c. 12–18 × 3.5cm; peduncle 10–50cm long; pedicels > flowers; rarely bracteose; flowers white; fruits red. Forest, forest-grassland transition; 1430–1500m.
Distr.: S Nigeria to Uganda [Lower Guinea & Congolian].
IUCN: LC
Bali Ngemba F.R.: Etuge 4712 11/2000; Ghogue 1070 11/2000; Pollard 932 fl., fr., 4/2002.

COSTACEAE

M. Etuge

Fl. Cameroun 4 (1965).

Costus sp. (Etuge 5432)
Herb, to 5m high; inflorescence terminal, green; flowers white. Secondary forest patch; 1400m.
Bali Ngemba F.R.: Etuge 5432 fl. 4/2004.
Note: field determination.

***Costus* sp.** (Etuge 5449)
Herb, to 1m high; leaves spiralling. Secondary forest patch; 1350m.
Bali Ngemba F.R.: Etuge 5449 4/2004.
Note: field determination.

CYPERACEAE

Y.B. Harvey (K)

Bulbostylis densa (Wall.) Hand.-Mazz. var. ***cameroonensis*** Hooper
Annual herb, to ± 30cm; stem deeply grooved, 0.2–0.4mm thick; leaves canaliculate, grooved, 0.2–0.3mm broad; inflorescence usually a compact umbel, somewhat contracted, with 3–8 shortly-pedicellate spikelets, each one 2–5 × 1.5–3mm; glumes few, each standing out from its neighbour, dark brown with conspicuous pale green or grey midrib. Grassland; 2100m.
Distr.: Mt Cameroon to Bali Ngemba F.R. [W Cameroon Uplands].
IUCN: VU
Bali Ngemba F.R.: Pollard 677 10/2001.

Bulbostylis pusilla (A.Rich.) C.B. Clarke subsp. ***congolensis*** (De Wild.) R.W.Haines

Syn. ***Bulbostylis congolensis*** De Wild.
Tufted annual, to 0.5m; culms densely-hairy, ridged; leaf-blades 10–40cm × 0.2–0.4mm, densely-hairy; sheaths straw-coloured or pinkish brown, long white hairs at throat; inflorescence an open anthela, 3–5cm long with 3–9 primary rays of unequal length and 10–40 spikelets; spikelets 3–5 × 1.5–2mm; nutlet 0.8–1 × 0.7–0.8mm, light brown. Grassland; 1300m.
Distr.: tropical Africa.
IUCN: LC
Bali Ngemba F.R.: Onana 1892 10/2001.

Cyperus flavescens L. subsp. ***flavescens***

Haines, R. & Lye, K. (1983). The Sedges and Rushes of East Africa: 281.
Syn. ***Pycreus flavescens*** (L.) Rchb.
Annual (rarely a stoloniferous perennial) herb; stems crowded, 6–50cm × 0.3–2.5mm; the base enclosed by reddish brown expanded leaf-bases; leaf-blades to 30 × 0.4cm, 2–4 per culm; inflorescence a dense anthela of clustered spikelets; involucral bracts 2–5, to 15cm; peduncles to 5cm, with a tubular basal purple prophyll; spikelets lanceolate to linear, 5–18 × 1.2–2.5mm, pale yellowish brown. Grassland; 1300m.
Distr.: pantropical and temperate regions.
IUCN: LC
Bali Ngemba F.R.: Kongor 52 10/2001; Onana 1898 10/2001.

Cyperus mannii C.B.Clarke

Stout perennial, to 2m; culms tufted or with a short rhizome; leaves often well-developed, to 1m or more; inflorescence anthelate, to c. 25 × 30cm, thrice-branched, bearing small clusters of spikelets; inflorescence branches grooved, each with a conspicuous basal prophyll, 1–2cm; involucral bracts leafy to 50 × 2cm; spikelets red-brown. Forest, forest edges, forest paths; 2000m.
Distr.: Sierra Leone to Bioko, W Cameroon [Upper & Lower Guinea].
IUCN: LC
Bali Ngemba F.R.: Tadjouteu 432 11/2000.

Cyperus renschii Boeck. var. ***renschii***

Large, robust perennial, to 1.5m; rhizome woody, 1.0–1.5cm thick; culms 50–150cm × 2–8mm; many large, basal leaves, 60–120 × 1–2.5cm; leaf-sheaths purplish near base; inflorescence a large anthela, 12–40 × 15–30cm, with 6–12 'pseudoumbels', which, in larger specimens bear further 'pseudoumbels' of the 3rd or 4th order; peduncles 0.5–30cm; involucral bracts leafy, to 90 × 4cm; spikelets 2–3 × 1–2mm, brown, 5–8-flowered. Forest; 1400–1790m.
Distr.: tropical Africa [Afromontane].
IUCN: LC
Bali Ngemba F.R.: Cheek 10495 11/2000; Onana 1920 10/2001; Tadjouteu 407 11/2000.

Cyperus sesquiflorus (Torr.) Mattf. & Kük. subsp. ***cylindricus*** (Nees) Koyama

Haines, R. & Lye, K. (1983). The Sedges and Rushes of East Africa: 241.
Syn. ***Kyllinga odorata*** Vahl subsp. ***cyllindrica*** (Nees) Koyama
Tufted perennial; rhizome creeping, to 4cm; culms crowded, 3–60cm, ridged; leaves 3–30cm (as long as culm or much shorter); inflorescence white, either a single cylindrical, or rarely globose, spike or a compound head of one larger cylindrical spike and several smaller lateral spikes; involucral bracts leafy, 2–4, to 12cm; spikelets 1.8–2.5mm, 1-flowered. Open grassland, especially in disturbed ground, heavily grazed or cut or burnt, often on sandy, or gravely soil; 1300–2250m.
Distr.: palaeotropical [Montane].
IUCN: LC
Bali Ngemba F.R.: Ghogue 1099 11/2000; Onana 1877 10/2001.

Cyperus tenuiculmis Boeck. var. ***tenuiculmis***

Medium-sized perennial with a thick creeping rhizome; culms, 30–80cm × 0.7–3mm, subglabrous; largest leaf-blade 10–40 × 0.2–0.4cm, scabrid on margin and major ribs; inflorescence, 5–20 × 3–15cm anthela consisting of one sessile, and 2–6 stalked, spikes; spikelets golden or yellowish brown, 6–20-flowered, linear-lanceolate; nutlets 1.6–1.8 × 1–1.2mm, blackish. Grassland; 1400m.
Distr.: old world tropics and to S Japan [Palaeotropics].
IUCN: LC
Bali Ngemba F.R.: Onana 1910 10/2001.

Lipocarpha nana (A.Rich.) Cherm.

Slender annual with crowded culms and very thin roots; stems 2–24cm × 0.2–0.5mm; leaves 2–8cm long, enrolled; leaf sheaths often tinged with violet-red; inflorescence of (1–)3(–4) spikes; involucre of (1–)2(–3) unequal leafy bracts to 40mm long, each subtending a lateral spike usually smaller than the central spike; bracts 1.5mm long; nutlet 0.6–0.9mm long. Grassland; 1300–1400m.
Distr.: tropical Africa to Transvaal [Tropical & subtropical Africa].
IUCN: LC
Bali Ngemba F.R.: Onana 1878 10/2001; 1911 10/2001.

Scleria afroreflexa Lye in press

Delicate annual; culms 10–50cm × 0.3–0.8mm; leaves 2–4 per stem, but only 1–3 perfecting leaf-blades; sheaths densely covered by retrorse white hairs, 0.2–0.4mm; blades 2–9cm × 0.8–1.8mm; inflorescence 3–9 × 1–2cm, appearing spike-like with sessile glomerules above, but, in fact, a narrow panicle with 1-several reflexed branches with 1–3 glomerules below; glomerules 4–5 × 3–6mm, consisting of 2–10 spreading spikelets. Grassland. 1300–1450m.
Distr.: Bakossi Mts, Bamenda Highlands [W Cameroon Uplands].
IUCN: EN
Bali Ngemba F.R.: Nkeng 194 fl., 10/2001; Onana 1880 10/2001.

Scleria foliosa Hochst. ex A.Rich.

Haines, R. & Lye, K. (1983). The Sedges and Rushes of E Africa: 344.
Robust, tufted annual with brown or reddish roots; stems 25–90cm × 1–4mm, thick, triangular, glabrous; leaves to 40cm × 2–6mm; sheath glabrous or hairy; inflorescences of 1–3 lateral, and 1 terminal, panicles; peduncles to 2cm; panicles 1–6 × 1–2cm, shorter than the leafy bracts; male spikelets 4–5mm; female spikelets 5–8mm long; nutlet ovoid, 2.7–3.5 × 1.8–2.5mm, white or greyish. Seasonally wet grassland; 1300m.
Distr.: tropical Africa and Madagascar [Palaeotropics].
IUCN: LC
Bali Ngemba F.R.: Kongor 49 fr., 10/2001; Onana 1876 10/2001.
Note: *Onana* 1876 & *Kongor* 49 are tentatively placed in this taxon since they were collected in montane grassland. In addition, *Onana* 1876 reached 2m tall.

Scleria melanotricha Hochst. ex A.Rich. var. *grata* (Nelmes) Lye

Haines, R. & Lye, K. (1983). The Sedges and Rushes of East Africa: 339.
A slender annual; stems 10–50cm; leaves 1–2mm wide, hairy; inflorescence spicate, 3–20cm; glomerules paired, shortly pedunculate; glumes straw-coloured to reddish brown, densely-hairy. Grassland; 1450m.
Distr.: Sierra Leone, Ivory Coast, N Nigeria, W Cameroon, scattered from Ethiopia to Zambia [Montane].
IUCN: LC
Bali Ngemba F.R.: Pollard 529 11/2000.

Scleria woodii C.B.Clarke var. *ornata* (Cherm.) W.Schultze-Motel

Haines, R. & Lye, K. (1983). The Sedges and Rushes of East Africa: 337.
Syn. *Scleria striatinux* De Wild.
Slender perennial, subglabrous; stems to 85cm long, slender, not dilated nor bulbous at the base; leaves to 30cm × 2.5mm; leaf sheaths hairy at the mouth; panicle lax, 5–14 × 2–5cm, occasionally branched twice; inflorescence branched; spikelets 4–5mm long, c. 1mm broad; nutlet to 2 × 1.3mm, ovoid to subglobose, surface white or greyish. In open woodland and grassland.
Distr.: Ghana, Nigeria & W Cameroon [Montane].
IUCN: LC
Bali Ngemba F.R.: Ujor 30321 5/1951.

DIOSCOREACEAE

P. Wilkin (K)

Dioscorea bulbifera L.

Herbaceous climber, glabrous, to 3–7m; stems left-twining (sinistrorse); leaves alternate with a pair of membranous, semicircular, lateral projections clasping stem at petiole base, apex short-acuminate, not thickened. Farmbush; 1400m.
Distr.: widespread in tropical Africa & Asia [Palaeotropics].
IUCN: LC
Bali Ngemba F.R.: Cheek 10742 10/2001.
Note: warty bulbils usually present.

Dioscorea praehensilis Benth.

Sturdy climber, to 8m; single tuber per growing season; stems right-twining, prickly, glabrous; leaves opposite, chartaceous, not leathery, shortly cordate, acuminate to long-acuminate. Forest & forest edge; 1200–1800m.
Distr.: Sierra Leone to Cameroon & E Africa, S to Zambia, Zimbabwe, Malawi & Mozambique [Tropical Africa].
IUCN: LC
Bali Ngemba F.R.: Onana 1811 10/2001; Tadjouteu 425 11/2000.

Dioscorea preussii Pax subsp. *preussii*

Robust, non-spiny climber, to 10m; stems often 6-winged, subglabrous, with a few ± caducous, medifixed (T-shaped) hairs, also on the inflorescence and leaf apices; leaves alternate, broadly ovate, obliquely acuminate, deeply cordate, 10–30 × 8–35cm, villous-tomentose beneath. Forest & farmbush; 1200m.
Distr.: Senegal to Uganda, Angola, Mozambique [Tropical Africa].
IUCN: LC
Bali Ngemba F.R.: Onana 1810 10/2001.

Dioscorea schimperiana Hochst. ex Kunth

Climber, 3–7m, twining, unarmed, pubescent; leaves subopposite, ovate, 10–15 × 10–15cm, acuminate, deeply cordate at the base, petiolate; male and female flowers on long, fascicled-spikes; capsule c. 4cm diam., glabrescent, 3 acute orbicular lobes and an axis c. 2.5cm long; seeds orbicular, winged. Forest; 1400–1560m.
Distr.: Nigeria to Ethiopia, southwards to Malawi and Angola [Tropical Africa].
IUCN: LC
Bali Ngemba F.R.: Biye 82 11/2000; Cheek 10740 10/2001; 10741 10/2001; Pollard 606 10/2001.

DRACAENACEAE

Y.B. Harvey (K) & I. Darbyshire (K)

Dracaena arborea Link

Tree, 10–20m, trunk 30cm, with aerial roots, several-branched; leaves in dense heads, sword-shaped, 50–120 × 4.5–6cm, widest above the middle, apex acute, mucro to 3mm, base clasping stem for ¾ circumference; inflorescence pendulous, to 1.5m; perianth cream-white, c. 1.5cm; fruit to 2cm, orange-red. Forest & planted.
Distr.: Sierra Leone to Angola [Guineo-Congolian].
IUCN: LC
Bali Ngemba F.R.: sight (& photographic) record 4/2004 (*Darbyshire*, pers. comm.).

Dracaena fragrans (L.) Ker-Gawl.

Syn. *Dracaena deisteliana* Engl.
Herb, to 3m, few-stemmed; stalk 1cm diam., leaves sword-shaped, to 70 × 9cm; inflorescence terminal, erect; flowers white with pink lines, very fragrant. Forest; 1420–2110m.
Distr.: tropical Africa [Afromontane].
IUCN: LC
Bali Ngemba F.R.: Darbyshire 368 fr. 4/2004; Etuge 5377 fr. 4/2004

Uses: ENVIRONMENTAL USES – Boundaries – used on farms for boundary markers (*Etuge* 5377).
Note: field determinations.

Dracaena phrynioides Hook.

Herb, to 0.7m, lacking aerial stem; leaves 4–6; blades lanceolate, 18 × 7cm, spotted yellow; petiole c. 1mm wide; inflorescence terminal; peduncle 5cm; fruit bilobed, orange, sessile, head c. 2.5cm diam. Forest; 1500m.
Distr.: Liberia to Bioko, Rio Muni & Gabon [Upper & Lower Guinea (montane)].
IUCN: LC
Bali Ngemba F.R.: Ghogue 1053 11/2000.

ERIOCAULACEAE

M. Cheek (K) & B.J. Pollard (K)

Eriocaulon asteroides S.M.Phillips

Kew Bull. 53: 943 (1998).
Annual rosulate herb, about 2–3cm diam.; leaves linear-subulate, 0.8–1.5cm, about 1mm wide; scapes up to 10, 1–2.5cm; capitula 5–7mm wide, few-flowered, star-like. Basalt pavement in thin peaty, seasonally waterlogged, soil; 1450–1500m.
Distr.: W Cameroon (Bamenda highlands) and S Nigeria (Mambilla Plateau) [Western Cameroon uplands (montane)].
IUCN: VU
Bali Ngemba F.R.: Pollard 525 11/2000; 700 10/2001.

GRAMINEAE

T.A. Cope (K) & Y.B. Harvey (K)

Andropogon schirensis A.Rich.

Syn. ***Andropogon dummeri*** Stapf
Erect, caespitose perennial, to 2m; leaf blades linear, to 45 × 1.4cm, mostly cauline; racemes 6–12cm; sessile spikelets, 5–7mm. Deciduous bushland and wooded grassland; 1400–2000m.
Distr.: tropical and subtropical Africa [Afromontane].
IUCN: LC
Bali Ngemba F.R.: Gosline 270 11/2000; Onana 1909 10/2001.

Arthraxon micans (Nees) Hochst

F.T.E.A. Gramineae: 742 (1982).
Annual, 10–60cm; culms slender, ascending, much-branched; leaf-blades 1–8cm × 3–20mm, lanceolate to ovate, amplexicaul at base, almost glabrous, rough at edges; sheaths swollen, pubescent; inflorescence (2–)8–14(–30) racemes, 2–6cm, digitate; rachis filiform;pedicelled spikelets 2.5–4mm, narrowly lanceolate, sessile; upper glume reddish purple; awns 4–8mm. Grassland; 2000m.
Distr.: W, Central and E Africa, tropical Asia, Australia, introduced in tropical America [Pantropical].
IUCN: LC
Bali Ngemba F.R.: Gosline 277 11/2000.

Ctenium ledermannii Pilg.

Perennial, 60–90cm with 2–5, conjugate racemes, ventrally yellow, dorsally grey. Steep grassland and rocks; 1500m.
Distr.: Nigeria, Cameroon, Burundi, Rwanda and CAR [Afromontane].
IUCN: NT
Bali Ngemba F.R.: Pollard 710 10/2001.
Note: although locally abundant in the Mambilla Plateau, Nigeria, this taxon is only known from about 12 localities, and each of these, only from one or two collections.

Ctenium newtonii Hack.

Caespitose perennial, up to 110cm tall; leaf laminas 5–30cm × up to 4mm, tightly involute; raceme 4–16cm long; upper glume 3.5–8mm long. Damp and swampy grassland; 1300m.
Distr.: from Senegal eastwards to Sudan and southwards into Angola [Tropical & subtropical Africa].
IUCN: LC
Bali Ngemba F.R.: Onana 1899 10/2001.

Digitaria abyssinica (Hochst. ex A.Rich.) Stapf

Perennial, 20–60cm, rhizomatous, glabrous to villous; leaf-blades linear to lanceolate, 5–15cm × 3–11mm; ligule 1.5–3mm, glabrous or with hairs; inflorescence 3–10 racemes, each 3–8cm, digitate or subdigitate, along 2–8cm rachis; racemes 2–9cm, very slender, rachis triquetrous; pedicels 0.3mm, spikelets in pairs; spikelets 1.5–2.4mm, about 3 times as long as broad, ovate-elliptic, acute at tip, glabrous, rarely pubescent; lower glume a green, ovate scale 0.3mm long; upper glume ⅔–¾ length of spikelet; back of upper lemma partly exposed; anthers and stigmas crimson. Grassland and forest edges; 2000m.
Distr.: Nigeria to Ethiopia, south to Congo (Kinshasa) and Zambia [Tropical & subtropical Africa].
IUCN: LC
Bali Ngemba F.R.: Gosline 275 11/2000.

Digitaria diagonalis (Nees) Stapf var. *diagonalis*

Robust, erect, perennial herb, 1–3m, with handsome, spreading inflorescence; spikelets glabrous. Open grassy places from waterlogged soils to rocky slopes; 1300–1500m.
Distr.: W Cameroon, Tanzania, Malawi, Zambia, Zimbabwe and South Africa [Afromontane].
IUCN: LC
Bali Ngemba F.R.: Onana 1896 10/2001; Pollard 711 fl., 10/2001.

Elymandra androphila (Stapf) Stapf

Coarse perennial, to 2.5m; racemes paired; pedicelled and homogamous spikelets glabrous. Fallow; 1300m.
Distr.: W and central Africa down to Angola [Tropical & subtropical Africa].
IUCN: LC
Bali Ngemba F.R.: Onana 1895 10/2001.

Eragrostis atrovirens (Desf.) Trin. ex Steud.

A very variable perennial; culms 0.45–1m, 1.5–3mm diam. at base, leafy; leaf-blades 15–30 × 0.2–0.4cm, flat or rolled; spikelets pallid to grey-purple. Swamps and wet places; 2200m.
Distr.: tropical Africa eastwards through India to the Philippines [Palaeotropics].
IUCN: LC
Bali Ngemba F.R.: Ghogue 1087 11/2000.

Eragrostis pobeguinii C.E. Hubb.

Wageningen Agric. Univ. Pap. 92.1: 120 (1992).
Densely caespitose perennial, about 30cm; panicle scantily branched, bearing to 15 spikelets, pallid to olive-grey; leaf-blades to 2mm wide, usually rolled and setaceous; basal sheaths bulbously swollen and hardened below. Grassland.
Distr.: Guinea (Conakry), Senegal, Ghana and Cameroon.
IUCN: LC
Bali Ngemba F.R.: Ujor 30323 5/1951.

Eragrostis volkensii Pilg.

Straggly perennial with a dense tussock; culms 30–90m, slender, wiry, many-noded; spikelets dark olive-green. Grassland and wet places; 2000m.
Distr.: tropical Africa southwards to South Africa (Transvaal) [Tropical & subtropical Africa].
IUCN: LC
Bali Ngemba F.R.: Gosline 271 11/2000.

Hyparrhenia diplandra (Hack.) Stapf

Coarse, caespitose perennial; culms up to 3m high, erect; leaf sheaths glabrous; leaf laminas 20–60cm × 3–10mm, subglabrous, usually with long, grey-hairs at base; false panicle 20–40cm long, narrow, usually purplish. Wet places, farmbush-grassland; 1400–2000m.
Distr.: tropical Africa west of (and including) Cameroon mountain chain to Indochina and Indonesia [Palaeotropics].
IUCN: LC
Bali Ngemba F.R.: Gosline 276 11/2000; Onana 1902 10/2001.

Hyparrhenia niariensis (Franch.) Clayton

Annual; culms up to 2m high, glabrous; leaf sheaths glabrous or pubescent along the margins; leaf laminas to 60cm × 15mm, subglabrous; false panicle 30–50cm long; homogamous spikelets 9–11mm long; sessile spikelets 7.5–11mm long; pedicelled spikelets 8–13mm long. Rocky grassland; 1500m.
Distr.: from Cameroon south to Angola and east to Uganda and Tanzania [Tropical Africa].
IUCN: LC
Bali Ngemba F.R.: Pollard 712 10/2001.

Hyparrhenia rufa (Nees) Stapf

Perennial or sometimes annual, 0.3–2.4m; leaf laminas 30–60 × 2–8mm; false panicle 5–8mm long; spikelets sessile, 3–5mm. Wet places, roadsides; 2100m.
Distr.: tropical Africa to South Africa, introduced in tropical America and Australia [Pantropical].
IUCN: LC
Bali Ngemba F.R.: Pollard 685 10/2001.
Note: more typically a savanna grass.

Hyparrhenia smithiana (Hook. f.) Stapf var. *major* Clayton

Syn. ***Hyparrhenia chrysangyrea*** (Stapf) Stapf
Caespitose perennial, 1.5–3m, basal pubescence dense, dark red, sometimes almost black; spikelet indumentum rufous or fulvous. Grassland, farmbush-grassland; 2000m.
Distr.: Guinea to Congo (Brazzaville) [Guineo-Congolian (montane)].
IUCN: LC
Bali Ngemba F.R.: Gosline 280 11/2000.

Hyparrhenia umbrosa (Hochst.) T.Anderson ex Clayton

Perennial; culms 1–2m; leaf-blades to 60cm × 12mm; ligule 2mm; spathes navicular, 1.5–2.5cm, glabrous or lightly pilose along edges, raceme base unappendaged; sessile spikelets 4mm, villous with white hairs, awn 0.7–1.3cm, pedicelled spikelets 5–6mm, villous with white hairs, muticous or mucronate. Meadows, roadsides, fallow; 1430m.
Distr.: N Nigeria, W Cameroon, widely scattered in tropical and South Africa, but not common [Tropical Africa].
IUCN: LC
Bali Ngemba F.R.: Etuge 4705 11/2000.

Isachne buettneri Hack.

Perennial, spreading; culms ascending, 30–50cm, rooting at nodes; leaf-blades 5–17cm × 5–20mm, linear-lanceolate, slender, glabrous or scabrous to pubescent; inflorescence an ovate panicle, 8–20cm, lax and open, reddish; branches up to 10cm, straight, extending obliquely; spikelets 0.8–1.4mm, light green; florets subrotund, indurated; lemmas similar in size, minutely pubescent; glumes slightly shorter than florets. Humid forests, often at edges of streams; 1500m.
Distr.: Guinea Bissau to Bioko, Cameroon, Gabon, Congo (Brazzaville) & (Kinshasa), Sudan, Uganda, Burundi, Zambia [Afromontane].
IUCN: LC
Bali Ngemba F.R.: Ghogue 1069 11/2000.

Loudetia arundinacea (Hochst. ex A.Rich.) Steud.

Tufted perennial, to 3m; leaf-blades 30–70cm; panicle robust, 20–60cm; branches spreading and conspicuously whorled; spikelets 6–13mm long. Grassland, rocky slopes and in swampy soils; 1300m.
Distr.: Senegal to Ethiopia and southwards to Angola [Tropical & subtropical Africa].
IUCN: LC
Bali Ngemba F.R.: Kongor 50 10/2001.

Loudetia kagerensis (K.Schum.) C.E.Hubb. ex Hutch.

Perennial; culms to 1m high, slender, darkened around the nodes; leaf blade linear, 2–15cm × 4–5mm; panicle loose, 2–25 × 8cm; spikelets lanceolate to lanceolate-oblong, to c. 1cm long, brown or greyish brown. Grassland; 1400–1500m.
Distr.: W Africa to E Africa (including Congo (Kinshasa) [Tropical Africa].
IUCN: LC
Bali Ngemba F.R.: Onana 1905 10/2001; Pollard 708 10/2001.

Melinis minutiflora P.Beauv. var. *minutiflora*

Perennial, erect or ascending from a prostrate base; culms to 2m; leaves 10–18cm × 8–14mm, covered in sticky hairs and smelling strongly of molasses or linseed oil; panicle to 15–30cm, pinky-lilac. Forest edge, grassland; 2000m.
Distr.: tropical Africa and introduced throughout the tropics [Pantropical].
IUCN: LC
Bali Ngemba F.R.: Gosline 278 11/2000.

Oplismenus hirtellus (L.) P.Beauv.

Variable perennial; culms slender, erect or climbing, rooting at nodes, 15–100cm; inflorescence 0.5–1.5cm; spikelets glabrous to pubescent; awns 2–4(–15)mm, viscid, smooth, stiff, the tips truncate. Shady places in forests and along roadsides; 1400–1500m.
Distr.: pantropical.
IUCN: LC
Bali Ngemba F.R.: Cheek 10460 9/2000; Ghogue 1013 11/2000; 1067 11/2000.

Panicum hochstetteri Steud.

Perennial, up to 1m; sheaths shorter than internodes; leaf-blades 7–10cm × 5–8mm, pubescent; inflorescence an ovate panicle, 7–10cm; lower branches ascending or patent, up to 7cm; inflorescence with or without long, glistening, white hairs on axis; spikelets up to 2.5mm, ovate-oblanceolate, purplish, glabrous or sometimes scabrid-hairy; nerves prominent and often forming raised ribs; lower glume 3-nerved, ⅔–¾ length of spikelet, lanceolate; lemma 1.5mm, white, tough, shiny; anthers yellow, 0.8mm long. Forest edges; 2000m.
Distr.: Cameroon, São Tomé, Congo (Kinshasa), Burundi, Kenya, Ethiopia [Afromontane].
IUCN: LC
Bali Ngemba F.R.: Gosline 272 11/2000; 274 11/2000.

Panicum nervatum (Franch.) Stapf

F.T.E.A. Gramineae: 475 (1982).
Syn. ***Panicum baumannii*** K.Schum.
Perennial, with weak scrambling culms, 30–130cm long; leaf lamina 5–15cm long, narrowly lanceolate to linear; panicle 6–18cm long, ovate-oblong; spikelets 1.5–2.5mm long. Wooded grassland and open hillsides; 1500m.
Distr.: from W Africa to Sudan and southwards to Angola and Zimbabwe [Tropical & subtropical Africa].
IUCN: LC
Bali Ngemba F.R.: Pollard 709 10/2001.

Pennisetum purpureum Schumach.

Perennial, stoloniferous, usually 1–3m, but can reach 6m; culm pubescent just below spike; leaf-blades 0.5–1m × 1–4cm, edges serrulate; ligules ciliate; sheath and leaf-blade scabrous, pubescent; inflorescence a terminal panicle, linear, 20–35cm, 15–30mm diam.; involucres subsessile; rachis densely-pubescent to loosely pilose; spikelets 4–7.5mm, glabrous, very short pedicels; lower floret variable; tuft of small hairs at tip of anthers. Streambanks in forest, river valleys, fallow, grassland; 1400m.
Distr.: tropical Africa, widely introduced elsewhere in tropics [Pantropical].
IUCN: LC
Bali Ngemba F.R.: Gosline 282 11/2000.

Pennisetum unisetum (Nees) Benth.

F.T.E.A. Gramineae: 681 (1982).
Syn. ***Beckeropsis uniseta*** (Nees) K.Schum.
Perennial, 1–4m, caespitose; culms 10–15mm diam.; nodes and basal sheath pubescent; apex of sheath usually bearded with a line of hairs on abaxial side; leaf-blades 60cm × 0.5–3cm, usually with slender petiole-like base; inflorescence a false panicle comprising numerous axillary racemes, 3.5–4cm; spikelets 2–3mm, elliptic, acute, overlapping by ¾ their own length or more; glumes subequal, up to 0.5mm long; lower lemma hispidulous, upper scaberulous; anthers orange-red. Shady humid places in woodlands and savanna; 2000m.
Distr.: tropical & southern Africa, Yemen [Afromontane & Arabian Peninsula].
IUCN: LC
Bali Ngemba F.R.: Gosline 281 11/2000.

Setaria poiretiana (Schult.) Kunth

Fl. Rwanda 4: 371 (1988).
Syn. ***Setaria caudula*** Stapf
Tufted perennial, 1.2–1.8m; leaf-blades broad, plicate; spikelets 3–4mm. Forest; 1500m.
Distr.: Nigeria east to Ethiopia and E Africa, introduced in tropical Asia and America [Montane].
IUCN: LC
Bali Ngemba F.R.: Ghogue 1043 11/2000.

Setaria sphacelata (Schumach.) Stapf & C.E.Hubb. ex M.B.Moss var. *sphacelata*

Syn. ***Setaria anceps*** Stapf ex Massey
Perennial, 0.5–2m, densely caespitose; culms erect, sometimes ascending, glabrous; leaf-blades linear, 15–50cm × 2–15mm; ligule short, ciliate; inflorescence a spike-like panicle, 5–40cm × 4–8mm; spikelets in glomerules or solitary; 6–10 hairs, 4–8mm, per glomerule, pale reddish or yellowish; spikelets subtended by bristle(s), 2–2.3mm, elliptic, glabrous, ± dorsally compressed; upper glume ¼–¾ length of spikelet, exposing conspicuously rugose upper lemma. Grassland; 2000m.
Distr.: tropical & southern Africa.
IUCN: LC
Bali Ngemba F.R.: Gosline 279 11/2000.
Note: species highly variable.

HYACINTHACEAE

Y.B. Harvey (K)

Drimia altissima (L.f.) Ker Gawl.

Bothalia 13: 452 (1981).
Syn. ***Urginea altissima*** (L.f.) Baker
Herb, to 2m; bulb globose, 10–15cm diam.; leaves 5–6, lorate-lanceolate, glabrous, 30–45cm long; peduncle to 2m; raceme cylindrical, 30–60cm long; perianth campanulate, c. 7mm long; capsule globose, c. 1cm diam.; seeds black. Grassland; 1450m.
Distr.: Sierra Leone down to South Africa [Tropical & subtropical Africa].
IUCN: LC
Bali Ngemba F.R.: Pollard 949 fr., 4/2002.

HYPOXIDACEAE

B.J. Pollard (K)

Fl. Cameroun 30 (1987).

Hypoxis camerooniana Baker

Fl. Cameroun 30: 37 (1987).
Syn. ***Hypoxis recurva*** Nel
Subterranean rhizome, c. 2–4 × 2–4cm when dried, bearing

succulent, white roots and masses of old leaf fibres; whole plant covered with golden hairs, especially on the nerves of the leaves, the scape and capsule; leaves to 50 × 0.5–2cm , recurved, prostrate to erect; inflorescences appearing during the dry season, before the leaves, which develop before flowering is completed; scape to 25cm, with c. 5–7 flowers; flowers yellow, frequented by bees; tepals 6, lanceolate, 6–12 × 4–5.5mm; fruit with c. 15 seeds. Grassland; 2100m.
Distr.: SE Nigeria, Cameroon, to Ethiopia [Afromontane].
IUCN: LC
Bali Ngemba F.R.: Pollard 680 fl., 10/2001; 961 fl., 4/2002.

IRIDACEAE

B.J. Pollard (K)

Gladiolus dalenii van Geel var. *andongensis* (Baker) Goldblatt

Taxonomania 6: 12 (2002).
Herb, to 60–90cm, often puberulous; corm (15–)20–30mm diam.; tunics of brittle, membranous layers, sometimes fibrous, reddish brown, usually with numerous tiny cormlets around the base; leaves not contemporaneous with the flowers, appearing on separate shoots after flowering, at least 2, narrowly lanceolate, 30–50cm × 4–16mm; spike (2–)3–9-flowered; flowers orange or rarely yellow; perianth tube 25–33(–40)mm, the dorsal sepal exceeding the laterals. Grassland, savanna, woodland; 1450–1680m.
Distr.: Guinea, Sierra Leone, Ivory Coast, Cameroon, CAR., Congo (Kinshasa) to Ethiopia, E Africa, south to Mozambique, Zimbabwe [Tropical Africa].
IUCN: LC
Bali Ngemba F.R.: Onana 2038 fl., 4/2002; Pollard 934 fl., 4/2002; 975 fl., 4/2002.

Gladiolus unguiculatus Baker

Slender herb, 30–60cm; corm 15–25(–35)mm diam., often dark red outside; cataphylls membranous, lower 2 red-brown, the upper green; leaves 2–3, short, similar to cataphylls, sheathing, 6–9cm; foliage leaves produced on separate shoots after blooming, (1–)2–3, linear to narrowly lanceolate, to 30–45cm × 4–8(–12)mm; spike 10–18-flowered; flowers cream to light purple; the lower tepals marked purple; tepals unequal. Seasonally wet sites, rock outcrops, savanna, woodland; 1450–1680m.
Distr.: tropical & subtropical Africa, but suprisingly mostly absent from E Africa, only occurring there in SW Tanzania.
IUCN: LC
Bali Ngemba F.R.: Onana 2037 fl., 4/2002; Pollard 937 fl., 4/2002; 974 fl., 4/2002.
Note: this species flowers early in the wet season in Cameroon.

Moraea schimperi (Hochst.) Pic.Serm.

Erect herb; corm 2cm diam.; leaves linear-lanceolate, about 1m, appearing after flowering; inflorescence of 1–3 flowers; peduncle about 30cm; perianth 4–4.5cm, mauve. Grassland, marshes, streamsides; 2100m.
Distr.: tropical Africa [Afromontane].
IUCN: LC
Bali Ngemba F.R.: Pollard 958 fl., 4/2002.

MARANTACEAE

B.J. Pollard (K)

Fl. Cameroun 4 (1965).

Hypselodelphys poggeana (K.Schum.) Milne-Redh.

Lianescent herb, to several metres long; leaves linear-oblong, 8–15 × 3–9cm, abruptly acuminate, subtruncate; calloused portion of petiole above point of articulation, 1–2cm; inflorescence, loose, little-branched spikes, nearly straight, 5–9cm; internodes c. 5mm; bracts 2–3.5cm; flowers violet and white; fruit muricate, 3-lobed, 4.5–5.0 × 3.0–3.5cm; tubercles short and dense, not curved, less than 2mm. Lowland forest; 1550m.
Distr.: Sierra Leone to Cameroon, Bioko, Gabon, Congo (Brazzaville) & (Kinshasa), Angola [Guineo-Congolian].
IUCN: LC
Bali Ngemba F.R.: Pollard 978 fl., 4/2002.

Hypselodelphys scandens Louis & Mullend.

Lianescent herb, to several metres long; branched; leaves elliptic or oblong-linear, 12–35 × 5–17cm, shortly acuminate, subtruncate; calloused portion of petiole above point of articulation, to 3cm, conspicuously beaked at junction with midrib adaxially; inflorescence pendulous with a number of bifurcations; branches usually zig-zag, c. 20cm long; internodes c. 1cm; bracts 3.5–4.5cm; flowers pale violet, white and brown; fruit muricate, 3-lobed, c. 5 × 2–3cm; tubercles long, often curved, to 5mm. Forest; 1430m.
Distr.: Ivory Coast to Cameroon, Gabon, Congo (Kinshasa) [Guineo-Congolian].
IUCN: LC
Bali Ngemba F.R.: Etuge 4709 fr., 11/2000.

Marantochloa leucantha (K.Schum.) Milne-Redh.

Erect or climbing herb, to 5m; leaves homotropic, 15–20(–40) × 7–12(–25)cm, abruptly long-acuminate, not pruinose; petiole with calloused portion, c. 1cm; inflorescence of long loose-panicles, much-branched, pendent, 30–40cm; flowers whitish; fruits smooth (before drying), spherical, 1cm diam., reddish, becoming yellowish on drying; perianth not persistent. Forest; 1350m.
Distr.: tropical Africa [Afromontane].
IUCN: LC
Bali Ngemba F.R.: Etuge 5426 fl., 4/2004.

ORCHIDACEAE

B.J. Pollard (K), D.L. Roberts (K) & P.J. Cribb (K)

Fl. Cameroun 34 (1998), 35 (2001) & 36 (2001).

The species names listed below are derived from threee main sources:
1. Names based on specimens collected by Ujor, or on our expeditions 2000–2002. These were determined at K by Cribb and Pollard.

2. Names taken from Richardiana 1(4): 153–186 (2001), mainly from DeMarco field dets. based on *Chiron* specimens collected earlier that year. These specimens have not yet been verified at K, so no descriptions or conservation assessments are given for these taxa unless collected by others at Bali Ngemba.
3. Names based on field dets. by Dave Roberts, mainly of specimens under *Simo* numbers, in April 2004.

Aërangis gravenreuthii (Kraenzl.) Schltr.

Epiphyte; stem woody, to 7 × 0.5cm; leaves distichous, oblanceolate, strongly falcate, to 15 × 1.5–3cm, narrowing basally; inflorescences pendent, axillary, racemose, 10–20cm, 2–5-flowered; flowers white, sometimes with an orange flush; dorsal sepal 20–32 × 5–7mm; spur reddish, 4–8(–12)cm. Forest, woodland; 1400–2075m.
Distr.: Bioko, Cameroon, Tanzania [Lower Guinea & Congolian (montane)].
IUCN: NT
Bali Ngemba F.R.: Simo 142 fr., 4/2004; 156 fl., 4/2004; 166 fl., 4/2004.
Note: *Simo* collections are field determinations. In addition, the following collections have been made by *Chiron*, 3–4/2001: 1229; 1236; 1252; 1255; 1264; 1265 (1470–1900m alt.). *Chiron* tentatively places his following specimens in this taxon: 1223; 1231 & 1243 (forest; 1700–1950m alt.).

Aërangis sp.

Secondary forest; 1800m.
Bali Ngemba F.R.: Chiron 1228 3–4/2001.

Ancistrorhynchus serratus Summerh.

Epiphyte; stem 5–10cm; leaves 5–9, 5–18 × 0.7–1.2cm, with parallel sides; apex irregularly serrate; inflorescence < 1cm, several–multi-flowered; flowers white; labellum 2.5–3 × 2.5–4.5mm; spur 3.5–4.5mm. Forest; 2075m.
Distr.: SE Nigeria, W Cameroon, Bioko, Rio Muni, São Tomé [Lower Guinea].
IUCN: NT
Bali Ngemba F.R.: Simo 160 fl., 4/2004.
Note: field determination.

Angraecopsis ischnopus (Schltr.) Schltr.

Epiphyte; stem to 5cm; leaves elliptic-oblong, elliptic-ligulate or ligulate, 2–11 × 0.4–1.3cm; inflorescence lax, 3–9cm, to 10-flowered; flowers small, resupinate, dull greenish white, fading brown; labellum to 8.5mm; spur 1.2–3.8cm, very slightly or not at all swollen apically. Forest, plantations; 1400–1950m.
Distr.: Guinea, Sierra Leone, Nigeria, Cameroon [Upper & Lower Guinea].
IUCN: LC
Bali Ngemba F.R.: Biye 128 11/2000; Onana 1814 fr., 11/2001; Simo 151 fl., 4/2004.
Note: *Simo* 151 is a field determination. In addition, the following collection has been made by *Chiron*, 3–4/2001: 1104 (forest; 1600m alt.).

Angraecopsis sp. vel *tenerrima* Kraenzl.

Epiphyte; inflorescence long; flowers white; labellum with 3 equal lobes; spur very long, with a wide mouth. Forest; 1200–1450m.
Distr.: Bali Ngemba F.R
Bali Ngemba F.R.: Onana 1813 fl., 10/2001; Simo 181 fl., 4/2004.
Note: *Onana* 1813 may well prove to represent an undescribed species, more material is required. *Simo* 181 is a field determination.

Angraecopsis sp. 1

Secondary forest near village; 1390m.
Bali Ngemba F.R.: Chiron 1013 3–4/2001.

Angraecopsis sp. 2

1500m.
Bali Ngemba F.R.: Chiron 1069 3–4/2001.

Brachycorythis kalbreyeri Rchb.f.

Terrestrial herb or epiphyte, 15–40cm; stem delicate; leaves up to 15, lanceolate to broadly-lanceolate, acute, to 11 × 2.5cm; inflorescence a spike, laxly up to 22-flowered, usually much less, to 17cm; flowers whitish, tinged mauve or lilac; labellum 2–3cm. Riverine forest, forest, plantations; 1450m.
Distr.: Guinea, Sierra Leone, Liberia, Cameroon, Congo (Brazzaville) & (Kinshasa), E Africa [Afromontane].
IUCN: LC
Bali Ngemba F.R.: Pollard 971 fl., 4/2002; 1036 fl., 4/2002; Simo 191 fl., 4/2004.
Note: *Simo* 191 is a field determination.

Brachycorythis ovata Lindl. var. *schweinfurthii* (Rchb.f.) Szlach. & Olszewski

Terrestrial herb, 40–100cm; leaves numerous, to c. 35, to 8 × 2.5cm, lanceolate to oblong-ovate; inflorescence 13–35cm, ± lax; bracts to 4cm; flowers violet or mauve, and somewhat whitish; labellum 12–16 × 6.5–10.5mm; spur absent. Grassland.
Distr.: Senegal, Ivory Coast, N Nigeria, Cameroon, Congo (Kinshasa), Sudan, Uganda, Kenya [Afromontane].
IUCN: LC
Bali Ngemba F.R.: Ujor 30324a fl., 5/1951.

Bulbophyllum calyptratum Kraenzl. var. *calyptratum*

Orchid Monogr. 2: 129 (1987).
Epiphyte; pseudobulbs bifoliate, 1.3–5cm apart; petiole 5–25mm; leaves linear-lanceolate to linear, 7.2–26 × 0.5–1.6cm; inflorescence 22–47cm, 8–50-flowered; peduncle 11–21cm; rachis greenish to ± white, terete or ± flattened; flowers white or greenish, ± suffused purple; labellum 1–2 × 0.4–0.8mm. Forest, secondary growth, plantations; 1440–1500m.
Distr.: Guinea to Cameroon, Gabon, Rio Muni, Congo (Brazzaville & Kinshasa) [Guineo-Congolian].
IUCN: LC
Bali Ngemba F.R.: Ghogue 1064 11/2000; Pollard 421 fr., 11/2000.

Bulbophyllum cochleatum Lindl. var. *bequaertii* (De Wild.) J.J.Verm.

Orchid Monogr. 2: 44 (1987); Fl. Rwanda 4: 530 (1988).
Epiphytic herb; pseudobulbs 1–6cm apart, 3–14 × 0.7–1.3cm; leaves linear-lanceolate, 6.5–24 × 0.6–2.0cm; inflorescence 13–45cm; flowers red or reddish brown. Forest; 1310–1400m.

Distr.: W Cameroon, Congo (Kinshasa), Rwanda, Uganda, Tanzania [Afromontane].
IUCN: LC
Bali Ngemba F.R.: Simo 152 fl., 4/2004; Zapfack 2006 fl., 4/2002.
Note: *Simo* 152 is a field determination. In addition, the following collections have been made by *Chiron*, 3–4/2001: 1025; 1058; 1094; 1269 (1385–1470m alt.).

Bulbophyllum cochleatum Lindl. var. *cochleatum*

Forest; 1400–1530m
Bali Ngemba F.R.: Chiron 1034 3–4/2001; 1075 3–4/2001; 1082 3–4/2001.
Note: *Chiron* tentatively places his following specimens in this taxon also: 1002, 1015 & 1090 (1420–1440m alt.).

Bulbophyllum cochleatum Lindl. var. *gravidum* (Lindl.) J.J.Verm.

Bull. Jard. Bot. Natl. Belg. 56: 230 (1986).
Syn. ***Bulbophyllum gravidum*** Lindl.
Epiphytic herb; pseudobulbs 0.7–4.5 × 0.4–1.3cm, bifoliate, 0.6–10cm apart; leaves 2.5–12 × 0.3–1.8cm; inflorescence 7–23cm; peduncle 6.2–19cm; rachis not thickened, 1.5–9.5cm, 10–40-flowered; tepals yellowish or greenish, often suffused with purple-red or entirely dark purple-red; labellum dark purple-red, 3.5–5.8 × 0.7–1.5mm, marginal hairs ≥ labellum width. Forest, woodland; 1500m.
Distr.: Cameroon, Bioko, Congo (Kinshasa), Tanzania, Zambia, Malawi [Afromontane].
IUCN: LC
Bali Ngemba F.R.: Zapfack 2125 fl., 4/2002.
Note: in addition, the following collections have been made by *Chiron*, 3–4/2001: 1042; 1043; 1088; 1247 (1400–1450m alt.).

Bulbophyllum cochleatum Lindl. var. *tenuicaule* (Lindl.) J.J.Verm.

Bull. Jard. Bot. Nat. Belg. 56: 230 (1986).
Syn. ***Bulbophyllum tenuicaule*** Lindl.
Epiphytic herb; pseudobulbs narrowly conical or cylindrical, 1–10 × 0.4–1.2cm, bifoliate, 1–9cm apart, leaves 1.8–16 × 0.4–1.5cm; inflorescence 6.5–22cm; peduncle 3–14.5cm; rachis slightly thickened, or not, 2–12cm, 8–60-flowered; flowers red-purple, whitish at base; labellum dark brown-red or purple-red, with marginal hairs ≥ labellum width. Forest, woodland; 1450–2100m.
Distr.: SW Nigeria, W Cameroon, Bioko, São Tomé, Congo (Kinshasa), Rwanda, Uganda, Kenya [Afromontane].
IUCN: LC
Bali Ngemba F.R.: Onana 1817 fl., 10/2001; Pollard 619 fl., 10/2001; 672 fl., 10/2001; Simo 180 fl., 4/2004.
Note: *Simo* 180 is a field determination.

Bulbophyllum encephalodes Summerh.

Bot. Mus. Leafl. 14: 228 (1951).
Epiphyte; pseudobulbs unifoliate, (1.5–)3–8cm apart, 1.2–3.5 × 1–2.2cm, c. sharply 4-angled; petiole 4–6mm; lamina lanceolate, 3–14.5 × 1.2–3.2cm; inflorescence 13–43cm, 14–36-flowered; rachis 4-angled in section, 2–11 × 0.2–0.35cm, reddish-brown; sepals greenish, suffused, spotted, striated or entirely purple; petals pale greenish or purple; lip very dark purple or white, marked purple. Forest, plantations; 1400m.
Distr.: W Cameroon, Congo (Kinshasa), Burundi, E Africa, Zambia, Malawi, Zimbabwe [Afromontane].
IUCN: LC
Bali Ngemba F.R.: Simo 198 fl., 4/2004.
Note: *Simo* 198 is a field determination, as yet unconfirmed. In addition, the following collection has been made by *Chiron*, 3–4/2001: 1048 (forest; 1390m alt.). *Chiron* tentatively places his following specimen in this taxon also: 1035 (1400m alt.).

Bulbophyllum intertextum Lindl. var. *intertextum*

Diminutive epiphyte; pseudobulbs unifoliate, 0.2–2.5(–4)cm apart, 0.4–1.0 × 0.3–0.7cm; leaves elliptic to linear-lanceolate, 0.7–10 × 0.3–1.1cm; inflorescence 2–30cm, 2–14(–20)-flowered; rachis arching, nodding, terete, usually zigzag-bent, 0.2–19cm; floral bracts 1.5–4 × 1–2mm; flowers very pale yellowish or greenish, often suffused red. Forest, forest patches in grassland; 1400m.
Distr.: tropical & subtropical Africa.
IUCN: LC
Bali Ngemba F.R.: Simo 146 fl., 4/2004.
Note: *Simo* 146 is a field determination. In addition, the following collections have been made by *Chiron*, 3–4/2001: 1080; 1249 (1530m alt.). *Chiron* tentatively places his following specimens in this taxon also: 1045 & 1078 (c. 1500m alt.).

Bulbophyllum josephi (Kuntze) Summerh. var. *josephi*

Orchid Monogr. 2: 68 (1987).
Epiphytic or epilithic herb; pseudobulbs unifoliate, 0.7–3cm apart, 1.5–4 × 0.6–2.4cm; lamina lanceolate, 4.5–28 × 0.9–3.2cm, slightly emarginate; petiole 2–35mm; inflorescence 8.5–40cm, 7–80-flowered; rachis arching to pendulous; flowers cream, tinged pink, pale green or yellowish; column without teeth along the adaxial margins. Forest; 1400m.
Distr.: Cameroon, Congo (Kinshasa), Rwanda, Burundi, Ethiopia, Kenya, Tanzania, Malawi, Mozambique [Afromontane].
IUCN: LC
Bali Ngemba F.R.: Simo 199 fl., 4/2004.
Note: *Simo* 199 is a field determination. In addition, the following collection has been made by *Chiron*, 3–4/2001: 1270 (1500m alt.). *Chiron* tentatively places his following specimens in this taxon also: 1067; 1238 & 1246 (1400–1900m alt.).

Bulbophyllum maximum (Lindl.) Rchb.f.

Syn. ***Bulbophyllum oxypterum*** (Lindl.) Rchb.f.
Epiphyte; pseudobulbs 2(–3)-leaved, 2–10cm apart, 3.5–10 × 1–3cm; leaves oblong to linear-lanceolate, maximum width usually just above middle, 3.8–20 × 1.3–5.cm; inflorescence 15–90cm; 16–120-flowered; rachis bladelike, 6–56 × 8–50mm; floral bracts spreading to reflexed, 2.5–7 × 2–4mm; flowers yellowish or greenish, spotted purple. Forest, savanna woodland, bush fallow; 1450m.
Distr.: tropical & subtropical Africa.
IUCN: LC
Bali Ngemba F.R.: Simo 188, 4/2004.

Note: *Simo* 188 is a field determination. In addition, the following collections have been made by *Chiron*, 3–4/2001: 1001; 1087 (1400m alt.).

Bulbophyllum nigericum Summerh.

Epiphyte; pseudobulbs bifoliate, ovoid, 0.8–2.5cm apart, 1.3–2.7 × 0.7–1.5cm, obtusely 4-angled; petiole 1–2mm; lamina lanceolate to linear-lanceolate, 3–7 × 0.5–1.2cm, tip obtuse; inflorescence 0.8–2.3cm, 7–30-flowered; peduncle 4–9.5cm × 3.5mm; rachis 3.5–14cm; floral bracts 2–12 × 4–6.5cm ; sepals pale green to yellow, lateral petals purple or yellow; labellum yellow or orange. Forest; 1700m.
Distr.: S Nigeria, W Cameroon [W Cameroon uplands].
IUCN: VU
Bali Ngemba F.R.: Plot voucher BAL25 fr., 4/2002.

Bulbophyllum oreonastes Rchb.f.

Syn. ***Bulbophyllum zenkerianum*** Kraenzl.
Epiphytic or epilithic herb; pseudobulbs bifoliate, 0.7–4.0cm apart, 0.4–3.5 × 0.4–1.2cm; leaves elliptic to linear-lanceolate, 0.6–8.2 × 0.4–2.0cm; inflorescence 1.5–17.5cm, 5–36-flowered; floral bracts yellow, often suffused red; tepals yellow, orange to dark red-purple, with conspicuous, dark longitudinal stripes; labellum 1.4–2.5 × 0.6–1.2mm. Forest; 1310–1580m.
Distr.: tropical and subtropical Africa [Afromontane].
IUCN: LC
Bali Ngemba F.R.: Pollard 951 fl., 4/2002; Simo 150 fl., 4/2004; Zapfack 2042 fl., 4/2002.
Note: *Simo* 150 is a field determination. In addition, the following collections have been made by *Chiron*, 3–4/2001: 1040; 1052; 1059; 1063; 1105 (1390–1500m alt.).

Bulbophyllum cf. *oxychilum* Schltr.

Orchid Monogr. 2: 76 (1987) (*B. oxychilum*).
Epiphyte; pseudobulbs unifoliate, ovoid to orbicular, 0.8–2cm apart, 0.6–2.3 × 0.5–1.6cm, obtusely 3–4-angled; petiole 1–11mm; lamina narrowly elliptic to broadly linear-lanceolate, 1.6–12 × 0.7–2cm, tip obtuse to acute; inflorescence 6.1–31cm, 15–100-flowered; peduncle 1.8–7.5cm × 1–2.5mm; rachis 4.5–24cm; flowers yellow to white. Bush fallow; 1450m.
Distr.: Liberia, Ivory Coast, Ghana, Nigeria, Cameroon, Gabon, Congo (Kinshasa), CAR, Uganda [Upper & Lower Guinea] (*B. oxychilum*).
Bali Ngemba F.R.: Simo 189 fl., 4/2004.
Note: field determination.

Bulbophyllum resupinatum Ridl. var. *filiforme* (Kraenzl.) J.J.Verm.

Forest; 1530m.
Bali Ngemba F.R.: Chiron 1081 3–4/2001.

Bulbophyllum resupinatum Ridl. var. *resupinatum*

Epiphyte; pseudobulbs bifoliate, narrowly ovoid to ellipsoid, 0.5–3.5cm apart, 0.8–4 × 0.5–1.4cm, obtusely to rather sharply 2–4-angled, green, often with purple spots; petiole 1–3cm; lamina oblong to linear-lanceolate, 1.3–5(–8.5) × 0.5–1.6cm, tip obliquely emarginate; inflorescence 1.8–8.6cm, 8–60-flowered; peduncle 0.9–4cm × 0.8–1.5mm; rachis 0.9–6.5cm; floral bracts 2–4 × 1.5–2.5cm ; flowers dark red. Forest; 1400–1450m.
Distr.: Ivory Coast, Ghana, Nigeria, Cameroon, São Tomé, Principle, Congo (Kinshasa).
IUCN: LC
Bali Ngemba F.R.: Simo 149, 4/2004; 170 fl., 4/2004.
Note: field determinations.

Bulbophyllum sandersonii (Hook.f.) Rchb.f. subsp. *sandersonii*

F.T.E.A. Orchidaceae: 320 (1984).
Syn. ***Bulbophyllum tentaculigerum*** Rchb.f.
Epiphyte; pseudobulbs bifoliate, 1.2–6.5cm apart, narrowly ovoid; leaves lanceolate to linear, 3.5–26 × 0.5–2.5cm, oblique; inflorescence 5.5–30cm; rachis 1.5–9 × 0.3–1.1cm; floral bracts distinctly narrower than the fully developed part of the rachis; flowers usually placed along an excentric line on the rachis, 3–10mm apart; flowers yellowish or greenish, suffused purple. Forest; 1310–1800m.
Distr.: Cameroon, Gabon, Congo (Kinshasa), E Africa, Zambia, Malawi, Zimbabwe, Mozambique, South Africa [Tropical & subtropical Africa].
IUCN: LC
Bali Ngemba F.R.: Pollard 955 fl., 4/2002; Simo 141 fl., 4/2004; Zapfack 2035 fl., 4/2002; 2059 fl., 4/2002.
Note: *Simo* 141 is a field determination. In addition, the following collections have been made by *Chiron*, 3–4/2001: 1064; 1092; 1258 (forest; 1460–1610m alt.). *Chiron* tentatively places his following specimen in this taxon also: 1091 (1450m alt.).

Bulbophyllum unifoliatum De Wild. subsp. *unifoliatum*

Orchid Monographs 2: 154 (1987).
Epiphyte; pseudobulb unifoliate; 5.5–9cm apart, 1.4–4cm, narrowly ellipsoid; leaves linear-lanceolate to linear, 6–17 × 0.6–1.4cm; rachis glabrous; flowers yellowish or brownish-orange; sepals abaxially glabrous or, at most, very finely papillose; labellum orange-red, basally purple, 1.8–3 × 1–1.6mm. Forest, farmland; 1400m.
Distr.: Cameroon, Congo (Kinshasa), Rwanda, Tanzania, Angola, Zambia [Afromontane].
IUCN: LC
Bali Ngemba F.R.: Simo 153 4/2004.
Note: *Simo* 153 has tentatively been placed in this taxon and is a field determination.

Bulbophyllum sp. nov..

Diminutive epiphytic herb; pseudobulbs bifoliate, c. 2–5cm apart, 0.5–1.3cm tall, ovoid; leaves elliptic to oblong-elliptic, c. 1.5–2.5 × 0.2–0.6cm, slightly unequally bilobed at apex; inflorescence very slender, peduncle erect, to 2.5cm × 0.5mm; rachis c. 1cm long, very congested, to c. 10-flowered; floral bracts to 3mm, conduplicate; labellum margins papillose. Forest; 1800m.
Distr.: Bali Ngemba F.R. [narrow endemic].
Bali Ngemba F.R.: Onana 1815 fl. fr. 10/2001.
Note: entire flowers have not been seen in our specimen, but the remains of a labellum, still attached to a mature fruit, clearly show the margins to be papillose, ruling out *B. stolzii* Schltr. and *B. inornatum* J.J.Verm. This collection appears most closely related to *B. bavonis* J.J.Verm., but with pseudobulbs 0.5–1.3cm tall (not 1.2–3.5cm); rachis congested, to c. 1cm (not lax, 2.3–7.4cm); floral bracts c.

3mm, conduplicate (not 3.2–4.5mm, recurved). More complete flowering material is required to investigate.

Bulbophyllum sp. 1
Forest; 1400m.
Bali Ngemba F.R.: Chiron 1037 3–4/2001.

Bulbophyllum sp. 2
Forest; 1400m.
Bali Ngemba F.R.: Chiron 1021 3–4/2001.

Bulbophyllum sp. 3
Forest; 1400m.
Bali Ngemba F.R.: Chiron 1019 3–4/2001.

Bulbophyllum sp. 4
Secondary forest; 1390m.
Bali Ngemba F.R.: Chiron 1050 3–4/2001.

Chamaeangis ichneumonea (Lindl.) Schltr.
Epiphyte; stem covered in persistent, sheathing leaf-bases; leaves arranged in a fan, oblanceolate, 17–43 × 2.5–5cm; inflorescence 20–50cm, many-flowered; flowers white, cream or greenish brown, fragrant - especially at night; spur 13–18mm. Forest; 1400–2075m.
Distr.: Sierra Leone, Liberia, Ghana, S Nigeria, W Cameroon, Gabon [Upper & Lower Guinea].
IUCN: LC
Bali Ngemba F.R.: Simo 143 4/2004; 157 4/2004.
Note: field determinations.

Chamaeangis lanceolata Summerh.
Forest; 1400m.
Bali Ngemba F.R.: Chiron 1028 3–4/2001.
Note: *Chiron* tentatively places his following specimens in this taxon also: 1027; 1029; 1030; 1041; 1056; 1070; 1072; 1073; 1226 (1400–1700m alt.).

Chamaeangis letouzeyi Szlach. & Olszewski
Epiphyte; leaves 18–20 × 1.3–3cm, linear-lanceolate, apex acute, obscurely bilobed; inflorescence 5.5–25cm, 13-flowered; flowers small, yellow-orange; labellum oblong-ovate with callus at base, 3.3 × 1.5cm; spur incurved, swollen, circular at apex, 6mm. Disturbed forest; 1400–1450m.
Distr.: Sierra Leone, Liberia, Ivory Coast, Cameroon.
IUCN: NT
Bali Ngemba F.R.: Simo 145, 4/2004; 190 fl., 4/2004.
Note: *Simo* collections are field determinations. Although occurring in four countries, this taxon has very few collections and is therefore considered 'Near Threatened'. In addition, the following collection has been made by *Chiron*, 3–4/2001: 1060.

Chamaeangis odoratissima (Rchb.f.) Schltr.
Epiphyte; stem 20–46 × –0.6cm; leaves well-separated, numerous, 10–24 × 1.8–3.1cm, oblanceolate to oblong-oblanceolate, slightly falcate; inflorescence dense, 8–28cm, many-flowered, pendent or spreading; bracts 1mm, amplexicaul; flowers fleshy, non-resupinate, 2–6 per node, small, yellow-green to yellow; labellum 1.7–2 × 1.3–1.5mm; spur 5–11mm. Forest; 1430–1450m.
Distr.: Sierra Leone to Cameroon, Congo (Kinshasa), Angola, CAR., Uganda, Rwanda, Malawi, E Africa [Afromontane].
IUCN: LC
Bali Ngemba F.R.: Etuge 4721 fr., 11/2000; Simo 173 fl., 4/2004; 179 fl., 4/2004.
Note: *Simo* 173 & 179 are field determinations. In addition, the following collections have been made by *Chiron*, 3–4/2001: 1106; 1079 (forest; 1450–1530m alt.).

Cyrtorchis cf. *arcuata* var. *variabilis* Summerh.
Epiphyte; 15–30cm; leaves 8–24 × 1.5–4cm, oblong-elliptic or oblanceolate, apex unequally bilobed; inflorescence 5–20cm, 5–15-flowered; flowers white; sepals 1.5–3.5cm; labellum lanceolate to subelliptic, 1.5–3.6 × 0.4–8.1cm; spur straight or slightly sigmoid, 3–5.5cm. Plantations; 1450m.
Distr.: Cameroon, E Africa & southern tropical Africa (for *C. arcuata* var. *variabilis*).
Bali Ngemba F.R.: Simo 167 fl., 4/2004.
Note: field determination.

Cyrtorchis belloneorum Chiron
Forest boundary; 1470m.
Bali Ngemba F.R.: Chiron 1254 3–4/2001.
Note: Chiron mentions (2001: 173) that J. DeMarco places this specimen in *C. aschersonii* (Kraenzl.) Schltr.

Cyrtorchis chailluana (Hook.f.) Schltr.
Epiphyte; stems long, becoming pendent; leaves 11–25 × 1.9–5.0cm, linear-oblanceolate; inflorescence 13–25cm, 7–10-flowered; flowers waxy white or cream, turning apricot; spur 9–16cm; fruits trilobed, to 4cm long. Forest; 1400–1450m.
Distr.: Sierra Leone, Nigeria to Uganda [Guineo-Congolian].
IUCN: LC
Bali Ngemba F.R.: Simo 147 fr., 4/2004; 183 fr., 4/2004; 184 fr., 4/2004.
Note: field determinations.

Cyrtorchis ringens (Rchb.f.) Summerh.
Epiphyte; stem arcuate, woody, to 30cm in old plants; leaves linear, usually 6–7, thick and leathery, 7.5–12.5 × 1.2–2.5cm; inflorescence 6–7(–16)cm, to 16-flowered; flowers closely-placed, creamy white, sweetly-scented; spur to 3cm. Forest; 1450m.
Distr.: Senegal, Sierra Leone to Cameroon, São Tomé, Congo (Kinshasa), Burundi, Uganda, Tanzania, Zambia, Malawi, Zimbabwe [Afromontane].
IUCN: LC
Bali Ngemba F.R.: Simo 187 fl., 4/2004.
Note: *Simo* 187 is a field determination. In addition, the following collections have been made by *Chiron*, 3–4/2001: 1232; 1233; 1234; 1256; 1262; 1263 (1400–1900m alt.). *Chiron* tentatively places his following specimens in this taxon also: 1093; 1221; 1255 1091 (1470–1620m alt.).

Cyrtorchis seretii (De Wild.) Schltr.
Epiphyte; stems 9–20cm; leaves 8–17 × 0.6–1.6cm, linear; inflorescence 5.5–12cm, 5–10-flowered; flowers white; spur 55–63mm. Bush fallow; 1450m.
Distr.: Cameroon, CAR, Congo (Kinshasa) Uganda.
IUCN: LC

Bali Ngemba F.R.: Simo 185 fl., 4/2004.
Note: field determination.

Cyrtorchis sp. 1
Near village; 1390m.
Bali Ngemba F.R.: Chiron 1098 3–4/2001.

Cyrtorchis sp. 2
Degraded forest; 1900m.
Bali Ngemba F.R.: Chiron 1237 3–4/2001.

Diaphananthe bueae (Schltr.) Schltr.
Forest; 1900m.
Bali Ngemba F.R.: Chiron 1235 3–4/2001.

Diaphananthe cf. *bueae* (Schltr.) Schltr.
Epiphyte; stem short, bearing leaves in a relatively dense tuft; leaves 4–16 × 0.5–2cm, lanceolate-ligulate, acute; inflorescence to 18cm, 5–10-flowered; flowers white-greenish to yellow-greenish; labellum narrowly ovate-lanceolate to oblong-elliptical, narrowing towards apex; spur 12–15.5mm. Forest opening; 1400m.
Distr.: Ivory Coast, Cameroon, Uganda.
Bali Ngemba F.R.: Simo 202 fl., 4/2004.
Note: field determination.

Diaphananthe polyantha (Kraenzl.) Rasm.
Forest; 1470m.
Bali Ngemba F.R.: Chiron 1251 3–4/2001.
Note: *Chiron* tentatively places his following specimens in this taxon also: 1260; 1242 (occurring at 1600–1950m alt.).

Diaphananthe pulchella Summerh.
Degraded forest & near village; 1390–1920m.
Bali Ngemba F.R.: Chiron 1222 3–4/2001; 1240 3–4/2001; 1257 3–4/2001.
Note: *Chiron* tentatively places his following specimen in this taxon also: 1014 (occurring at 1390m alt.).

Diaphananthe rohrii (Rchb.f.) Summerh.
Forest; 1415m.
Bali Ngemba F.R.: Chiron 1007 3–4/2001.
Note: *Chiron* tentatively places his following specimen in this taxon also: 1065 (occurring at 1550m alt.).

Diaphananthe rutila (Rchb.f.) Summerh.
Forest; 1400m.
Bali Ngemba F.R.: Chiron 1024 3–4/2001.

Diaphananthe sp.
Epiphyte; stem 15–20cm; leaves 4–16 × 0.5–2cm, lanceolate-ligulate, apex subacute unequally bilobed; inflorescence to 20cm; 20–25-flowered; Bush fallow, plantations; 1400–2075m.
Bali Ngemba F.R.: Simo 148 fl., 4/2004; 159 fr., 4/2004; 169 fr., 4/2004; 204 fl., 4/2004.
Note: field determinations.

Disa equestris Rchb.f.
Terrestrial herb, 20–70cm; leaves 6–19, the lowest 1–3 sheath-like, all tinged wine-purple, 5–15(–22)cm × 4–12mm; inflorescence 3–17(–23) × 1.5–3.5cm, lax, 4–40-flowered; bracts 7–17mm; flowers mauve, violet or purple; dorsal sepal 9–12mm; labellum 4.5–6 × ± 2mm. Wet grassland; 2100m.
Distr.: Nigeria, Cameroon, CAR, Congo (Brazzaville & Kinshasa), Angola, Malawi, Zambia, Zimbabwe, Mozambique [Afromontane].
IUCN: LC
Bali Ngemba F.R.: Pollard 963 fl., 4/2002.

Disa aff. *nigerica* Rolfe
A slender, terrestrial herb, 20–30cm; leaves on stem narrowly lanceolate; inflorescence rather lax, about 10cm; flowers small, white, tinged purple through pink to purple; dorsal sepal with spur ± swollen. Grassland; 1600m.
Distr.: N Nigeria, Cameroon (Bamenda Highlands) [Western Cameroon Uplands (montane)] (*D. nigerica*).
Bali Ngemba F.R.: Darbyshire 409 fl., 4/2004.
Note: field determination.

Disa welwitschii Rchb.f. subsp. *occultans* (Schltr.) Linder
An erect, terrestrial herb, 0.3–1m; leaves 10–18, linear, 8–36 × 0.5–4cm; inflorescence 5–18(–21) × 2–5cm, 20–100-flowered; flowers red or purple; dorsal sepal oblong-obovate, rounded at apex; lateral sepals 5–9mm; labellum 4–12 × 0.7–2mm; spur 3–10mm. Swampy or stony grassland.
Distr.: Guinea (Conakry), Liberia, Ivory Coast, Nigeria, Cameroon, Gabon, CAR., Congo (Kinshasa), E Africa, Zambia, Malawi [Afromontane].
IUCN: LC
Bali Ngemba F.R.: Ujor 30324 fl., 5/1951.

Epipogium roseum (D.Don) Lindl.
Forest; 1400m.
Bali Ngemba F.R.: Chiron 1008 3–4/2001.

Eulophia guineensis Lindl.
Terrestrial or, occasionally, lithophytic herb, up to 65cm; leaves 3–4, plicate, elliptic, acute, 10–35 × 3–9.5cm; inflorescence laxly few- to several-flowered; flowers showy; sepals 16–30 × 3–6m, greenish to purple-brown; petals 15–27 × 4–7mm, greenish to purple-brown; labellum 20–30 × 13–32mm, pale to mid purple-pink, sometimes with a magenta blotch in the throat. In riverine forest and thicket shade; 1430–1560m.
Distr.: tropical Africa, southern tropical Africa, N Yemen and Oman [Palaeotropics].
IUCN: LC
Bali Ngemba F.R.: Etuge 4723 11/2000; Pollard 614 fl., 10/2001.

Eulophia odontoglossa Rchb.f.
Linnea 19: 373 (1846).
Syn. ***Eulophia shupangae*** (Rchb.f.) Kraenzl.
Terrestrial herb, 0.6–1.0m; leaves 5–6, erect, plicate, oblanceolate, acuminate, 40–70 × 1–2.1cm, basal 3 sheathing; inflorescence densely many-flowered; flowers yellow; labellum with yellow, orange or red papillae. Grassland, bushland, rocky areas; 1500m.
Distr.: tropical and subtropical Africa [Afromontane].
IUCN: LC
Bali Ngemba F.R.: Pollard 704 fr., 10/2001.

Graphorkis lurida (Sw.) Kuntze
Robust epiphyte; pseudobulbs cylindrical-fusiform or conical ovoid, 3–9 × 1–3cm, 4–6-leaved, yellowish and ribbed;

inflorescence appearing before the leaves, erect, paniculate, 15–50cm; flowers yellowish with brown sepals and petals; sepals about 5–6mm; spur sharply bent forward, ± as long as labellum; mature fruits yellow. Forest; 1450m.
Distr.: Senegal to Bioko, Cameroon, Gabon, Congo (Kinshasa), Burundi, Uganda, Tanzania [Guineo-Congolian].
IUCN: LC
Bali Ngemba F.R.: Simo 165 fr., 4/2004.
Note: field determination.

Habenaria malacophylla Rchb.f. var. *malacophylla*

A slender, terrestrial herb, 0.3–1m; stem leafy in centre part, bare below; inflorescence a long, lax raceme, 8–34 × 2.5–3.5cm; bracts lanceolate, 0.9–2.0cm; flowers numerous, small, green; petals bipartite near the base, anterior lobe 5–9.5 × 0.3–0.8mm, posterior lobe 4–7.3 × 0.4–1.5mm; labellum trilobed, median lobe 4.5–8 × 0.8–1mm; spur 9–18mm. Woodland and grassland; 1560m.
Distr.: Sierra Leone, Nigeria, Cameroon, Ethiopia, E Africa, Malawi, Zambia, Zimbabwe, South Africa, Oman [Afromontane & Arabia (montane)].
IUCN: LC
Bali Ngemba F.R.: Pollard 615 fl., 10/2001.

Liparis nervosa (Thunb.) Lindl. var. *nervosa*

Gen. & Sp. Orch.: 26 (1830).
Syn. ***Liparis guineensis*** Lindl.
Syn. ***Liparis rufina*** (Ridl.) Rchb.f. ex Rolfe
Terrestrial, epilithic or, rarely, epiphytic herb to 70cm; stem basally swollen; leaves 2–5, petiolate, sheathing, lanceolate, to 35 × 7.5cm; peduncle to 55cm; rachis many-flowered, to 15cm; flowers green or yellow to reddish or purplish brown. Forest, marshy-grassland, or stony grassland and on wet rock outcrops; 1300–1560m.
Distr.: tropics and subtropics of the World, including Africa, India to Japan, Philippines, Costa Rica, W Indies, S America [Pantropical].
IUCN: LC
Bali Ngemba F.R.: Etuge 4702 fr., 11/2000; 4261r fr., 10/2001; Ghogue 1018 fr., 11/2000; Onana 1887 fr., 10/2001.
Note: in addition, the following collections have been made by *Chiron*, 3–4/2001: 1046 & 1085 (1500m alt.).

Malaxis weberbaueriana (Kraenzl.) Summerh.

Terrestrial herb, to 25cm; rhizome creeping, slender; pseudobulbs 2.5–8 × 0.1–0.2cm, fusiform; leaves thin-textured, ovate or elliptic-ovate, to 5.5 × 3cm; inflorescence lax, 5–20-flowered, 4–17cm; flowers small, flat, purple or green; labellum 2–2.6 × 2–3mm. Deciduous woodland, forest, plantations.
Distr.: Cameroon, Bioko, Congo (Kinshasa), Kenya, Tanzania, Zambia, Malawi, Zimbabwe [Afromontane].
IUCN: LC
Bali Ngemba F.R.: Ujor 30424 fl., 6/1951.

Microcoelia globulosa (Hochst. ex A.Rich.) L.Jonss.

Epiphyte; aphyllous; roots 3.5–6.5cm; inflorescence 2cm, 15–26-flowered; flowers white; labellum yellow; spur orange to brown; spur to 3.7mm. Plantations; 1450m.
Distr.: Nigeria, Cameroon, CAR, Congo (Kinshasa), Rwanda, Burundi, Sudan, Ethiopia, Uganda, Kenya, Tanzania, Angola, Zambia, Zimbabwe.
IUCN: LC
Bali Ngemba F.R.: Simo 164 fl., 4/2004.
Note: *Simo* 164 is a field determination. In addition, the following collections have been made by *Chiron*, 3–4/2001: 1018; 1022; 1026 (1380–1400m alt.).

Nervilia bicarinata (Blume) Schltr.

Fl. Zamb. 11(1): (1995).
Erect, terrestrial herb, 17–75cm, glabrous except for labellum and subterranean parts; leaf solitary, appearing after flowering, to 22.5 × 26.5cm, orbicular, apiculate, cordate, heavily pleated; inflorescence 1–12-flowered; flowers greenish; sepals and petals subequal; labellum greenish white with purple or green venation, 20–31 × 17–25mm, ovate, obscurely 3-lobed. Riverine and waterfall-spray forest and in *Syzygium* thicket; 1400–1700m.
Distr.: tropical Africa, south to South Africa, Comoro Is, Madagascar, Mascarenes, Oman & Yemen [Palaeotropics].
IUCN: LC
Bali Ngemba F.R.: Darbyshire 405 fl., 4/2004; Pollard 1043 fl., 4/2002; Simo 205 fl., 4/2004.
Note: the 2004 collections cited above are field determinations.

Podangis dactyloceras (Rchb.f.) Schltr.

Epiphyte; stem leafy, 1–8(–11)cm; leaves 4–8, linear-ligulate, often falcate, 4–10(–16) × 0.5–1.2cm; inflorescence to 6cm, borne laterally in leaf axil, few–20-flowered; peduncle and rachis 1–4cm; flowers glistening-white, semi-transparent; labellum 4–6 × 5mm; spur 9–11 × 2–3mm, inflated into 2 apical lobules. Forest; 1310–1450m.
Distr.: Guinea to Cameroon, Congo (Kinshasa), Uganda, Tanzania, Angola, Madagascar [Afromontane].
IUCN: LC
Bali Ngemba F.R.: Simo 171 fl., 4/2004; Zapfack 2007 fr., 4/2002.
Note: *Simo* 171 is a field determination. In addition, the following collection has been made by *Chiron*, 3–4/2001: 1011 (1390m alt.).

Polystachya adansoniae Rchb.f.

Forest; 1400m.
Bali Ngemba F.R.: Chiron 1101 3–4/2001.

Polystachya albescens Ridl. subsp. *albescens*

Epiphyte or epilith, 15–25(–40)cm; pseudobulbs absent; leaves linear-lanceolate, 10–13 × 1.1–1.3cm; inflorescence racemose or weakly once-branched, 2–5cm; flowers non-resupinate, greenish or whitish, sometimes with reddish veins; labellum tinged or veined red, indistinctly trilobed, 6–6.25 × 5mm; spur to 8mm, sacciform-conical. Forest, riverine forest, sometimes on rocks, farmland; 1450m.
Distr.: SE Nigeria, W Cameroon, Bioko, São Tomé, Príncipe Annobon [W Cameroon Uplands].
IUCN: NT
Bali Ngemba F.R.: Simo 178 4/2004.
Note: *Simo* 178 is a field determination. In addition, the following collections have been made by *Chiron*, 3–4/2001: 1031; 1036; 1230 (forest; 1400–1900m alt.). *Chiron*

tentatively places his following specimen in this taxon also: 1006 (1145 [1415]m alt.).

Polystachya anthoceros la Croix & P.J.Cribb

Kew Bull. 51(3): 571 (1996).

Epiphyte; pseudobulbs densely clustered, slightly superposed, ± fusiform, 2–4 × 0.4–0.5cm, bifoliate; leaves 7–16(–28) × 0.4–0.5cm, somewhat canaliculate; inflorescence slender, 4–6cm, laxly 4–6(–9)-flowered; peduncle with one loose sheath; bracts lanceolate-subsetose, 2–2.5mm; flowers white, with a remarkable, conspicuous, elongate, erect, spur-like mentum to 14mm, curving forwards with maturity; labellum 3-lobed with a very long, slender claw, 12 × 3mm; claw canaliculate, 10mm; type specimen apparently abnormal, with 3 anthers. Forest; 2100m.

Distr.: SE Nigeria, W Cameroon (Bali Ngemba F.R. & Baba II) [W Cameroon Uplands].

IUCN: EN

Bali Ngemba F.R.: Pollard 696 fl., 10/2001; Simo 158 fl, 4/2004.

Note: previously known only from the type. This taxon was also collected recently from the nearby forests above the village of Baba II (*Etuge* 4302r). *Simo* 158 is a field determination.

Polystachya* cf. *bertauxiana Szlach. & Olszewski

Small epiphyte; pseudobulbs 1.3 × 0.4cm; 1–2 sheathing leaves, 2–3 leaves at apex; leave oblong-lanceolate, 4 × 0. 6cm; inflorescence simple, pubescent, 2.5cm, 5–8-flowered; flowers non-resupinate, white with violet coloration; labellum 8 × 3mm; spur cylindrical, 4mm. Secondary forest & plantations; 1450m.

Distr.: Cameroon.

Bali Ngemba F.R.: Simo 177 fl., 4/2004.

Note: field determination.

Polystachya billietiana Geerinck

Forest; 1700m.

Bali Ngemba F.R.: Chiron 1227 3–4/2001.

Note: *Chiron* tentatively places his following specimens in this taxon also: 1261 & 1266 (1600–1710m alt.).

Polystachya calluniflora Kraenzl.

Epiphyte, 2–30cm; stems arching; pseudobulbs superposed, fusiform, (2.5–)4–7(–11.5) × 0.3–0.5cm; bifoliate; leaves linear, grass-like, 4.5–20 × 0.5–0.8cm; inflorescence erect, racemose, 4–10cm, 5–30-flowered; flowers white, non-resupinate; labellum and dorsal sepal purple, but whole flower drying orange; labellum 2.5–4 × 1.8–2mm; spur 0.5–1mm, sacciform. Forest, farmland; 1400m.

Distr.: Nigeria, Cameroon, Bioko, Rwanda, Uganda [Lower Guinea & Congolian].

IUCN: LC

Bali Ngemba F.R.: Simo 207 4/2004.

Note: *Simo* 207 is a field determination. In addition, the following collection has been made by *Chiron*, 3–4/2001: 1016 (forest; 1400m alt.). *Chiron* tentatively places his following specimen in this taxon also: 1005 (1415m alt.).

Polystachya elegans Rchb.f.

Epiphyte, 20–30(–40)cm; pseudobulbs superposed, 7–15 × 0.5–0.7cm, 3–5-leaved; leaves strap-shaped or lanceolate-oblong, 5–22 × 0.5–1.6(–2.5)cm, borne apically; inflorescence 6–15cm, branched basally, many-flowered; flowers non-resupinate, greenish white or yellowish white, tinged purple; labellum 4.6–5.9 × 2.5mm; spur 2.8mm, clylindric-sacciform. Forest, plantations; 1450m.

Distr.: Nigeria, W Cameroon, Bioko, Congo (Kinshasa) [Lower Guinea & Congolian].

IUCN: LC

Bali Ngemba F.R.: Simo 176 fl., 4/2004.

Note: *Simo* 176 is a field determination. In addition, the following collection has been made by *Chiron*, 3–4/2001: 1016 (forest; 1500m alt.).

Polystachya fusiformis (Thouars) Lindl.

Suberect or pendent epiphyte or epilith, to ± 60cm; stems (pseudobulbs) cylindrical to fusiform, superposed, longitudinally ridged, drying yellow, to 22 × 0.3–0.4cm; leaves 3–7, oblong-lanceolate, lanceolate to oblanceolate, 5–16 × 0.6–1.6(–3.2)cm, largest borne apically; inflorescence terminal, paniculate, 3–8(–15)cm, densely 20–80-flowered; peduncle pubescent; flowers miniscule, non-resupinate, persistent on developed ovary, cream, yellow-green tinged or entirely purple or mauve; pedicel and ovary 4mm, glabrous; labellum 2–2.5 × 2.5mm; spur 1mm, sacciform. Forest; 1400m.

Distr.: Ghana, Cameroon, Bioko, Congo (Kinshasa), Rwanda, Burundi, E Africa, Zambia, Malawi, Zimbabwe, South Africa, Mascarene Is. [Afromontane].

IUCN: LC

Bali Ngemba F.R.: Simo 206 fl., 4/2004.

Note: *Simo* 206 is a field determination. In addition, the following collection has been made by *Chiron*, 3–4/2001: 1253 (1470m alt.).

Polystachya galeata (Sw.) Rchb.

Forest; 1600m.

Bali Ngemba F.R.: Chiron 1103 3–4/2001.

Polystachya odorata Lindl. var. ***odorata***

Epiphyte or rarely epilith, 20–60cm; pseudobulbs 2–4.5 × 0.6–1.5cm; leaves 4–8, oblanceolate to oblong-elliptic, 8–26 × 2.8–5.5cm, prominently veined beneath; inflorescence paniculate, 10–30cm, conspicuously branched; branches 6–15, puberulous, many-flowered, to 8cm; flowers non-resupinate, white, yellow, dull red-brown or pale green; labellum 6–7(–8) × 5–7.5mm; spur 4–5mm, sacciform. Forest openings and farmland; 1400–1450m.

Distr.: Ivory Coast, Burkina Faso, Ghana, Nigeria to Tanzania, Angola [Guineo-Congolian].

IUCN: LC

Bali Ngemba F.R.: Simo 140 fl., 4/2004; 175 fl., 4/2004; 201 fl., 4/2004.

Note: *Simo* collections are field determinations. In addition, the following collections have been made by *Chiron*, 3–4/2001: 1047; 1074; 1268 (1390–1500m alt.).

Polystachya tessellata Lindl.

Syn. *Polystachya concreta* (Jacq.) Garay & H.R.Sweet

Epiphyte, (10–)20–60cm; pseudobulbs 15 × 0.5–0.7cm, 3–5-foliate; leaves oblanceolate or elliptic, (3–)10–30 × 0.8–6.0cm; inflorescence paniculate, 10–50cm; branches secund, distant, densely 20–200-flowered; rachis and peduncle covered in sheaths; flowers small, non-resupinate, cream,

yellow, clear green or red-purple. Forest, savanna or woodland; 1400–2075m.
Distr.: tropical & subtropical Africa.
IUCN: LC
Bali Ngemba F.R.: Simo 155, fl., 4/2004; 174 4/2004; 203 fl., 4/2004.
Note: field determinations.

Polystachya sp. 1
Natural/degraded forest zone; 1920m.
Bali Ngemba F.R.: Chiron 1241 3–4/2001.

Polystachya sp. 2
Forest; 1470m.
Bali Ngemba F.R.: Chiron 1095 3–4/2001.

Polystachya sp. 3
Near village; 1390m.
Bali Ngemba F.R.: Chiron 1071 3–4/2001; 1083 3–4/2001.

Polystachya sp. 4
Forest; 1700m.
Bali Ngemba F.R.: Chiron 1244 3–4/2001.

Rangaeris muscicola (Rchb.f.) Summerh.
Epiphyte or epilith; stem very short, 1–6cm; leaves arranged in a fan, conduplicate, 5–11, linear, 6.5–20 × 0.6–1.3cm; inflorescence 5–22(–42)cm, 5–16-flowered; bracts amplexicaul; flowers fragrant, white; labellum 6.7–8.5 × 4–7mm; spur pinkish-olive, 5.5–7cm. Forest; 1450m.
Distr.: Guinea (Conakry) to Cameroon, Congo (Kinshasa), E Africa to South Africa [Tropical & subtropical Africa].
IUCN: LC
Bali Ngemba F.R.: Simo 162 fl., 4/2004.
Note: *Simo* 162 is a field determination. In addition, the following collection has been made by *Chiron*, 3–4/2001: 1010 (forest; 1390m alt.). *Chiron* tentatively places his following specimen in this taxon also: 1255A (1380m alt.).

Satyrium volkensii Schltr.
Terrestrial herb; sterile stems to 5cm, with 3–4 leaves, 5–18 × 2–6cm, lanceolate to oblong-elliptic; fertile stems 20–110cm, entirely covered with imbricate cauline bracts; cauline bracts 7–15, 2.5–11 × 1–3cm; inflorescence lax, 5–42cm, several-flowered; flowers non-resupinate, green to yellowish green; spurs 2, 11–23mm. Grassland; 2100m.
Distr.: Nigeria, Cameroon, Congo (Kinshasa), Kenya, Tanzania, Malawi, Zambia, Zimbabwe [Afromontane].
IUCN: LC
Bali Ngemba F.R.: Pollard 960 fl., 4/2002; Ujor 30324b fl., 5/1951.

Stolzia repens (Rolfe) Summerh. var. *repens*
Creeping dwarf epiphytic herb, to 1cm high; pseudobulbs prostrate except at apex, widely spaced, 1.3–4cm apart, to 3 × 0.3cm, with 2 leaves near insertion of next pseudobulb; leaves subsessile, 0.5–1.4 × 0.3–0.8cm; peduncle with a single flower; flowers yellow, brown or reddish, striped red or brown. Forest, woodland; 1310m.
Distr.: Ghana, S Nigeria, W Cameroon, Congo (Kinshasa), Ethiopia, E Africa, Malawi, Zambia, Zimbabwe [Afromontane].
IUCN: LC
Bali Ngemba F.R.: Zapfack 2034 fl., 4/2002.

Note: in addition, the following collections have been made by Chiron, 3–4/2001: 1012; 1084 (1390m alt.).

Tridactyle anthomaniaca (Rchb.f.) Summerh.
Near village; 1390m.
Bali Ngemba F.R.: Chiron 1097 3–4/2001.

Tridactyle bicaudata (Lindl.) Schltr.
Near village; 1390m.
Bali Ngemba F.R.: Chiron 1009 3–4/2001.

Tridactyle filifolia (Schltr) Schltr
Forest; 1400m.
Bali Ngemba F.R.: Chiron 1023 3–4/2001.

Tridactyle gentilii (De Wild.) Schltr.
Near village; 1390–1450m.
Bali Ngemba F.R.: Chiron 1038 3–4/2001; 1039 2–4/2001; 1049 3–4/2001.

Tridactyle cf. *gentilii* (De Wild.) Schltr.
Epiphyte; stems 30–85cm; leaves 6–21 × 0.8–2.1cm, ligulate; inflorescence 3.5–15cm, 7–15-flowered; flowers cream-colored; labellum 0.7–1.4cm, trilobed, lateral lobes longer than mid-lobe, fimbriate at apex, mid-lobe triangular-lanceolate; spur 35–85mm. Bush fallow; 1450m.
Distr.: Ghana, Nigeria, Cameroon, Congo (Kinshasa), Uganda, Zambia (*T. gentilii*).
Bali Ngemba F.R.: Simo 186 fl., 4/2004.
Note: *Simo* 186 is a field determination. In addition, the following collections have been made by *Chiron*, 3–4/2001:1033; 1055 (1390–1400m alt.).

Tridactyle tridactylites (Rolfe) Schltr.
Epiphyte or epilith; stems pendent, 0.4 to 1.6m, robust; leaves numerous, 6–21 × 0.6–1.3cm, linear or linear-lanceolate; inflorescence lax, 1.7–10cm, to 18-flowered; flowers small, resupinate, yellow, orange or brownish-orange, sometimes tinged green, fragrant; labellum 4–5 × 6.5mm, side lobes ± equal in length to mid-lobe, entire or rarely slightly bifid; spur 6–11mm. Forest and farmland; 1400m.
Distr.: Sierra Leone to Congo (Kinshasa), E Africa, S to Mozambique, Malawi, Zambia, Angola, Zimbabwe [Afromontane].
IUCN: LC
Bali Ngemba F.R.: Simo 144 fl., 4/2004.
Note: *Simo* 144 is a field determination. In addition, the following collection has been made by *Chiron*, 3–4/2001: 1102 (1400–1500m alt.).

Tridactyle tridentata (Harv.) Schltr.
Epiphyte or epilith; stem 10–50 × 0.2–0.5cm, pendent or arcuate, usually branched; leaves numerous, 6–10.5 × 0.1–0.35cm; inflorescence lax, (0.5–)1.8–2.7cm, 4–5-flowered; flowers small, whitish, pale ochre-yellow or salmon-pink; labellum 2.1–5 × 1.4–4mm, auriculate at base, weakly 3-lobed; spur long, 6–18mm, filiform, straight or incurved. Forest, secondary forest, bush fallow; branches of exposed trees; 1450m.
Distr.: Cameroon, CAR, Congo (Brazzaville) & (Kinshasa), Uganda, Tanzania, Mozambique, Malawi, Zimbabwe, S Africa [Afromontane]

IUCN: LC
Bali Ngemba F.R.: Simo 168 4/2004; 182 4/2004.
Note: field determination.

PALMAE

I. Darbyshire (K) & M. Cheek (K)

Phoenix reclinata Jacq.

Tree or shrub; stems clustered; fruiting when 1–10m tall; leaves pinnately compound, leaflet apices often spine-like; fruits ellipsoid, fleshy, 2 × 1cm, orange, ripening brown. Forest, farms & farmbush; 1650m.
Distr.: tropical and subtropical Africa.
IUCN: LC
Bali Ngemba F.R.: Darbyshire 419 4/2004.
Note: field determination.

Raphia mambillensis Otedoh

J. Niger. Inst. Oil Palm Res. 6(22) 145–189 (1982).
Herb, to 7m, lacking aerial stem; leaves 5–8m, marginal fibres 20 × 3–4cm, red-brown, leathery; leaflets 60–90 on each side, with only a few weak spines on margins or upper surface midveins; petioles 1.5–2m, to 5cm in diam.; inflorescence axis to 1.5m, 1.2cm diam., slightly curved, primary branches 3–5, 25–120cm long, largely wrapped in papery sheathing, second order bracts, spikes crowded with flowers in two ranks; fruit subellipsoid, broader in distal half, 3.5–7.5 × 2.5–3.5cm, apical beak sharp, 3–4mm, scales dark brown, with a central furrow, in 8–11 rows. Forests, sometimes near streams (usually cultivated); 1250–1600m alt.
Distr.: Cameroon Highlands (Bamboutos, Bamenda Highlands, to Mambilla Plateau (Nigeria), CAR and SW Sudan.
IUCN: LC
Bali Ngemba F.R.: sight record 4/2004 (*Darbyshire*, pers. comm.).

SMILACACEAE

Y.B. Harvey (K) & M. Cheek (K)

Smilax anceps Willd.

Meded. Land. Wag. 82(3): 219 (1982).
Syn. ***Smilax kraussiana*** Meisn.
Climber, to 7m; stem spiny; leaves coriaceous, alternate, elliptic, c. 14 × 8cm, mucron 0.5cm, base obtuse, nerve palmate, 3–5; petiole c. 2cm; inflorescence terminal, umbellate, 5cm diam. Forest; 1400–1450m.
Distr.: Senegal to South Africa [Tropical & subtropical Africa].
IUCN: LC
Bali Ngemba F.R.: Etuge 4714 11/2000; Nkeng 199 fr., 10/2001.

XYRIDACEAE

B.J. Pollard (K) & J.M. Lock (K)

Xyris sp.

Erect herb, to 40cm; leaves to 30cm × 2–3mm; capitula 7–9mm, spherical; flowers yellow. Rocky grassland; 1450m.
Distr.: Bali Ngemba F.R.
Bali Ngemba F.R.: Pollard 528 11/2000.
Note: material not matched.

ZINGIBERACEAE

M. Cheek (K)

Fl. Cameroun 4 (1965).

Aframomum cf. *zambesiacum* (Baker) K.Schum.

Herb, 1–3m, rhizome with purple, papery, sheathing scales, to 2 × 2cm, internodes 2.5cm; leaves to 30 × 6.5cm, acumen ligulate 1cm, base obtuse, green below; petiole 1cm, ligule bilobed; peduncle 7cm, erect; bracts 4, resembling rhizome scales; 10-flowered; capitulum 4 × 6cm; flowers white and red; fruit spherical, red, ridged. Secondary forest; 1500–2110m.
Distr.: Bali Ngemba F.R.
Bali Ngemba F.R.: Darbyshire 367 fl., 4/2004; Tadjouteu 424 11/2000; Zapfack 2123 fl., 4/2002.
Uses: FOOD ADDITIVES – Seeds - savoury preparations - seeds used in 'sauce jaune comme condiment' (*Tadjouteu* 424).
Note: see note with *Aframomum* sp. 1 of Bali Ngemba. *Darbyshire* 367 is a field determination.

Aframomum sp. 1 of Bali Ngemba

Herb, 4m, rhizome not seen, glabrous; leaves 40 × 8cm, apex ligulate, 2.5cm, twisted, base cuneate, sessile, ligule slightly bilobed, blade dull-white below; flowers not seen; fruit narrowly ovoid, 13 × 3cm, including entire calyx, 4cm, brick-red when ripe, when unripe with deep ridges. Disturbed riverine forest; 1450–1830m.
Distr.: Bali Ngemba F.R.
Bali Ngemba F.R.: Biye 101 11/2000; Cheek 10505 11/2000.
Local name: Nchou (Bali, *Cheek* 10505). **Uses:** FOOD – infructescence; MATERIALS – Other materials – Tools – leaves used to wrap and cook ground nut paste & stems or leaves used for giving enemas to small children; MEDICINES (*Biye* 101).
Note: essential reference material for genus at K is on loan to E (27.i.2004).

Aframomum sp.

Herb with leafy stems, to 2m tall; flowers borne at ground level from rhizome, to 3m long; bracts red-brown; corolla cream, lip with yellow-cream centre, rolled inwards; fertile parts cream-white. Open forest with burnt areas and some cultivation; 1570m.
Bali Ngemba F.R.: Darbyshire 348 fl. 4/2004.
Note: field determination, as yet unconfirmed.

GYMNOSPERMAE

PINOPSIDA

CUPRESSACEAE

B.J. Pollard (K)

Cupressus lusitanica Mill.
Tree, to 30–35m, evergreen, monoecious; trunk monopodial, large trees buttressed, up to 2m dbh; leaves scale-like; seed cones solitary or in groups near the upper ends of lateral branches, terminal on short,leafy branchlets, maturing in 2 growing seasons, persistent. From cultivation (forming reserve boundary); 1450m.
Distr.: pantropical and temperate regions.
IUCN: LC
Bali Ngemba F.R.: Darbyshire 447 5/2004.
Uses: MATERIALS; SOCIAL USES (*fide* Pollard); ENVIROMENTAL USES – marking boundaries.
Note: field determination.

PODOCARPACEAE

I. Darbyshire (K)

Podocarpus milanjianus Rendle
Dioecious shrub or tree, to 35m; bark exfoliating in papery flakes; slash pale brown; stems much-branched, sympodial; leaves spreading, alternate, linear-lanceolate, (5–)10–15 × 0.5–1.5cm, stomata on lower side only, midrib prominent and raised below; male cones solitary or paired, flesh-pink, c. 3cm; female cones solitary; fruit green, obovoid to subglobose, c. 1cm long; receptacle well-developed, obconical to subglobose, fleshy, red; seeds 1–2, subglobose, 8–9mm. Forest; 900–2000m.
Distr.: Cameroon, Congo (Kinshasa), Angola, Sudan to Zimbabwe [Afromontane].
IUCN: LC
Bali Ngemba F.R.: Ujor 30326 5/1951.

PTERIDOPHYTA

Fl. Cameroun 3 (1964).

LYCOPSIDA

P.J. Edwards (K)

LYCOPODIACEAE

Huperzia brachystachys (Baker) Pic.Serm.
Webbia 23: 162 (1968); Acta Bot. Barcinon. 31: 9 (1978).
Syn. *Lycopodium brachystachys* (Baker) Alston
Pendulous epiphyte; stems to 1m long; leaves c. 1cm long; strobili lax, 9–20cm long; sporophylls c. 5cm long. Forest-grassland transition; 1700–2100m.
Distr.: Guinea to SW Cameroon and Bioko [Guineo-Congolian (montane)].
IUCN: LC
Bali Ngemba F.R.: Pollard 697 10/2001; Zapfack 2049 4/2002.

SELAGINELLACEAE

Selaginella vogelii Spring
Terrestrial in deep shade; stolons long-creeping, pink; stems erect, at least partly pink when dry, pubescent; frond-like portion triangular, 15–40cm long and wide; branches pubescent on underside; lateral microphylls oblong-lanceolate, with entire margins; median leaves with basal auricles, long-acuminate-aristate. Forest; 1400m.
Distr.: tropical Africa.
IUCN: LC
Bali Ngemba F.R.: Cheek 10455 11/2000.

FILICOPSIDA

P.J. Edwards (K)

ASPLENIACEAE

Asplenium aethiopicum (Burm.f.) Bech.
Terrestrial, epilithic or epiphytic fern; rhizome erect; fronds oblong-lanceolate; stipe black or very dark brown, glossy; lamina 1–2-pinnate pinnatisect; scales deciduous; pinnae very acute to long-caudate; sori extending along pinna-midrib (costa) and for c. ⅔ pinna length. Forest-grassland transition; 1310–1700m.
Distr.: pantropical [Montane].
IUCN: LC
Bali Ngemba F.R.: Onana 1819 10/2001; Zapfack 2033 4/2002; 2051 4/2002.

Asplenium anisophyllum Kunze
Fl. Zamb. Pteridophyta: 170 (1970).
Syn. *Asplenium geppii* Carruth.
Terrestrial or epilithic fern, to 1m; rhizome short, erect; scales brown, broad; lamina linear-lanceolate, pinnate; larger fronds 60 × 18cm; pinnae linear-oblong, apex acuminate, base unequally cuneate, bluntly serrate; sori elongate, short, at c. 45°; bud/plantlet at base of terminal pinna. Forest; 1310–1600m.
Distr.: Guinea to Bioko, Cameroon, Gabon and Angola [Guineo-Congolian].
IUCN: LC
Bali Ngemba F.R.: Kongor 41 10/2001; Pollard 423 11/2000; Zapfack 2011 4/2002.
Note: *Asplenium elliottii* C.H.Wright is included here. According to C.H.Wright the base of the lower side of the pinnae in *A. elliottii* is much more obtuse than in *A.*

anisophyllum, and is sometimes almost parallel to the rachis. The specimens cannot be clearly divided on the basis of this single character. Further study of specimens from this complex (including *A. boltonii*) is required to clarify the specific delimitations.

Asplenium dregeanum Kunze

Epiphyte, to 25cm; rhizome short-creeping; fronds densely tufted; lamina ovate-lanceolate in outline; pinnae dimidiate, deeply dissected, one linear sorus on many lobes. Forest and *Aframomum* thicket; 1440–1560m.
Distr.: tropical and subtropical Africa [Afromontane].
IUCN: LC
Bali Ngemba F.R.: Cheek 10465 11/2000; Pollard 424 11/2000; 603 10/2001.

Asplenium jaundeense Hieron.

Terrestrial or on rotting logs; rhizome short-creeping; stipe 4–13cm; lamina 8 × 5 to 30 × 9cm; 1-pinnate, pinnae sessile, irregular trapezoid, acute, cuneate at base; terminal pinna acutely trilobed. Forest; 1500m.
Distr.: Guinea, Cameroon and Gabon [Guineo-Congolian].
IUCN: LC
Bali Ngemba F.R.: Ghogue 1045 11/2000.

Asplenium mannii Hook.

Very small epiphyte, to 6cm; rooting runners present; lamina 2-pinnate, one linear sorus per lobe. Forest-grassland transition; 1500m.
Distr.: tropical Africa [Afromontane].
IUCN: LC
Bali Ngemba F.R.: Zapfack 2124 4/2002.

Asplenium preussii Hieron.

Terrestrial or epilithic fern, to 60cm ; rhizome short; lamina linear-lanceolate in outline, 1-pinnate pinnatisect, larger fronds c. 40 × 12cm; pinnae ± pectinate; basal acroscopic lobe very enlarged and almost stipitate, most lobes with a single elongated sorus along ½ to ¾ their length. Forest; 1440–1500m.
Distr.: tropical and subtropical Africa [Afromontane].
IUCN: LC
Bali Ngemba F.R.: Ghogue 1022a 11/2000; 1022b 11/2000; Pollard 426 11/2000.

Asplenium sp. aff. *staudtii* Hieron.

Epiphyte, to 20cm; stipe purplish green. Damp places in forest understorey; 1440m.
Distr.: Bali Ngemba F.R.
Bali Ngemba F.R.: Pollard 425 11/2000.
Note: too little material, which is juvenile, to be able to identify.

Asplenium theciferum (Kunth) Mett. var. *cornutum* (Alston) Benl

Acta Bot. Barcinon. 40: 32 (1991).
Syn. ***Asplenium cornutum*** Alston
Small epiphyte, to 15cm tall; rhizome short-creeping; lamina 2-pinnate, with a rounded sorus embedded at the tip of many of the lobes. Forest, grassland and *Aframomum* thicket; 1700–1800m.
Distr.: W Cameroon, Bioko, Uganda and Kenya [Tropical Africa].
IUCN: LC
Bali Ngemba F.R.: Onana 1818 10/2001; Zapfack 2050 4/2002.
Note: smaller plants are similar to *Asplenium mannii*, but runners are absent.

Asplenium unilaterale Lam.

Rhizome creeping; stipe 10–20cm, black, shiny; fronds lanceolate in outline, 10–20cm long, pinnate; pinnae subopposite, narrowly oblong, dimidiate, margin dentate, tips pinnatifid without a terminal pinna; sori short, thick. Shady forest understorey and streambanks; 1400–1700m.
Distr.: Africa and tropical Asia [Palaeotropical].
IUCN: LC
Bali Ngemba F.R.: Etuge 4796 11/2000; Plot voucher BAL43 4/2002.

CYATHEACEAE

Cyathea camerooniana Hook. var. *camerooniana*

Syn. ***Alsophila camerooniana*** (Hook.) R.M.Tryon var. ***camerooniana***
Tree fern, to 4m tall; trunk slender, 0.6–3m tall, not spiny; fronds 1.2 × 0.4 to 2.5 × 0.55m; stipe c. 10cm; pinnae sessile, gradually reducing in size in lower ¼ of frond; sori near costules, at forking of nerve. Forest and streambanks; 1200–1450m.
Distr.: Guinea to Bioko, Cameroon and Gabon [Guineo-Congolian (montane)].
IUCN: LC
Bali Ngemba F.R.: Cheek 10507 11/2000; Onana 1812 10/2001; Zapfack 2021 4/2002.

DENNSTAEDTIACEAE

Lonchitis occidentalis Baker

Fl. Zamb. Pteridophyta: 86 (1970).
Syn. ***Anisosorus occidentalis*** (Baker) C.Chr.
Terrestrial; rhizome short-creeping, fleshy, with many long, brown-hairs; stipe 30–80cm; lamina pinnate-pinnatifid to bipinnate-tripinnatisect, oblong to ovate-deltate, 40 × 28 to 180 × 36cm. Forest; 1400m.
Distr.: tropical Africa [Afromontane].
IUCN: LC
Bali Ngemba F.R.: Etuge 4803 11/2000.

Pteridium aquilinum (L.) Kühn subsp. *aquilinum*

Terrestrial fern; thicket forming; rhizome long-creeping, subterranean; fronds, to 1.5m tall; stipe erect, the base black (remainder brown); lamina 3-pinnate pinnatisect; sori marginal with fimbriate indusia on both sides. Forest and grassland; 1700m.
Distr.: cosmopolitan [Montane].
IUCN: LC
Bali Ngemba F.R.: sight record 4/2004 (*Darbyshire* pers. comm.).

DRYOPTERIDACEAE

Didymochlaena truncatula (Sw.) J.Sm.
Terrestrial, erect, tufted fern, to 2m tall; stipe and rachis with many, broad, dark brown scales; lamina 2-pinnate; pinnules dimidiate, trapeziform. Forest; 1700m.
Distr.: pantropical [Montane].
IUCN: LC
Bali Ngemba F.R.: Plot voucher BAL75 4/2002.

Tectaria fernandensis (Baker) C.Chr.
Terrestrial fern; rhizome erect; fronds tufted to 1m; lamina mostly 1-pinnate, pinnatisect, with 3–4 pairs of pinnae, but the basal pair with much enlarged basal pinnules. Forest, on rocky ground; 1440m.
Distr.: Guinea to Bioko, Cameroon, Gabon, Congo (Kinshasa) and Rwanda [Guineo-Congolian (montane)].
IUCN: LC
Bali Ngemba F.R.: Pollard 418 11/2000.

HYMENOPHYLLACEAE

Vandenboschia radicans (Sw.) Copel.
Acta Bot. Barcinon. 32: 14 (1980).
Syn. Trichomanes giganteum Bory ex Willd.
Syn. Trichomanes radicans Sw.
Trailing/climbing epiphyte; rhizome long-creeping, 1–3mm diam., densely black-hairy, bearing remotely large 2–3 pinnate pinnatisect pinnae, 18 × 3 to 40 × 23cm, all axes in the lamina green winged. Beside stream; 1400m.
Distr.: pantropic and in N temperate region.
IUCN: LC
Bali Ngemba F.R.: Etuge 4801 11/2000.

LOMARIOPSIDACEAE

Bolbitis acrostichoides (Afzel. ex Sw.) Ching
Terrestrial or on rocks; rhizome long-creeping; stipe 13–50cm (longest on fertile fronds); lamina 10 × 4 to 50 × 35cm, 1-pinnate, margins slightly crenate; terminal pinna often with a scaly bud, some lateral pinnae may have buds too; fertile pinnae narrower than sterile pinnae, completely covered in sori on undersurface. Forest, sometimes on rocks; 1400m.
Distr.: tropical Africa.
IUCN: LC
Bali Ngemba F.R.: Etuge 4802 11/2000.

Bolbitis fluviatilis (Hook.) Ching
Terrestrial on soil or rocks; rhizome long-creeping; stipe 10–40cm; sterile lamina entire to pinnatipartite, 10 × 2 to 40 × 25cm, with venation conspicuously reticulate; fertile fronds often smaller and on longer stipes, entire to pinnatipartite; sori covering all of underside of lamina. Forest, sometimes on rocks; 1400m.
Distr.: Liberia to Bioko, São Tomé and Príncipe, Cameroon, Gabon, Congo (Brazzaville), Congo (Kinshasa) [Guineo-Congolian].
IUCN: LC
Bali Ngemba F.R.: Etuge 4716 11/2000.

Elaphoglossum kuhnii Hieron.
Small epiphyte, to 30cm; stipe c. ½ to ¾ length of the narrowly oblong-elliptic lamina; fronds to 50cm or more; stipe and sterile lamina densely covered on both surfaces with lanceolate, appressed, red-brown, ciliate scales. Forest; 1700m.
Distr.: Sierra Leone, Liberia, Bioko and Cameroon [Guineo-Congolian].
IUCN: NT
Bali Ngemba F.R.: Zapfack 2068 4/2002.
Note: known from very few collections made at c. 8 different localities.

MARATTIACEAE

Marattia fraxinea J.Sm. var. ***fraxinea***
Very large terrestrial fern; rhizome erect, to 40 × 30cm; fronds tufted to 4m, stiff, fleshy; stipe with brown flushing and long white- or green-streaks; swollen base with a pair of green to dark brown, thick, fleshy stipules; lamina ovate in outline, 2-pinnate, to 2 × 1m. Forest; 1450m.
Distr.: palaeotropical [Montane].
IUCN: LC
Bali Ngemba F.R.: Pollard 931 4/2002.

OLEANDRACEAE

Nephrolepis undulata (Afzel. ex Sw.) J.Sm. var. ***undulata***
Rhizome vestigial, <1cm; stolons numerous, long and wiry (many bearing scaly tubers); fronds tufted, erect; lamina 1-pinnate, with 30–100 pairs of slightly crenate, elliptic-lanceolate pinnae; bases auricled; sori semi-circular. Forest, plantations and roadsides; 1300–2000m.
Distr.: tropical and subtropical Africa.
IUCN: LC
Bali Ngemba F.R.: Etuge 4817 11/2000; Kongor 51 10/2001; Onana 1840 10/2001; Pollard 427 11/2000.

POLYPODIACEAE

Drynaria volkensii Hieron.
Epiphytic fern; rhizome 1–1.5cm diam., long-creeping; "nest-leaves" brown, rigid, erect, 12 × 6 to 23 × 16cm, deeply lobed; green fronds 24 × 8 to 95 × 34cm, ovate-oblong, pinnatisect almost to rachis; stipe 8–25cm; sporangia in a single row on each side of midrib (costa). Forest and forest-grassland transition; 1500m.
Distr.: Bioko and Cameroon to E Africa [Afromontane].
IUCN: LC
Bali Ngemba F.R.: Zapfack 2118 4/2002.

Lepisorus excavatus (Bory ex Willd.) Ching
Zink, M. (1993). Systematics of *Lepisorus*: 37.
Syn. *Pleopeltis excavata* (Bory ex Willd.) T.Moore
Epiphytic fern; rhizome long-creeping; stipe short; lamina very thin, glabrous, 15–30 × 2–3.5cm; sori circular, large, in a single series each side of the midrib. Forest-grassland transition; 1800m.

Distr.: Guinea to Bioko and Cameroon, east to Sudan & Ethiopia, south to South Africa [Tropical and subtropical Africa].
IUCN: LC
Bali Ngemba F.R.: Onana 1820 10/2001.

Loxogramme abyssinica (Baker) M.G.Price
Amer. Fern. J. 74(2): 61 (1984).
Syn. ***Loxogramme lanceolata*** (Sw.) C.Presl
Epiphytic fern; rhizome wide-creeping, 1–2mm diam.; fronds mostly with short stipes; lamina entire, glabrous, very leathery, 5–30 × 1–2.5cm; sori very elongated, forming oblique, parallel lines near midrib. Forest-grassland transition; 1440–1600m.
Distr.: tropical Africa [Afromontane].
IUCN: LC
Bali Ngemba F.R.: Kongor 39 10/2001; Pollard 422 11/2000; Zapfack 2122 4/2002.

Pleopeltis macrocarpa (Bory ex Willd.) Kaulf. var. *macrocarpa*
Fl. Zamb. Pteridophyta: 152 (1970).
Syn. Pleopeltis lanceolata (L.) Kaulf.
Small leathery epiphytic fern similar to *Loxogramme abyssinica*, but lower surface with many scattered, small, brown scales; sori as a single series of large oval spots each side of the midrib. Forest-grassland transition, sometimes on rocks; 1700m.
Distr.: pantropical [Montane].
IUCN: LC
Bali Ngemba F.R.: Zapfack 2058 4/2002.

PTERIDACEAE

Pteris prolifera Hieron.
Terrestrial, usually near watercourses; stem short,erect; lamina 30 × 26 to 70 × 55cm, ovate-oblong, pinnate-pinnatisect; 3–9 pairs of pinnae; a single, black, scaly bud near base of an apical lateral pinna; stipe approx. same length as lamina. Forest and forest-grassland transition; 1400m.
Distr.: Liberia, SW Cameroon, Bioko, CAR, Congo (Brazzaville), Congo (Kinshasa), Uganda and Sudan [Guineo-Congolian (montane)].
IUCN: LC
Bali Ngemba F.R.: Etuge 4799 11/2000.

THELYPTERIDACEAE

Pneumatopteris afra (Christ) Holttum
Bull. Jard. Bot. Natl. Belg. 53: 283 (1983); Acta Bot. Barcinon. 38: 57 (1988).
Syn. ***Cyclosorus afer*** (Christ) Ching
Terrestrial; rhizome wide-creeping; fronds 3 × 18 to 120 × 36cm; stipe c. ¼ length of the lamina, 1-pinnate; pinnae crenate, 1–2 pairs of basal pinnae much-reduced; sori medial; indusia hairy. Plantations and open areas in forest; 1400m.
Distr.: tropical Africa.
IUCN: LC
Bali Ngemba F.R.: Etuge 4804 11/2000.

VITTARIACEAE

Antrophyum mannianum Hook.
Epitphytic fern; trunks/lower branches epilithic; rhizome short-creeping; fertile fronds 8 × 5 to 25 × 17cm, entire, obovate-orbicular, acuminate; stipe length similar to frond length; sori elongate along the veins all over undersurface. Forest, sometimes on rocks; 1310m.
Distr.: tropical Africa [Afromontane].
IUCN: LC
Bali Ngemba F.R.: Zapfack 2025 4/2002.

Vittaria guineensis Desv. var. *camerooniana* Schelpe
Acta Bot. Barcinon. 33: 8 (1982).
Pendulous epiphyte, very grass-like; rhizome short-creeping; fronds 18–30cm × 4–10mm; base of stipe black; spores monolete. Forest-grassland transition; 1400m.
Distr.: Cameroon and Bioko [Lower Guinea (montane)].
IUCN: VU
Bali Ngemba F.R.: Cheek 10464 11/2000